Computational Intelligence for Missing Data Imputation, Estimation, and Management:
Knowledge Optimization Techniques

Tshilidzi Marwala
University of Witwatersrand, South Africa

INFORMATION SCIENCE REFERENCE

Hershey · New York

Director of Editorial Content:	Kristin Klinger
Senior Managing Editor:	Jamie Snavely
Managing Editor:	Jeff Ash
Assistant Managing Editor:	Carole Coulson
Typesetter:	Sean Woznicki
Cover Design:	Lisa Tosheff
Printed at:	Yurchak Printing Inc.

Published in the United States of America by
    Information Science Reference (an imprint of IGI Global)
    701 E. Chocolate Avenue, Suite 200
    Hershey PA 17033
    Tel: 717-533-8845
    Fax: 717-533-8661
    E-mail: cust@igi-global.com
    Web site: http://www.igi-global.com/reference

and in the United Kingdom by
    Information Science Reference (an imprint of IGI Global)
    3 Henrietta Street
    Covent Garden
    London WC2E 8LU
    Tel: 44 20 7240 0856
    Fax: 44 20 7379 0609
    Web site: http://www.eurospanbookstore.com

Library of Congress Cataloging-in-Publication Data

Marwala, Tshilidzi, 1971-
  Computational intelligence for missing data imputation, estimation and management : knowledge optimization techniques / by Tshilidzi Marwala.

    p. cm.

  Includes bibliographical references and index.
  Summary: "This book is for those who use data analysis to build decision support systems, particularly engineers, scientists and statisticians"--Provided by publisher.

  ISBN 978-1-60566-336-4 (hardcover) -- ISBN 978-1-60566-337-1 (ebook)
  1. Statistical decision--Data processing. 2. Decision support systems. 3. Computational intelligence. 4. Missing observations (Statistics) I. Title.
  QA279.4.M37 2009
  519.5'42--dc22
               2008041549

British Cataloguing in Publication Data
A Cataloguing in Publication record for this book is available from the British Library.

All work contributed to this book is new, previously-unpublished material. The views expressed in this book are those of the authors, but not necessarily of the publisher.

Table of Contents

Nomenclature

AANN	Autoassociative Neural Network
AANN-HSAGA	Hybrid of Autoassociative Neural Network, Simulated Annealing and Genetic Algorithm
AIDS	Acquired Immunodeficiency Syndrome
ANN	Artificial Neural Network
ANNGA	Combination of Neural Network and Genetic Algorithm
ARCH	Autoregressive Conditional Heteroskedasticity
ART	Adaptive Resonance Theory
CART	Classification and Regression Trees
CI	Computational Intelligence
DMS	Dimethyl Sulfide
dom	Domain Attribute
EM	Expectation Maximization
ENN	Extension Neural Network
FSA	Fast Simulated Annealing
GA(s)	Genetic Algorithm(s)
GMM(s)	Gaussian Mixture Model(s)
GMM-EM	Gaussian Mixture Models Trained Using Expectation Maximization
H	Hessian or Hamiltonian
HC	Jill Climbing
HGA	Hybrid Genetic Algorithm
HIV	Human Immunodeficiency Virus
HGAPSO	Hybrid Genetic Algorithm and Particle Swarm Optimization
HSAGA	Hybrid Simulated Annealing and Genetic Algorithm
ID3	Iterative Dichotomizer 3
IND	Indiscernibility
INVNN	Inverse Neural Networks
K	Kernel
KTT	Karush-Kuhn Tucker
MAR	Missing At Random
MBND	Missing by Natural Design
MCAR	Missing Completely At Random
MDEEF	Missing Data Estimation Error Function
MI	Multiple Imputation
ML	Maximum Likelihood

MLP(s)	Multi-layer Perceptron(s)
MNAR	Missing Not At Random
MSE	Mean Square Error
MTBF	Mean Time Before Failure
NN-FSA	Neural Networks and Fast Simulated Annealing
NNH	Non-stationary Non-linear Heteroskedasticity
NN-HGA	Neural Networks and Hybrid Genetic Algorithm
NN-PSO	Neural Network and Particle Swarm Optimization Method
OC1	Oblique Classifier 1
PC	Principal Component
PCA	Principal Component Analysis
PCA-ANN(s)	Hybrid Principal Component Analysis and Artificial Neural Network(s)
PCANNGA	Hybrid of Principal Component Analysis, Neural Network and Genetic Algorithm
PCA-NN-HSAGA	Hybrid of Principal Component Analysis, Neural Network, Simulated Annealing and Genetic Algorithm
PE-LE	Present Extract Minus Limit Extract
PSO	Particle Swarm Optimization
RBF(s)	Radial Basis Function(s)
RED	Reducts
rel	Relevant Attribute
SA	Simulated Annealing
SCG	Scaled Conjugate Gradient
SE	Standard Error
SSE	Sum of Square Error
SVM	Support Vector Machine
SVR	Support Vector Regression
SVRGA	Combined Support Vector Regression and Genetic Algorithm

Foreword

Information is unquestionably a critical resource in several organizations, in the fields of engineering, computer science, political sciences, social sciences, health sciences, finance and many more. Most of this information is obtained from data that need to be managed for an effective use in planning and decision making. Through an increasing use of computers in data processing and information retrieval, many organizations are now able to cope with large quantities of data. It is unfortunate that there is also a growing loss of information due to ill-management of data and due to the inabilities of the processing tools to process data that are not presented well or that are incomplete.

The book *Computational Intelligence for Missing Data Imputation, Estimation and Management: Knowledge Optimization Techniques* by Tshilidzi Marwala, PhD is breadth of fresh air in the field of missing data estimation. This book addresses a problem that has largely been avoided in engineering, mainly because of its difficulty. To circumvent this problem, when a signal is required from a source, engineers will use two sensors rather than one in case the sensor fails. We call this a safety factor of two. This is mainly because the area of missing data estimation is not a mature field within the engineering paradigm.

Methods that have been developed thus far are mainly applicable to surveys where computational time and the concept of real time are not strictly applicable. However, in engineering, when sensors fail and a decision has to be made based on the information from that sensor and quickly then that information must be as accurate as possible. This book covers the use of appropriate information technology for missing data.

The book introduces computational intelligence to achieve this goal. The methods covered are neural networks, support vector machines, decision trees, genetic algorithms, particle swarm optimization and many more. Applications in engineering and medicine are considered. This book makes an interesting read and it will open new avenues in the use of advanced information technology techniques to the problem of missing data estimation.

Fulufhelo Vincent Nelwamondo, PhD
Harvard University, Cambridge, Massachusetts

Fulufhelo Vincent Nelwamondo *is currently a Research Fellow at Harvard University Graduate School of Arts and Sciences. Concurrently, he is a Senior Research Leader at the Council for Scientific and Industrial Research responsible for Biometrics. An electrical engineer by training Nelwamondo holds a Bachelor of Science and a PhD in Electrical Engineering specializing in the area of Missing Data Using Computational Intelligence both from the University of the Witwatersrand. He furthermore has interests in areas of software engineering including data mining as well as modelling of complex systems using neural networks, support vector machines, Markov models and other evolutionary computing techniques. A prolific researcher, Nelwamondo has edited The Proceedings of the 12th World Multi-Conference on Systematics, Cybernetics and Informatics, Orlando, Florida. He has written over 30 publications in journals and proceedings in the area of missing data imputation. He has won many accolades including the SAIEE Premium Best Paper Award in 2008, International Institute of Informatics and Systemics Best Paper Award in 2008, an award from the Microsoft Research Limited in Cambridge and the best presenter award at the 2006 IEEE World Congress on Computational Intelligence in Vancouver, Canada. Nelwamondo has reviewed papers for many prestigious journals, conferences and organizations such as the International Journal of Systems Science, IEEE Transactions on Neural Networks, Neural Networks journal and the National Research Foundation. He has served on number of influential bodies including the Technical Committee of the IEEE World Congress on Computational Intelligence that was held in Hong Kong.*

Preface

In real life, a set of data invariably contains missing data. The problem then is to reconstitute the most probable values through processes such as interpolation and extrapolation before using that set.

Methods for resolving the problem of missing data have been extensively explored in statistical texts (Abdella, 2005; Little & Rubin, 1987). The initial work on compensating for missing data was focused on improving survey data. In this book, missing data interpolation is called *imputation* to distinguish it from the statistical approach. Imputation is viewed as an alternative approach to deal with missing data. There are two ways to deal with missing data: these are either to estimate the missing data or to delete any vector (data set) with missing value(s). This book focuses on methods that estimate the missing values.

Of particular importance to the area of missing data interpolation is to analyze the nature of the missing data, and this is termed the *missing data mechanism*. Little and Rubin (1987) categorized three missing data mechanisms, namely: Missing At Random (MAR), Missing Completely At Random (MCAR) and a non-ignorable case also known as Missing Not At Random (MNAR).

In the first case, MAR occurs when the probability that variable X is missing depends on other variables, but not on X itself. An example of this is the case where two variables: the vibration level of a machine and its temperature, X are measured. If a very high vibration level causes the temperature sensor to fall off and thus high and subsequently low values of X become missing because of the other variable *vibration level,* this is termed MAR.

MCAR occurs when the probability that variable X is missing is unrelated to the value of X itself or to any other variable in the data set. This refers to data sets where the absence of data does not depend on the variable of interest or of any other variable in the data set (Rubin, 1978).

MNAR occurs when the probability of variable X missing is related to the value of X itself even if the other variables are controlled in the analysis (Allison, 2000). An example of this is when in a survey of weights of candidates, a person omits mentioning his or her weight because its value is very high. In analyzing survey data, these mechanisms are very powerful and useful. Knowing these mechanisms assists one in choosing which missing data imputation method is best to use.

However, in many engineering problems, where on-line decision support tools are becoming widely used, these mechanisms are proving to be insignificant (Marwala & Hunt, 1999). For example, if an aircraft is flying over the Atlantic Ocean and one of its critical sensors fails, there is simply no time to investigate why that particular sensor has failed and, thereby, indentify its missing value mechanism. What ought to be done in this situation is to quickly estimate the sensor's value, so that an on-line auto-pilot system can continue to operate.

In using decision support tools, if data become missing, it is extremely important, particularly for critical applications, that the missing data estimation technique is accurate. The methods introduced in this book are *computational intelligence methods* and have proven to be very successful in modeling complex problems such as speech recognition (Nelwamondo, Mahola, & Marwala, 2006). In this book, many methods are considered. These include:

- The multi-layer perceptron model (Marwala, 2000),
- Radial basis functions (Bishop, 1995),
- Gaussian mixture models (Chen, Chen, & Hou, 2004),
- Rough sets (Wu, Mi, & Zhang, 2003),
- Support vector machines (Drezet & Harrison, 2001),
- Decision trees (Ssali, & Marwala, 2008),
- Fuzzy ARTMAP (Carpenter et al., 1992) and extension neural networks (Mohamed, Tettey, & Marwala, 2006).

Descriptions and implementations for using these missing data estimation process follow (Bishop, 1995; Marwala, 2007). These methods are implemented in both the Bayesian and maximum-likelihood framework (Marwala, 2001).

It is still very difficult to know beforehand which of these computational intelligence methods are ideal for missing data imputation. For this reason, hybrid methods are also introduced and implemented in this book for missing data imputation. In particular, the ensemble methods that use more than one learning algorithm are considered (Perrone & Cooper, 1993). Some of these methods are computationally intensive, and as a result, the book introduces methods that are computationally efficient, such as the principal component analysis (Adams et al., 2002) and dynamic programming method (Bellman, 1957; Bertsekas, 2000).

In this book, many optimization methods are used. For example, to train multi-layer perceptrons, a scaled conjugate gradient optimization method (Møller, 1993) is used. Other optimization methods used are:

- The expectation maximization algorithm (Dempster, Laird, & Rubin, 1977)
- Genetic algorithms (Goldberg, 1989)
- Particle swarm optimization (Poli, Langdon, & Holland, 2005)
- Hill climbing (Tanaka, Toumiya, & Suzuki, 1997)
- Simulated annealing (Tavakkoli-Moghaddam, Safaei, & Gholipour, 2006)

It is difficult to know in advance which optimization method to use for missing data estimation process and, therefore, this book also explores various hybrid optimization techniques. Some of the hybrid optimization techniques that are considered in this book include the hybrid of genetic algorithms and particle swarm optimization.

Traditional missing data imputation methods have been largely based on static models. Even computational intelligence methods are traditionally constructed in a static manner. These methods are static in the sense that they are the same over time. For situations where the concepts are drifting and, therefore, the data are non-stationary, these methods fail (Kubat, & Widmer, 1996). Many engineering problems have to model systems that are continuously changing because of aging. Therefore, for many engineering problems, data imputation methods are required that are immune or at best can handle these changes in the character of the systems. This problem, therefore, requires missing data models that evolve with the systems on which they are based. Evolutionary methods have been successful in designing learning machines that evolve with systems. These evolutionary methods include genetic algorithms (Goldberg, 2002), fuzzy maps (Carpenter et al., 1992), particle swarm optimization (Kennedy & Eberhart, 1995) and are described in detail in this book.

Throughout this book, examples from the literature and case studies are used to illustrate the effectiveness of the presented missing data estimation methods. Some of the case studies used include the artificial taster, HIV and a mechanical system.

SUMMARY OF THE BOOK

In Chapter I, traditional missing data issues, such as missing data patterns and mechanisms, are described. Attention is paid to the best models to deal with particular missing data mechanisms. A review of traditional missing data imputation methods is conducted, and the methods reviewed include case deletion and prediction rules (Acork, 2005). The *case deletion* methods reviewed are list-wise and pair-wise deletion. The *prediction rule* imputation techniques reviewed are mean substitution, hot-deck, regression and decision trees. Two missing data examples are studied, namely, the Sudoku puzzle and a mechanical system.

Missing data estimation processes require mathematical models that capture interrelationships amongst the variables. In Chapter II, a method is presented that is aimed at approximating missing data and, thereby, capturing variables' interrelationships by combining genetic algorithms and autoassociative neural networks. The neural network architectures implemented are the multi-layer perceptron and the radial basis function neural networks (Russell & Norvig, 1995). The proposed procedures are tested and then compared for missing data imputation.

The ability to identify a model which captures the interrelationships between the variables is very important. Different models bring unique perspectives to the missing data problem and one way to maximize the performance of the missing data procedure is to hybridize different methods. In Chapter III, hybrid autoassociative neural networks models are developed and used in conjunction with genetic algorithms (Goldberg, 2002) to estimate missing data. One hybrid technique combines three neural networks to form a hybrid autoassociative network, while the other merges principal component analysis and neural networks. These procedures are compared to the Bayesian auto-associative neural network (Bishop, 1995) and the genetic algorithm approach.

In Chapter IV, two techniques, i.e., Gaussian mixture models trained using the Expectation Maximization (EM) algorithm (Dempster, Laird, & Rubin, 1977) and the combined auto-associative neural networks and particle swarm optimization methods are implemented for missing data estimation and then compared. Of a particular interest is the nature of the data in the analysis that suits each of these methods.

Chapter V investigates an imputation technique based on rough sets computation (Wu, Mi, & Zhang, 2003). The characteristic relations are introduced to describe incompletely specified decision tables and then used for missing data estimation. Empirical results obtained using real data are given and insights into the problem of missing data are derived.

In Chapter VI, autoassociative neural networks, principal components analysis and support vector regression (Marivate, Nelwamondo, & Marwala, 2008) are all combined with genetic algorithms, and then used to impute missing variables. The impact of using the principal component analysis on the overall performance of the autoassociative network and support vector regression is then assessed.

In Chapter VII, a committee of networks is introduced for missing data estimation. This committee of networks consists of a multi-layer perceptron, support vector machines and radial basis functions. It is constructed through a weighted combination of the three networks. The networks committee is implemented collectively with a hybrid of the genetic algorithm and the particle-swarm optimization method for missing data estimation, and is then tested and assessed. Furthermore, evolutionary methods are used to evolve a committee of networks. The results of this committee are compared to the results from a traditional committee and stand-alone networks.

The use of inferential sensors is common in on-line fault detection systems in various control applications. A problem arises when sensors fail while the system is designed to make a decision based on the data from those sensors. Various techniques to handle missing data are discussed in Chapter VIII. First, a novel algorithm that classifies and regresses in the presence of missing data is proposed. The algorithm is tested for both classification and regression problems. Second, an estimation algorithm that uses an ensemble of regressors within the context of the boosting mechanism is proposed. Hybrid genetic algorithms and fast simulated annealing are used to predict missing values and the results are compared.

In Chapter IX, a classifier method is presented that is based on a missing data estimation framework, and which uses auto-associative multi-layer perceptron neural networks and genetic algorithms. The method is tested and compared to conventional feed-forward neural network using classification accuracies and the area under the receiver operating characteristics curve.

In Chapter X, various optimization methods are compared with the aim of optimizing the missing data estimation equation, which is made out of the autoassociative neural networks with missing values as design variables. These optimization techniques are the genetic algorithm, particle swarm optimization, hill climbing and simulated annealing. They are tested and the results obtained are compared.

In implementing solutions to the missing data estimation problem, using optimization techniques, the definition of variable bounds is of critical importance. Chapter XI introduces a novel paradigm to impute missing data that combines decision trees with an auto-associative neural network and principal component analysis. This is designed to answer the crucial question on whether the optimization bounds actually matter in the estimation of missing data. In the model, a decision tree is used to predict search bounds for a hybrid simulated annealing and genetic algorithm that minimizes an error function derived from the respective models. The results obtained are compared.

Chapter XII presents a control mechanism to assess the effect of a demographic variable, *education level*, on the HIV risk of individuals. This is intended to assist for understanding the extent to which the spread of HIV can be controlled by using the variable *education level*. This control mechanism is based on missing data frameworks where the missing data are the set points for control. An inverse neural network model and a missing data approximation model, based on an auto-associative neural network and the genetic algorithm, are used for the control mechanism and the results obtained are then compared.

In Chapter XIII, a computational intelligence approach to predicting missing data in the presence of concept drift is presented, using an ensemble of multi-layered feed-forward neural networks. An algorithm that detects concept drift is presented. Six instances prior to the occurrence of missing data are used to approximate the missing values. The algorithm is applied to simulated time-series data set resembling the non-stationary data from a sensor. Second, an algorithm that uses dynamic programming and neural networks to solve the problem of missing data imputation is presented, tested and the results are assessed. Third, the impact of missing data estimation on fault classification in mechanical systems is studied. The missing data estimation method is based on auto-associative neural networks where the network is trained to recall the input data through some non-linear neural network mapping using genetic algorithm. The classification methods used are extension neural networks and Gaussian mixture models.

TARGET AUDIENCE OF THIS BOOK

This book is intended for researchers and practitioners who use data analysis to build decision support systems. In particular the target audience includes engineers, scientists and statisticians. The areas of engineering where decision support tools are becoming widely used (the target audience of this book) are aerospace, mechanical, civil, biomedical and electrical engineering. Furthermore, researchers in statistics and social science will also find the techniques introduced in this book to be highly applicable to their work. This book is carefully written to give a good balance between theory and application of various missing data estimation techniques. The applications selected reflect the target audience of this book and include examples from various branches of engineering.

Tshilidzi Marwala

REFERENCES

Abdella, M. (2005). *The use of genetic algorithms and neural networks to approximate missing data in database.* Unpublished master's thesis, University of the Witwatersrand, Johannesburg.

Acork, A. C. (2005). Working with missing values. *Journal of Marriage and Family, 67,* 1012–1028.

Adams, E., Walczak, B., Vervaet, C., Risha, P. G., & Massart, D. L. (2002). Principal component analysis of dissolution data with missing elements. *International Journal of Pharmaceutics, 234* (1-2), 169-178.

Allison, P. (2000). Multiple imputation for missing data: A cautionary tale. *Sociological Methods and Research, 28,* 301-309.

Bellman, R. (1957). *Dynamic programming.* Princeton, NJ: Princeton University Press.

Bertsekas, D. P., (2000). *Dynamic programming and optimal control.* New York, NY: Athena Scientific.

Bishop, C. M. (1995). *Neural networks for pattern recognition.* Oxford, UK: Oxford University Press.

Carpenter, G. A., Grossberg, S., Markuzon, N., Reynolds, J. H., & Rosen, D. B. (1992). Fuzzy ARTMAP: A neural network architecture for incremental supervised learning of analog multidimensional maps. *IEEE Transactions on Neural Networks, 3,* 698-713.

Chen, C. T. Chen, C., & Hou, C. (2004). Speaker identification using hybrid Karhunen–Loeve transform and Gaussian mixture model approach. *Pattern Recognition, 37*(5), 1073-1075.

Dempster, A. P, Laird, N. M., & Rubin, D. B. (1977). Maximum likelihood for incomplete data via the EM algorithm. *Journal of the Royal Statistical Society, B39,* 1-38.

Drezet, P. M. L., & Harrison, R. F. (2001). A new method for sparsity control in support vector classification and regression. *Pattern Recognition, 34*(1), 111-125.

Goldberg, D. E. (1989). *Genetic algorithms in search, optimization, and machine learning.* Reading, MA: Addison-Wesley.

Goldberg, D. E. (2002). *The design of innovation: Lessons from and for competent genetic algorithms.* Reading, MA: Addison-Wesley.

Kennedy, J. E, & Eberhart, R. C. (1995). Particle swarm optimization. In *Proceedings of the IEEE International Conference on Neural Networks,* (pp. 942-1948).

Kubat, M., & Widmer, G. (1996). Learning in the presence of concept drift and hidden contexts. *Machine Learning, 23,* 69–101.

Little, R., & Rubin, D. (1987). *Statistical analysis with missing data.* New York: John Wiley and Sons.

Marivate, V. N., Nelwamondo, V. F., & Marwala, T. (2008). Investigation into the use of autoencoder neural networks, principal component analysis and support vector regression in estimating missing HIV data. In *Proceedings of the 17th World Congress of the International Federation of Automatic Control* (pp. 682-689).

Marwala, T. (2000). On damage identification using a committee of neural networks. *American Society of Civil Engineers, Journal of Engineering Mechanics, 126,* 43-50.

Marwala, T. (2001). Scaled conjugate gradient and Bayesian training of neural networks for fault identification in cylinders. *Computers and Structures 79/32,* 2793-2803.

Marwala, T. (2007). *Computational intelligence for modelling complex systems.* India, New Delhi: Research India Publications.

Marwala, T., & Hunt, H. E. M. (1999). Fault identification using finite element models and neural networks. *Mechanical Systems and Signal Processing, 13,* 475-490.

Mohamed, S., Tettey, T., & Marwala, T. (2006). An extension neural network and genetic algorithm for bearing fault classification. In *Proceedings of the IEEE International Joint Conference on Neural Networks* (pp. 7673-7679).

Møller, A. F. (1993). A scaled conjugate gradient algorithm for fast supervised learning. *Neural Networks, 6,* 525-533.

Nelwamondo, F. V., Mahola, U., & Marwala, T. (2006). Multi-scale fractal dimension for speaker identification system. *Transactions on Systems 5(5),* 1152-1157.

Perrone, M. P., & Cooper, L. N. (1993). When networks disagree: Ensemble methods for hybrid neural networks. In R. J. Mammone (Ed.), *Artificial neural networks for speech and vision* (pp. 126-142). London: Chapman and Hall.

Poli, R., Langdon, W. B., & Holland, O. (2005). Extending particle swarm optimization via genetic programming. *Lecture Notes in Computer Science, 3447,* 291-300.

Rubin, D. B. (1978). Multiple imputations in sample surveys: A phenomenological Bayesian approach to non-response. In *Proceedings of the Survey Research Methods Section of the American Statistical Association* (pp. 20-34).

Russell, S., & Norvig, P. (1995). *Artificial intelligence: A modern approach.* Englewood Cliffs, NJ: Prentice Hall.

Ssali, G., & Marwala, T. (2008). Estimation of missing data using computational intelligence and decision trees. In *Proceedings of the IEEE International Joint Conference on Neural Networks* (pp. 201-207).

Tanaka, T., Toumiya, T., & Suzuki, T. (1997). Output control by hill-climbing method for a small scale wind power generating system. *Renewable Energy, 12*(4), 387-400.

Tavakkoli-Moghaddam, R., Safaei, N., & Gholipour, Y. (2006). A hybrid simulated annealing for capacitated vehicle routing problems with the independent route length. *Applied Mathematics and Computation, 176*(2), 445-454.

Wu, W., Mi, J., & Zhang, W. (2003). Generalized fuzzy rough sets. *Information Sciences, 151*, 263-282.

Acknowledgment

I would like to thank the following institutions for financially contributing towards the writing of this book: Harvard University through the Harvard/SA Fellowship, Wolfson College (University of Cambridge) for a Visiting Fellowship that I was granted for a year and the University of the Witwatersrand.

I also would like to thank my following people: Dr Vincent Fulufhelo Nelwamondo, Busisiwe Vilakazi, Shakir Mohamed, Dalton Lunga and Mussa Abdella for their assistance in developing this manuscript. Furthermore, I thank the following former students for assisting me in running the simulations: George Sssali, Vukosi Marivate, A.K. Mohamed, Jaisheel Mistry, Dr. Brain Leke and Dan Golding. In particular, I thank Dr Ian Kennedy and Lesedi Masisi.

I dedicate this book to the schools that gave me the foundation to always seek excellence in everything I do and these are: Mbilwi Secondary School, Case Western Reserve University, University of Pretoria, University of Cambridge (St. John' College) and Imperial College (London). I also thank my supervisors who played pivotal roles in my education and these are: Professor P.S. Heyns, Dr. H.E.M. Hunt and Professor Philippe de Wilde. I also thank the valuable comments of the three anonymous reviewers of this book.

This book is dedicated to the following people: Nhlonipho Khathutshelo Marwala, Jabulile Vuyiswa Manana, Mrs Reginah Marwala and Mr. Shavhani Marwala.

Professor Tshilidzi Marwala, PhD
Johannesburg

Chapter I
Introduction to Missing Data

ABSTRACT

In this chapter, the traditional missing data imputation issues such as missing data patterns and mechanisms are described. Attention is paid to the best models to deal with particular missing data mechanisms. A review of traditional missing data imputation methods, namely case deletion and prediction rules, is conducted. For case deletion, list-wise and pair-wise deletions are reviewed. In addition, for prediction rules, the imputation techniques such as mean substitution, hot-deck, regression and decision trees are also reviewed. Two missing data examples are studied, namely: the Sudoku puzzle and a mechanical system. The major conclusions drawn from these examples are that there is a need for an accurate model that describes inter-relationships and rules that define the data and that a good optimization method is required for a successful missing data estimation procedure.

INTRODUCTION

Datasets are frequently characterized by their incompleteness. There are a number of reasons why data become missing (Ljung, 1989). These include sensor failures, omitted entries in databases and non-response in questionnaires. In many situations, data collectors put in place firm measures to circumvent any incompleteness in data gathering. Nevertheless, it is unfortunate that despite all these efforts, data incompleteness remains a major problem in data analysis (Beunckens, Sotto, & Molenberghs, 2008; Schafer, 1997; Schafer & Olsen, 1998). The specific reason for the incompleteness of data is usually not known in advance, particularly in engineering problems. Consequently, methods for averting missing data are normally not successful. The absence of complete data then hampers decision-making processes because of the dependence of decisions on *full* information (Stefanakos & Athanassoulis, 2001; Marwala, Chakraverty, & Mahola, 2006).

In one way or another, most scientific, business and economic decisions are related to the information available at the time of making such decisions. For example, many business decisions are dependent on the availability of sales data and other information, while progresses in research are based on discovery of knowledge from various experiments and measured parameters. For example, in aerospace engineering, there are many fault detection mechanisms where the measured data are either partially corrupted or otherwise incomplete (Marwala & Heyns, 1998). In many applications, merely ignoring the incomplete record is not an optimal option because this may lead to biased results in statistical modeling resulting in, for example, a breakdown in machine automation or control. For this reason, it is essential to make decisions based on available data.

Most decision support systems such as the commonly used neural networks, support vector machines and many other computational intelligence techniques are predictive models that take observed data as inputs and predict an outcome (Bishop, 1995; Marwala & Chakraverty, 2006). Such models fail when one or more inputs are missing. Consequently, they cannot be used for decision-making purpose if the data variables are not complete. The end goal of the missing data estimation process is usually to make optimal decisions. To achieve this goal, appropriate approximations to the missing data need to be found. Once the missing variables values have been estimated, then pattern recognition tools for decision-making can be used.

The problem that missing data poses to a decision making process is more apparent in online applications where data have to be used nearly instantly after being obtained. In a situation where some variables are not available, it becomes difficult to carry on with the decision making process thereby stopping the application all together. In essence, the major challenge is that the standard computational intelligence techniques are not able to process input data with missing values. They cannot perform classification or regression if one of the variables is missing. Another major issue that is of concern here is that many missing data imputation techniques developed thus far are mainly suited for survey datasets. In this case, data analysts do have adequate time to study the reasons why data components are missing. However, in many engineering problems, missing data are usually required in real-time. Therefore, there is no time to understand why data components are missing. This calls for a development of robust methods that are effective for missing data estimation regardless of the cause of why the data are missing.

It is important to differentiate between missing data estimation and imputation. Missing data imputation essentially means *dealing* with missing data. This can include either by deleting that set with missing values or by using techniques such as list-wise deletion or estimating the missing values. Therefore, in this chapter missing data estimation is viewed as a sub-set of missing data imputation. It has generally been observed that in many data sets from the social sciences that missing data imputation is a valid way of dealing with missing data. This is because in social science the goal of the statistical analysis is to estimate statistical parameters such as averages and standard deviations. However, in the engineering field, where data are usually needed for manual decision support or automated decision support, the deleting of the entry is usually not an option. Therefore, in most cases, an estimation of the missing values has to be made. For that reason, in engineering problems, the term "missing data estimation" is more valid than the term "missing data imputation". Therefore, this book concentrates on the specific problem of missing data estimation rather than the general term missing data imputation.

This chapter discusses general approaches that have been used to deal with the problem of estimating missing values in an information system. Their advantages and disadvantages are discussed and some missing data imputation theory is discussed. This chapter concludes by discussing two classical missing data problems, which are the Sudoku puzzle and a problem that has been confronting engineers for some time, particularly in the aerospace sector. In discussing these two problems, key issues that are the subject of this book are identified.

THE HISTORICAL EVOLUTION OF MISSING DATA ESTIMATION METHODS

Before the 1970s, missing data problems were resolved through editing, whereby a missing variable was logically inferred from additional data that have been observed. A method for inference from incomplete data was only developed in 1976. Immediately afterwards, Dempster, Laird, and Rubin (1977) invented the Expectation Maximization (EM) algorithm that resulted in the use of the Maximum Likelihood (ML) methods for missing data estimation. Barely a decade later, Little and Rubin (1987) did acknowledge the limitations of *Case Deletion* and *Single Imputations* and then introduced Multiple Imputations (MI). *Multiple Imputations* would not have been achievable without parallel progress in computational power because generally they are computationally expensive (Ho, Silva, & Hogg, 2001; Faris et al., 2002; Hui *et al.*, 2004; Sartori, Salvan, & Thomaseth, 2005).

From the 1990s until today, many methods for estimating missing data have been developed and applied to diverse applications. In recent times, researchers are beginning to study the sensitivity of the missing data estimation results on the outcome of the decision process that use these estimated variables. Research has been conducted to determine new approaches to approximating missing variables. As an example, computational intelligence techniques such as neural networks and optimization techniques such as evolutionary computing are some of the methods that are increasingly used in a number of missing data estimation problems (Dhlamini, Nelwamondo, and Marwala, 2006; Nelwamondo, Mohamed, & Marwala, 2007; Nelwamondo & Marwala, 2007a, 2008). Some of the methods used for missing data imputation include the semi-hidden Markov models (Yu & Kobayashi, 2003), rough sets (Nelwamondo & Marwala, 2007b), fuzzy approaches (Gabrys, 2002; Nelwamondo & Marwala, 2007c), Hopfield neural networks (Wang, 2005) and genetic algorithms (Junninen et al., 2004; Abdella, 2005; Abdella & Marwala, 2005).

MISSING DATA PATTERNS

Missing data can be characterized into a number of patterns, as is shown in Figure 1.1. In this figure, the rows match up to observational units whereas the columns correspond to different data variables. A Univariate pattern is the case where data in an observational unit are only missing in one variable, indicated by Y in Figure 1.1(a). A Monotone pattern occurs when data are missing from a number of data variables and, in addition, missing data entries pursue a particular pattern that can easily be noticed as is shown in Figure 1.1(b). Finally, an arbitrary pattern occurs when data are missing in accordance with some random pattern as is shown in Figure 1.1(c). The pattern that the data follow depends on the manner in which data become missing and on the particular application in question. For example, sensor failure is more likely to follow the pattern in Figure 1(a) or (b) whereas in databases where information is recorded by different individuals, as in medical databases, the pattern in Figure 1.1(c) is the one most likely to be observed.

Wasito and Mirkin (2006) used nearest-neighbors least-squares data imputation methods to analyze three missing data showing different patterns and observed that the nearest neighbors method performed well on data that are missing in a random pattern. Yang and Shoptaw (2005) assessed the impact of missing data in longitudinal studies of smoking cessation. Gad and Ahmed (2006) used a stochastic EM algorithm to analyze longitudinal data with intermittent missing values, which are missing values that are followed by observed values.

Figure 1.1. Patterns of missing data in rectangular databases: (a) univariate pattern, (b) monotone pattern, and (c) arbitrary pattern (Schafer & Graham, 2002)

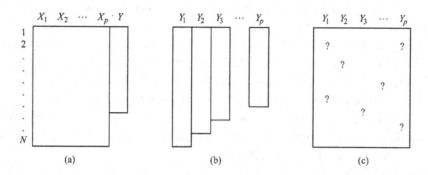

The next section describes what is termed the *mechanisms* of missing data. This is done, because an understanding of the mechanism by which data become missing can assist in choosing which missing data estimation method to use.

MISSING DATA MECHANISMS

As stated before, it is very vital to identify the reason why data are missing. When the explanation is known, a suitable method for missing data imputation may then be chosen or derived, resulting in higher effectiveness and prediction accuracy. In many situations, data collectors may be conscious of such reasons, whereas statisticians and data users may not have that information available to them when they perform the analysis. In such scenarios, data users may have to use other techniques that can assist in data analysis to comprehend how missing data are related to observed data and as a result, possible reasons may be derived.

A variable or a feature in the dataset is viewed as a *mechanism*, if it assists in explaining why other variables are missing or not missing. In datasets collected through surveys, variables that are mechanisms are frequently associated with details that people are embarrassed to divulge. However, such information can often be derived from the data that have been given. As an example, low-income people may be embarrassed to disclose their income but may disclose their highest level of education. Data users may then use the supplied educational information to acquire an insight into the income. If Y is the complete dataset then:

$$Y = \{Y_o, Y_m\} \tag{1.1}$$

Here Y_o is the observed component of Y while Y_m is the missing component of Y. Little and Rubin (1987) differentiate among three missing data mechanisms. These types are Missing at Random (MAR), Missing Completely at Random (MCAR) and Missing Not at Random (MNAR) and these are explained in the next sections. In this chapter, a data mechanism called Missing by Natural Design (MBND) is also introduced.

Missing at Random

Missing at random (MAR) is the case where the explanation for a variable entry being missing is not related to the missing variables themselves. Nonetheless, the cause may be related to other observed variables. MAR is also known as the *ignorable case*. It arises when cases with missing data are different from the observed cases, but the pattern of missing data is predictable from other observed variables and thus can be mathematically written as follows (Little & Rubin, 1987):

$$P(M \mid Y_o, Y_m) = P(M \mid Y_o) \tag{1.2}$$

In equation 1.2, M indicates missing value indicator and is equal to 1 if Y is observed and 0 if Y is missing. Simply put, in MAR the behavior of two observations that share observed variables have identical statistical behavior to the other observations, whether they are observed or not. Donders et al. (2006) observed that in the MAR case, both single and multiple imputations result in an unbiased estimate of missing values. Shih, Quan, and Chang (1994) studied the effect of MAR on the estimated mean. They illustrated this by using the asymptotic bias of the mean based on observed data, while Liang (2008) proposed merging a local linear regression and a local quasi-likelihood method to estimate the parameters and applied the model to study the relationship between virologic and immunologic responses in AIDS clinical trials. For this mechanism, computational intelligence methods, the subject of this book, may be used.

Missing Completely at Random

Missing Completely At Random (MCAR) refers to a condition where the probability of data missing is not related to the values of any other variables, whether missing or observed. In this mechanism, cases with complete data are not discernible from cases with incomplete data and can be mathematically written as follows (Little & Rubin, 1987):

$$P(M \mid Y_o, Y_m) = P(M) \tag{1.3}$$

In this case, neither the missing values, Y_m, nor the observed value, Y_o, can assist in predicting the missing values. Therefore, for MCAR, the analysis of only those observations with complete data offers legitimate inferences. Donders et al. (2006) observed that in the MCAR case, both single and multiple imputations result in unbiased estimates. If data are MCAR, then methods such as the list-wise or pair-wise deletion may be implemented. Otherwise, if not MCAR, missing values should be imputed. In this book, deletion is not an option because the missing values are intended for decision-making purposes whether a specific variable is available or not. Therefore, for this case, computational intelligence techniques that deal with discernibility such as rough sets method (Grzymala-Busse, 1992), which is explained in Chapter V, should be used (Nelwamondo & Marwala, 2008).

Missing Not at Random

Missing Not At Random (MNAR) is the case where the missing data mechanism is related to the missing values themselves. An example of data missing not at random is the case where databases for two cities are merged and one database is short of some features that have been measured in the other database. In this situation, the explanation on why some data are missing can be made clear. Nonetheless, this reason is merely dependent on the same variables that are missing. It cannot be explained in terms of any other variables in the database. Data that cannot be classified into the MAR or MCAR are generally classified into this class. Consequently, for this case, M can be predicted from Y_m and as a result, the expression for $P(M|Y_o, Y_m)$ cannot be computed.

Recently, there have been some different views from various researchers when data are not missing, but are simply not answered e.g. 'not applicable' or 'don't know'. Acork (2005) defines this case as missing by definition of the sub-population. Jansen et al. (2006) studied the sensitivity of MNAR in order to explore how inferences vary under assumptions of MAR, MNAR, and under various models. One approach to MNAR 'missingness' is to impute values based on data otherwise external to the research design and then use the case hot-deck models or likelihood methods.

Missing By Design

Missing by natural design (MBND) is a missing data mechanism where the values are missing because they cannot be measured physically (or naturally). This is a relatively new missing data mechanism and includes problems in mechanical engineering such as rotational degrees of freedom, which even though they are useful are almost impossible to measure. In the MBND mechanism, what is normally done is to model the missing value mathematically. This is usually possible if the physical characteristics of the missing values can be described. One example considered in this chapter includes the MBND mechanism.

A REVIEW OF COMMON MISSING DATA IMPUTATION TECHNIQUES

Many methods have been used because of their accessibility and computational efficiency. These techniques include list deletion, pair-wise deletion, simple-rule prediction, mean substitution, hot-deck imputation, cold-deck imputation, imputation using regression, regression-based nearest neighbor hot-decking, tree-based imputation and stochastic imputation. A concise account of these methods follows in the next sub-sections.

Case Deletion

The most used method to deal with missing data is merely to leave out those cases with missing data and to run analyses on what remains. There are two popular methods that can be used for deleting data with missing entries, namely: list-wise deletion and pair-wise deletion. Zhu et al. (2001) proposed case-deletion measures for models with incomplete data measures for evaluating the impact of an observation in complex models with real or assumed missing data corresponding to latent random variables. On the other hand, Velilla (1993) proposed an approximation for evaluating the impact of deleting an observation on the eigenvalues of the correlation matrix of a multiple linear regression model.

List-Wise Deletion

List-wise deletion is a technique whereby only the complete data are retained (Tsikritis, 2005). In the cases of missing entries in the respective observations, the entire observations are eliminated from the database. The major problem with this technique is the amount of information lost in the process. A further weakness with list-wise deletion is that it assumes that the observations with missing values are not significant and, therefore, can be ignored. As a consequence, the condition of a system being monitored, such as in aerospace engineering (Marwala, 2001), cannot be detected should one or more sensors fail. The impact of list-wise deletion can be seen in Tables 1.1 and 1.2 where this process leads to a loss of half of the observations.

Pair-Wise Deletion

Pair-wise deletion uses the incomplete record only if the missing variable is not needed in the calculation under consideration. In this case, the record with missing values is only used in the analyses that do not require the missing variable. Even though this technique appears to be better than list-wise deletion, it has been criticized by researchers such as Allison (2000) who noted that only if the data are MCAR, does the pair-wise deletion method give biased estimates.

Prediction Rules

There are a variety of analyses that are needed prior to performing the missing data imputation. A great deal of information, which varies from one application to another, is required. For example, in some applications what must be identified is the significance of the missing data, whereas in some applications the basis, for the data being missing, is required. This information may be gained by examining the dynamics of all the other records relative to the missing attribute.

Table 1.1. (left) An example of a table with missing values. The question mark denotes the missing values and D is the decision attribute. Table 1.2 (right) Data from Table 1 after cases with missing data have been deleted.

Instances	x_1	x_2	x_3	D
1	1	?	0.2	B
2	1	2	0.3	A
3	0	1	0.3	B
4	0	?	0.3	B
5	0	3	0.4	A
6	0	?	0.2	B
7	1	4	?	A
8	1	4	0.3	A

Instances	x_1	x_2	x_3	D
2	1	2	0.3	A
3	0	1	0.3	B
5	0	3	0.4	A
8	1	4	0.3	A

In some other cases, it is vital to learn the relationship between the missing variables and the observed ones while another approach is to search the records for similar cases and study how the missing attributes behave given the other observed data. Prediction rule based approaches to handle missing data rely mostly on very limited variables. Approaches to predicting missing data in this class can be classified into two groups, namely: simple-rule prediction, i.e., single imputations and multivariate rule prediction i.e., multiple imputations (Harel, 2007; Jerome, 2008).

Mean Substitution

When mean substitution is implemented, the missing entry is simply substituted by the historical mean of the observed data. This substitution, nevertheless, results in biased variances, which intensifies with an increase in the number of missing data records.

Hot / Cold-Deck Imputation

Hot-deck imputation is a method in which missing data are substituted with data from similar cases that have been observed from the same dataset. Hot-deck imputation has two major steps:

- Records are divided into classes using techniques such as the nearest neighbor clustering techniques.
- Incomplete entries are then imputed with values that fall within the same class.

The benefit of hot-deck imputation is that to estimate missing values, there is no need for robust model assumptions. Hot-deck imputations are also chosen for their theoretical simplicity. However, implementing them is difficult because of the complexity of characterizing the concept of similarity due to its subjectivity. In addition, in many cases, more than one value will be found suitable for imputing the missing one. A criterion for picking the most suitable value must be specified. Some approaches merely choose one value randomly or employ an average of the possible values. *Cold-deck imputation* is a case where missing values are substituted by constant values that are obtained from external sources such as previous observation or a modal value.

Linear Regression Imputation

Linear regression imputations substitute the missing values with values predicted from regression models constructed from the available data. Choosing a suitable regression imputation method is very reliant on the missing variable. Some regression imputation models begin by calculating estimates of the mean vector and co-variance matrix of the data from the sub-matrix from data with no missing values. Linear regression of the missing variables is then applied, based on the observed data. Missing data are then replaced by the values obtained from the linear regression. This method, nevertheless, underrates the variance and the covariance of the data.

Regression-Based Nearest Neighbor

The *regression-based nearest neighbor* hot-decking method (Huang & Zhu, 2002) is a mixture of the nearest neighbor hot-decking technique and the multivariate regression model. If the missing data are on a continuous covariate, then the missing value is imputed as the mean of the covariate values of the nearest neighbors, otherwise the majority of the 'votes' establishes the class of missing observation on the basis of nearest available data.

Decision Tree Based Imputation

Decision tree based imputation methods are classified into two groups: (1) the classification tree models, and (2) the regression tree models. In the classification tree models, it is pre-supposed that the response variable is *categorical* while regression tree models presume that the response variable can be represented using *numeric* values. Imputing the missing values using tree-based methods entails measuring the response variables and the independent variables.

Stochastic Imputation

Stochastic imputation is a type of regression imputation that replicates the uncertainty of the predicted values. It can be viewed as an extension of the simple-rule prediction. The distinction between this class and the preceding one is that, in place of using one variable, the multivariate rule prediction employs more than one variable.

Neural networks imputation can be viewed as one example of this class (Nelwamondo, 2008). This technique is also known as Multiple Imputation (MI), and was introduced by Rubin (1987). It merges statistical techniques by producing a maximum-likelihood based covariance matrix and a vector of means. Multiple Imputations involve drawing missing values from the posterior distribution of the missing values, given the observed values and is attained by averaging the posterior distribution for the complete data over the predictive distribution of the missing data. MI is related to hot-deck imputations but its advantage over hot-deck imputation is that it creates more than one imputation value. Its drawback is that it is computationally expensive.

TOWARDS A GENERALIZED MISSING DATA ESTIMATION METHOD: LESSONS FROM SUDOKU PUZZLE

Sudoku is a puzzle played on a partially filled 9x9 grid that became popular in Japan in 1986 and became internationally popular in 2005 (Perez & Marwala, 2008). The goal of this puzzle is to fill in the missing values in the grid. Therefore, this puzzle is essentially a missing data imputation problem.

A Sudoku puzzle is shown in Figure 1.2 and its solution is in Figure 1.3. Missing data in the Sudoku puzzle are completed by following these simple rules:

1. Only integers from 1 to 9 may be used to complete the puzzle.
2. Each cell has at least one value that satisfies the rules.
3. Each row must contain all the integers from 1 to 9.
4. Each column must contain all the integers from 1 to 9.
5. Each 3x3 sub-grid must contain all the integers from 1 to 9.

Figure 1.2. Sudoku puzzle

1							2	
		8			9		3	7
7			5	3			8	
	8			7	3		5	4
		6	4		2	7		
9	7		8	5			1	
	1			8	7			9
3	4		6			8		
8								1

This problem has been solved as a constrained problem by Simonis (2005). In this chapter, it is successfully solved using a constrained optimization approach. In Figure 1.2, there are 47 missing values which need to be estimated given the 34 values that have been observed. Therefore, the 47 missing values can thus be treated as design variables in an optimization problem.

Building a Model Describing the Inter-Relationships

The first general step to do is to build a model that describes all the inter-relationships in the Sudoku puzzle. From rules 1, 2 and 3, the following equation can be written as follows:

$$s_i = \sum_{j=1}^{9} x_{ij} = 45$$

(1.4)

In equation 1.4, s_i is the sum of entries x_{ij} in row i. From rules 1, 2 and 4 the following equation can be written as follows:

$$s_j = \sum_{i=1}^{9} x_{ij} = 45$$

(1.5)

In equation 1.5, s_j is the sum of entries x_{ij} in column j. From rules 1, 2 and 5, the following equation can be written as follows:

$$g_{nm} = \sum_{i=3n-2}^{3n} \sum_{j=3m-2}^{3m} x_{ij} = 45$$

(1.6)

In equation 1.6, g_{nm} is the sum of each 3x3 grid in 3x3 grid row n and column m and n and m are 1, 2 and 3.

Figure 1.3. Sudoku puzzle solution

1	5	3	7	6	8	9	4	2
4	6	8	1	2	9	5	3	7
7	2	9	5	3	4	1	8	6
2	8	1	9	7	3	6	5	4
5	3	6	4	1	2	7	9	8
9	7	4	8	5	6	2	1	3
6	1	5	3	8	7	4	2	9
3	4	2	6	9	1	8	7	5
8	9	7	2	4	5	3	6	1

Posing the Inter-Relationships as Optimization Problems

By re-arranging equations 1.4 to 1.6, squaring the results and summing all of them the following objective is obtained:

$$E = \sum_{i=1}^{9}(s_i - 45)^2 + \sum_{j=1}^{9}(s_j - 45)^2 + \sum_{n=1}^{3}\sum_{m=1}^{3}(g_{nm} - 45)^2$$

(1.7)

The fact that each row, column and 3x3 grid should have all integers from 1 to 9 can be enforced by ensuring that each of the design variables can only assume certain values that do not violate this rule. This in essence represents the bounds in an integer programming problem. The solution to the Sudoku puzzle is obtained when equation 1.7 is set equal to zero and this is successfully solved in this chapter using genetic algorithm (Goldberg, 1989). More details on genetic algorithms will be discussed in later chapters. In solving this puzzle, simple crossover, binary mutation and roulette selection are used (Goldberg, 1989). The crossover rate is set to 60% and the mutation rate is set to 0.3%. The population size is set to 500 and the genetic algorithm is run for 1000 generations. This problem has an exact global solution defined by the error in equation 1.7 being equal to zero. In other words, the solution to equation 1.7 should over-fit the observed values, a situation that is not necessarily desirable for real world problems where the observed data are measured and, therefore, are characterized by measurement errors. Finding a global optimum solution for a case where the observed values are measured means that the model is fitting the noise in the data as is discussed in the next example. Using this procedure the solution in Figure 1.3 is obtained.

The missing data are estimated correctly despite the fact that we do not assume a missing data mechanism. It can thus be concluded that if rules that govern inter-relationships amongst the variables in the system are identified, then the missing data estimation process can be conducted irrespective of the reason or the mechanism in which data are missing. Therefore, the condition that the rules that govern inter-relationships are known essentially renders the missing data mechanism irrelevant. This is a powerful factor because in mechanical systems where missing data are required to enable automated decisions, which are usually online and instantaneous, there is simply no time to investigate why the data are missing. In this situation, regardless of the missing data mechanism, an estimation of the missing data has to be made to enable a decision to be made.

In this book, methods that seek to identify rules that govern inter-relationships amongst the variables are proposed, developed and tested. The methods are based on computational intelligence. It is demonstrated that functionally correct estimates of missing data are obtained regardless of the knowledge of the missing data mechanism.

Lessons from Sudoku

From Sudoku puzzle the following lessons are learned:

- For a successful missing data estimation process, a correct model must describe the inter-relationships that exist amongst the data variables and define the rules that govern the data. This, therefore, requires a robust model selection process.
- To estimate the missing data, the correct estimated data are those that obey all the inter-relationships that exist amongst the data and the rules that govern the data. To ensure that the estimated values obey the inter-relationships in the data, the optimization method must treat the missing data as design variables and define the inter-relationships amongst the data and rules that govern the data for the objective to be reached. Therefore, the missing data estimation problem is fundamentally an optimization problem where the objective is to estimate missing data by ensuring that the rules and inter-relationships governing the data are maintained.
- If the rules that govern the inter-relationships are correctly identified, then the knowledge of the missing data mechanism is irrelevant for missing data imputation.
- Missing data mechanisms are valid for missing data *imputation* and not necessarily missing data *estimation*.
- It is vital to have a good idea of the bounds that govern the variables to be estimated to aid the optimization process.
- In the Sudoku puzzle example, the concept of a correct solution exists and can be found. Therefore, the issue of global optimization becomes very significant. This is because there is no contradiction in the observed data that can come from problems such as measurement errors, incorrect measurements, etc. This, however, is not the case for practical problems where the measurements fundamentally contain measurement errors.

MISSING DATA ESTIMATION IN MECHANICAL SYSTEMS

In this section a missing data mechanism, MBND, is illustrated. In this mechanism, data variable is missing because it is naturally un-measurable. Despite this fact, these data variable is still essential for the analysis of the problem in question. Many mechanical structures may be expressed using equations of motion as follows (Ewins, 1986):

$$[M]\{\ddot{x}\} + [C]\{\dot{x}\} + [K]\{x\} = \{F\} \qquad (1.8)$$

where $[M]$ is the mass matrix, $\{\ddot{x}\}$, is the acceleration vector, $[C]$ is the damping matrix, $\{\dot{x}\}$ is the velocity vector, $[K]$ is the stiffness matrix, $\{x\}$ is the displacement vector and $\{F\}$ is the force vector. Equation 1.8 can be transformed into a modal domain and, therefore, can be rewritten as follows (Ewins, 1986; Marwala & Hunt, 1999):

$$\left(-\omega_i^2[M] + j\omega[C] + [K]\right)\{\phi\}_i = \{0\} \tag{1.9}$$

In equation 1.9, $\{\phi\}_i$ is the ith mode's shape vector and ω_i is its corresponding natural frequency. The parameter ω_i and some components of $\{\phi\}_i$ can be measured. In structural mechanics, it is easier to measure the translational components of $\{\phi\}_i$ but extremely difficult to measure its rotational components. In this chapter, it is assumed that the un-measurable data are rotational degrees of freedom. To perform tasks such as model-based condition monitoring of structures complete (Doebling et al., 1996; Marwala & Chakraverty, 2006), mode shapes are required. It should be noted that in equation 1.9, the mass and stiffness matrices can be obtained from a finite element model (Bathe, 1982) and the damping matrix can be derived from the mass and stiffness matrices for lightly damped structures (Maia & Silva, 1997). Therefore, equation 1.9 can be broken down into two components (Guyan, 1965), i.e., the measured and un-measured components as follows:

$$\left(-\omega_i^2 \begin{bmatrix} M_{mm} & M_{um} \\ M_{mu} & M_{uu} \end{bmatrix} + j\omega \begin{bmatrix} C_{mm} & C_{um} \\ C_{mu} & C_{uu} \end{bmatrix} + \begin{bmatrix} K_{mm} & K_{um} \\ K_{mu} & K_{uu} \end{bmatrix}\right)\begin{Bmatrix} \phi_m \\ \phi_u \end{Bmatrix}_i = \{0\} \tag{1.10}$$

In equation 1.10, given the measured mode shapes $\{\phi_m\}$ and their respective natural frequencies as well as the finite element model, then the missing un-measurable modes $\{\phi_u\}$ can be estimated by reformulating the problem as a least squares optimization problem as follows:

$$E = \sum_{j=1}^{Q} \sum_{i=1}^{N} \left\{ \left(-\omega_i^2 \begin{bmatrix} M_{mm} & M_{um} \\ M_{mu} & M_{uu} \end{bmatrix} + j\omega \begin{bmatrix} C_{mm} & C_{um} \\ C_{mu} & C_{uu} \end{bmatrix} + \begin{bmatrix} K_{mm} & K_{um} \\ K_{mu} & K_{uu} \end{bmatrix}\right)\begin{Bmatrix} \phi_m \\ \phi_u \end{Bmatrix}_i \right\}^2 \tag{1.11}$$

In equation 1.11, Q is the size of the mode shape vectors and N is the total number of modes under consideration.

This missing data estimation method was tested on an asymmetrical structure whose details are described in Marwala and Heyns (1998). Measurements of modes $\{\phi_m\}$ and their respective natural frequencies were conducted. Using these measurements and the mass, stiffness and damping matrices from the finite element model, the unmeasured modes are calculated using a genetic algorithm. Then the asymmetric structure was damaged at three locations as described in Marwala and Heyns (1998) and the complete mode shapes and finite element model are used to locate these damages (Marwala and Heyns, 1998). The method is able to locate damages at these three locations and, therefore, implies that the missing mode shapes are estimated sufficiently accurately for the damage identification exercise.

From this example several lessons can be drawn:

- In missing data estimation problems, in some cases, the only way in which we can ever know that the missing estimates are correct is when these data can perform their intended task sufficiently well. In this example, the task is for damage to be identified.

- It is important to select models that can relate measurements to missing values. In this example, the model was derived using Newton's Laws of motion (Ewins, 1986). This book focuses more on the identification of such models instead of selecting models that have been derived from the Laws of Physics. As is the case in this chapter, computational intelligence is used to achieve this task.
- This model must be robust even in the presence of uncertainties in the measurements, which is a fundamental property of any measured data.
- As the model describing the data may be adapting because the system in question might be aging etc., there may be a great need to evolve the model so that it remains a fair reflection of the system.

CHALLENGES TO MISSING DATA IMPUTATION

Most of the methods described above do not provide an optimal solution to the problem of missing data. Nevertheless, they result in biases excluding few exceptional and specialized cases. List-wise deletion method has been found to perform well for the MCAR mechanism and especially for large databases. Nonetheless, establishing that the dataset observe the MCAR mechanism is not easy and MCAR is frequently unreasonable in rectangular survey data. The use of the mean substitution method cannot be justified for any condition as it distorts the variance. The next sub-sections describe a number of issues relevant to missing data estimation.

Several problems arise during missing data imputation. These include knowing which method to use, given a particular imputation problem. For example, does the choice of a data modeling method matter? A literature review demonstrates that the Expectation Maximization (EM) algorithm is still widely used (Shinozaki & Ostendorf, 2008; Stolkin et al., 2008) and therefore, a relevant question that may be posed is whether the EM algorithm is the state of art. Recently, systems that merge computational intelligence with optimization techniques have been reported in the literature (Abdella, 2005; Abdella & Marwala, 2006) particularly for cases that are highly nonlinear.

The question that needs to be asked is: Which computational intelligence approach is ideal for missing data estimation? Furthermore, which optimization technique is best for missing data estimation? To avoid some of these questions, researchers such as Twala (2005) have adopted a multiple models approach, particularly in cases where it is not clear which approach to use. The question then becomes: What is the optimal manner in which different models can be combined for effective missing data estimation solution? If models will be deployed and used over an extended period of time, how do we design them with inherent adaptive properties so that they remain valid for the period of use?

The models built for missing data estimation normally assume the traditional manner of quantifying uncertainty using probability models. Recently, possibility and plausibility models have been developed using fuzzy logic and rough sets. Therefore, how do we use these developments to tackle difficult problems that contain subjective observations? How do we exploit statistical machine learning tools such as support vector machines to deal with large multivariate data? How do we merge computational intelligence with decision-trees to decrease the search space in pursuit for the missing values? On designing our models, do we use the Bayesian or the maximum-likelihood approaches? How do we know that our missing data estimates are correct? What is the impact of these on the decision making process? Do the new missing data approaches offer us new insights into classification? How do concepts such as heteroskedasticity and dynamic programming assist us in designing better missing data estimation methods? Many of these questions will be answered in this book.

CONCLUSION

In this chapter, various missing data imputation methods, missing data patterns and mechanism were reviewed. Furthermore, conventional missing data imputation methods were also studied. Two missing data examples were then studied namely: the Sudoku puzzle and a mechanical system. For a successful missing data estimation procedure, it is concluded that an accurate model is required which describes the rules that define inter-relationships in the variables. In addition, if a model that characterizes inter-relationships in the variables is identified, then the knowledge of the missing data mechanism becomes irrelevant. Furthermore, a good optimization method, which has global optimum convergence characteristics, is also required.

REFERENCES

Abdella, M. (2005). *The use of genetic algorithms and neural networks to approximate missing data in database.* Unpublished master's thesis, University of the Witwatersrand, Johannesburg.

Abdella, M., & Marwala, T. (2005). Treatment of missing data using neural networks. In *Proceedings of the IEEE International Joint Conference on Neural Networks*, (pp. 598-603).

Abdella, M., & Marwala, T. (2006). The use of genetic algorithms and neural networks to approximate missing data in database. *Computing and Informatics, 24,* 1001–1013.

Acork, A. C. (2005). Working with missing values. *Journal of Marriage and Family, 67,* 1012–1028.

Allison, P. (2000). Multiple imputation for missing data: A cautionary tale. *Sociological Methods and Research, 28,* 301-309.

Bathe, K. J. (1982). *Finite element procedures in engineering analysis.* Englewood Cliffs: Prentice-Hall, Inc.

Beunckens, C., Sotto, C., & Molenberghs, G. (2008). A simulation study comparing weighted estimating equations with multiple imputation based estimating equations for longitudinal binary data. *Computational Statistics & Data Analysis, 52*(3), 1533-1548.

Bishop, C. M. (1995). *Neural networks for pattern recognition.* Oxford: Oxford University Press.

Dempster, A. P., Laird, N. M., & Rubin, D. B. (1977). Maximum likelihood for incomplete data via the EM algorithm. *Journal of Royal Statistic Society, B39,* 1–38.

Dhlamini, S. M., Nelwamondo, F. V., & Marwala, T. (2006). Condition monitoring of HV bushings in the presence of missing data using evolutionary computing. *Transactions on Power Systems, 1*(2), 280–287.

Doebling, S. W., Farrar, C. R., Prime, M. B., & Shevitz, D. W. (1996). *Damage identification and health monitoring of structural and mechanical systems from changes in their vibration characteristics: A literature review* (Los Alamos Tech. Rep. LA-13070-MS). New Mexico, USA: Los Alamos National Laboratory.

Donders, R. T., van der Heijden, G. J. M. G., Stijnen, T., & Moons, K. G. M. (2006). Review: A gentle introduction to imputation of missing values. *Journal of Clinical Epidemiology, 59*(10), 1087-1091.

Ewins, D. J. (1986). *Modal testing: Theory and practice.* Letchworth, UK: Research Studies Press.

Faris, P. D., Ghali, W. A., Brant, R., Norris, C. M., Galbraith, P. D., & Knudtson, M. L. (2002). Multiple imputation versus data enhancement for dealing with missing data in observational health care outcome analyses. *Journal of Clinical Epidemiology, 55*(2), 184-191.

Gabrys, B. (2002). Neuro-fuzzy approach to processing inputs with missing values in pattern recognition problems. *International Journal of Approximate Reasoning, (30),* 149–179.

Gad, A. M., & Ahmed, A. S. (2006). Analysis of longitudinal data with intermittent missing values using the stochastic EM algorithm. *Computational Statistics & Data Analysis, 50*(10), 2702-2714.

Goldberg, D. (1989). *Genetic algorithms in search, optimization, and machine learning.* Reading, MA: Addison-Wesley.

Grzymala-Busse, J. W. (1992). *LERS: A system for learning from examples based on rough sets. Handbook of Applications and Advances of the Rough Sets Theory.* Dordrecht: Kluwer Academic Publishers.

Guyan, R. J. (1965). Reduction of stiffness and mass matrices. *American Institute of Aeronautics and Astronautics Journal, 3*(2), 380.

Harel, O. (2007). Inferences on missing information under multiple imputation and two-stage multiple imputation. *Statistical Methodology, 4*(1), 75-89.

Ho, P., Silva, M. C. M., & Hogg, T. A. (2001). Multiple imputation and maximum likelihood principal component analysis of incomplete multivariate data from a study of the ageing of port. *Chemometrics and Intelligent Laboratory Systems, 55*(1-2), 1-11.

Huang, X., & Zhu, Q. (2002). A pseudo-nearest-neighbor approach for missing data recovery on Gaussian random data sets. *Pattern Recognition Letters, 23,* 1613–1622.

Hui, D., Wan, S., Su, B., Katul, G., Monson, R., & Luo, Y. (2004). Gap-filling missing data in eddy covariance measurements using multiple imputation (MI) for annual estimations. *Agricultural and Forest Meteorology, 121*(1-2), 93-111.

Jansen, I., Hens, N., Molenberghs, G., Aerts, M., Verbeke, G., & Kenward, M. G. (2006). The nature of sensitivity in monotone missing not at random model. *Computational Statistics & Data Analysis, 50*(3), 830-858.

Jerome, P. (2008). Reiter Selecting the number of imputed datasets when using multiple imputation for missing data and disclosure limitation. *Statistics & Probability Letters, 78*(1), 15-20.

Junninen, H., Niska, H., Tuppurainen, K., Ruuskanen, J., & Kolehmainen, M. (2004). Methods for imputation of missing values in air quality data sets. *Atmospheric Environment, 38*(18), 2895-2907.

Liang, H. (2008). Generalized partially linear models with missing covariates. *Journal of Multivariate Analysis, 99*(5), 880-895.

Little, R. J. A., & Rubin, D. B. (1987). *Statistical analysis with missing data.* New York: Wiley.

Ljung, G. M. (1989). A note on the estimation of missing values in time series. *Communications in Statistics, 18,* 459–465.

Maia, N. M. M., & Silva, J. M. M. (1997). *Theoretical and experimental modal analysis.* Letchworth, U.K: Research Studies Press.

Marwala, T. (2001). Scaled conjugate gradient and Bayesian training of neural networks for fault identification in cylinders. *Computers and Structures, 79/32,* 2793-2803.

Marwala, T., & Chakraverty, S., (2006). Fault classification in structures with incomplete measured data using autoassociative neural networks and genetic algorithm. *Current Science, 90*(4), 542-548.

Marwala, T., Chakraverty, S., & Mahola, U. (2006). Fault classification using multi-layer perceptrons and support vector machines. *International Journal of Engineering Simulation, 7*(1), 29-35.

Marwala, T., & Heyns, P. S. (1998). A multiple criterion method for detecting damage on structures. *American Institute of Aeronautics and Astronautics Journal, 195,* 1494-1501.

Marwala, T., & Hunt, H. E. M. (1999). Fault identification using finite element models and neural networks. *Mechanical Systems and Signal Processing, 13,* 475-490.

Nelwamondo, F. V. (2008). *Computational intelligence techniques for missing data imputation.* Unpublished doctoral dissertation, University of the Witwatersrand, Johannesburg.

Nelwamondo, F. V., & Marwala, T. (2007a). Handling missing data from heteroskedastic and non-stationary data. *Lecture Notes in Computer Science, 4491*(1), 1297-1306.

Nelwamondo, F. V., & Marwala, T. (2007b). Rough set theory for the treatment of incomplete data. In *Proceedings of the IEEE Conference on Fuzzy Systems,* London, UK 338-343

Nelwamondo, F. V., & Marwala, T. (2007c). Fuzzy ARTMAP and neural network approach to online processing of inputs with missing values. *SAIEE Africa Research Journal, 98*(2), 45-51.

Nelwamondo, F. V., & Marwala, T. (2008). Techniques for handling missing data: applications to online condition monitoring. *International Journal of Innovative Computing, Information and Control, 4*(6), 1507-1526.

Nelwamondo, F.V., Mohamed, S., & Marwala, T. (2007). Missing data: A comparison of neural network and expectation maximization techniques. *Current Science, 93*(11), 1514-1521.

Perez, M., & Marwala, T. (2008). Stochastic optimization approaches for solving Sudoku. *ArXiv: 0805.0697.*

Rubin, D. B. (1987). *Multiple imputation for nonresponse in surveys*. New York: Wiley.

Sartori, N., Salvan, A., & Thomaseth, K. (2005). Multiple imputation of missing values in a cancer mortality analysis with estimated exposure dose. *Computational Statistics & Data Analysis, 49 (3)*, 937-953.

Schafer, J. L. (1997). *Analysis of incomplete multivariate data*. New York: Chapman & Hall.

Schafer, J. L., & Graham, J. W. (2002). Missing data: Our view of the state of the art. *Psychological Methods, 7*(2), 147–177.

Schafer, J. L., & Olsen, M. K. (1998). Multiple imputation for multivariate missing-data problems: A data analysts perspective. *Multivariate Behavioral Research, 33*(4), 545–571.

Shih, W. J., Quan, H., & Chang, M. N. (1994). Estimation of the mean when data contain non-ignorable missing values from a random effects model. *Statistics & Probability Letters, 19*(3), 249-257.

Shinozaki, T., & Ostendorf, M. (2008). Cross-validation and aggregated EM training for robust parameter estimation. *Computer Speech & Language, 22*(2), 185-195.

Simonis, H. (2005). Sudoku as a constraint problem. In *Proceedings of the CP Workshop on Modeling and Reformulating Constraint Satisfaction Problems* (pp. 13-27).

Stefanakos, C., & Athanassoulis, G. A. (2001). A unified methodology for analysis, completion and simulation of non-stationary time series with missing values, with application to wave data. *Applied Ocean Research, 23,* 207–220.

Stolkin, R., Greig, A., Hodgetts, M., & Gilby, J. (2008). An EM/E-MRF algorithm for adaptive model based tracking in extremely poor visibility. *Image and Vision Computing, 26*(4), 480-495.

Tsikriktsis, N. (2005). A review of techniques for treating missing data in OM survey research. *Journal of Operations Management, 24*(1), 53-62.

Twala, B. E. T. H. (2005). *Effective techniques for handling incomplete data using decision trees*. Unpublished doctoral dissertation, The Open University, UK.

Velilla, S. (1993). On eigenvalues, case deletion and extremes in regression. *Computational Statistics & Data Analysis, 16*(3), 299-309.

Wang, S. (2005). Classification with incomplete survey data: a Hopfield neural network approach. *Computers & Operations Research, 24,* 53–62.

Wasito, I., & Mirkin, B. (2006). Nearest neighbors in least-squares data imputation algorithms with different missing patterns. *Computational Statistics & Data Analysis, 50*(4), 926-949.

Yang, X., & Shoptaw, S. (2005). Assessing missing data assumptions in longitudinal studies: an example using a smoking cessation trial. *Drug and Alcohol Dependence, 77*(3), 213-225.

Yu, S., & Kobayashi, H. (2003). A hidden semi-Markov model with missing data and multiple observation sequences for mobility tracking. *Signal Processing, 83*(2), 235–250.

Zhu, H., Lee, S-Y., Wei, B-C., & Zhou, J. (2001). Case-deletion measures for models with incomplete data. *Biometrika, 88*(3), 727-737.

Chapter II
Estimation of Missing Data Using Neural Networks and Genetic Algorithms

ABSTRACT

Missing data creates various problems in analyzing and processing data in databases. In this chapter, a method aimed at approximating missing data in a database that uses a combination of genetic algorithms and neural networks is introduced. The presented method uses genetic algorithms to minimize an error function derived from an auto-associative neural network. The Multi-Layer Perceptron (MLP) and Radial Basis Function (RBF) networks are employed to form an auto-associative network. An investigation is undertaken into using the method to predict missing data accurately as the number of missing cases within a single record increases. It is observed that there is no significant reduction in the accuracy of the results as the number of missing cases in a single record increases. It is also found that results obtained from using the MLP are better than from the RBF for the data used.

INTRODUCTION

Inferences made from available data for many applications depend on the completeness and the quality of the data being used in the analysis. Therefore, inferences made from complete data are most likely to be more accurate than those made from incomplete data are. However, there are time critical applications that necessitate estimation or approximation of the values of some missing variables, which have to be supplied with the values of other corresponding variables. Such situations may appear in a system that uses a number of instruments, where one or more of the sensors used in the system fail. In such a situation, the values from the missing sensor have to be estimated within a short time and with great precision by taking into account the values of other sensors in the system. In such situations, an approximation for the missing values involves estimating missing values while taking into account the inter-relationships that exists between the values of other variables.

Missing data in a database may arise for various reasons. They can arise from data entry errors, respondents' non-response to some items in the data collection process, failure of instruments and other reasons. In Table 2.1 presents a database consisting of five variables, namely x_1, x_2, x_3, x_4, and x_5, where the values of some variables are missing. If we assume that the observations from some variables in various records are not available, it becomes critical that a technique be formulated for estimating the values for the missing entries.

This is possible if there are techniques to approximate the missing data from the observed values that exploit the inter-relationships existing between the variables in the database. Therefore, the aim of this chapter is to use neural networks and genetic algorithms to approximate missing data in such situations. In this chapter, neural networks are used to identify the inter-relationships that exist amongst the variables while a genetic algorithm is used to identify the missing variables, given the inter-relationships that exist amongst the variables that have been identified by the neural network.

This chapter introduces the combination of neural networks and genetic algorithms for missing data estimation. An issue of the criticality of identifying the correct model that describes inter-relationships and rules that govern the data is solved by using the scaled conjugate gradient optimization method and model selection process through cross-validation (Bishop, 1995).

BACKGROUND

Missing data generates various problems in numerous applications that depend on access to accurate and complete data.

Missing Data

Methods that handle missing data have been areas of research in statistics, mathematics and other disciplines (Yuan, 2000; Allison, 2000; Rubin, 1978; Roth, 1994; Abdella, 2005; Abdella & Marwala, 2005; Nelwamondo & Marwala, 2007a; Nelwamondo & Marwala, 2008). The reasonable way to handle missing data depends upon how data points become missing.

As described earlier, according to Little and Rubin (1987), there are three types of missing data mechanisms. These are MCAR, MAR and the non-ignorable case, which is also called MNAR. MCAR arises if the probability of missing value for variable X is unrelated to the value X itself or to any other variable in the data set. This refers to data where the absence of data does not depend on the variable of interest or any other variable in the dataset (Rubin, 1978). MAR arises if the probability of missing

Table 2.1. Database with missing values

x_1	x_2	x_3	x_4	x_5
25	3.5	?	5000	-3.5
?	6.9	5.6	?	0.5
45	3.6	9.5	1500	46.5
27	9.7	?	3000	?

data on a particular variable X depends on other variables, but not on X itself and MNAR arises if the probability of missing data X is related to the value of X itself, even if the other variables in the analysis are controlled (Allison, 2000; Nelwamondo and Marwala, 2007b).

Depending on the mechanism of missing data, currently various methods are being used to handle missing data. For detailed discussions on the various missing data imputation methods used to handle missing data refer to Hu, Savucci, and Choen (1998); Mohamed and Marwala (2005); Leke and Marwala (2006); Mohamed, Nelwamondo, and Marwala (2007); Nelwamondo and Marwala (2007c); Nelwamondo, Mohamed, and Marwala (2007) as well as Nelwamondo (2008).

In this chapter, the first task is to capture the model that characterizes the rules governing the inter-relationships in the data. To identify these rules, auto-associative neural networks are used. Once this model is identified, it is used to estimate missing data given the observed data, and to achieve this task an optimization method is used. In this chapter, genetic algorithms are used. It should be noted here, that it is argued in this book, that once the model that defines inter-relationships in the data variables is identified, and then used with an optimization method, then it does not matter what missing data mechanism is responsible for the missing data. An accurate estimation of these missing data values can still be made. As indicated before, this is advantageous for an online system that requires fast estimation of missing data to continue to function.

Neural Networks

A neural network is an information-processing paradigm that is inspired by the way biological nervous systems, like the human brain, process information. It is a computer-based model of the way the brain performs a particular function of interest (Haykin, 1999). It is an exceptionally powerful instrument that has found successful application in mechanical engineering (Marwala & Hunt, 1999; Vilakazi & Marwala, 2007), civil engineering (Marwala, 2000; Marwala, 2001a), aerospace engineering (Marwala, 2001b; Marwala, 2003), biomedical engineering (Mohamed, Rubin, & Marwala, 2006; Mohamed, Tettey, & Marwala, 2006; Marwala, 2007a), finance (Patel & Marwala, 2006) and political science (Lagazio & Marwala, 2005). In this chapter, neural networks are viewed as generalized regression models that can model the data, which can be either linear or nonlinear. A neural network consists of four main parts (Haykin, 1999) and these are:

1. The processing units u_j, where each u_j has a certain activation level $a_j(t)$ at any point in time.
2. Weighted inter-connections between various processing units. These inter-connections determine how the activation of one unit leads to the input for another unit.
3. An activation rule, which acts on the set of input signals at a unit to produce a new output signal.
4. The learning rule that specifies how to adjust the weights for a given input / output pair (Haykin, 1999; Freeman & Skapura, 1991).

Due to their ability to gain meaning from complicated data, neural networks are employed to extract patterns and detect trends that are too complex to be noticed by many other computer techniques (Hassoun, 1995). A trained neural network can be considered as an expert in the category of information it has been given to analyze (Yoon & Peterson, 1990; Lunga & Marwala, 2006). This expert can then be used to provide predictions given new situations. Because of their ability to adapt to a nonlinear data, neural networks have been used to model various nonlinear applications (Haykin, 1999; Hassoun, 1995; Leke, Marwala, & Tettey, 2007).

The configuration of neural processing units and their inter-connections can have a profound impact on the processing capabilities of neural networks (Haykin, 1999). Consequently, there are many different connections defining how the data flows between the input, hidden and output layers. The next section gives details on the architecture of the two neural networks employed in this chapter.

Multi-Layer Perceptrons (MLP)

The first architecture considered in this chapter is the *Multi-Layer Perceptron* (MLP). The MLP can be defined as a feed-forward neural network model that approximates a relationship between a set of input data and a set of appropriate output data. Its foundation is the standard linear perceptron and it makes use of three or more layers of neurons (nodes) with nonlinear activation functions and is more powerful than the perceptron. This is because it can distinguish data that are not linearly separable, or separable by a hyperplane. The multi-layer perceptron has been used to model many complex systems in areas such as mechanical engineering (Marwala, 2001c).

The MLP neural network consists of multiple layers of computational units, usually inter-connected in a feed-forward way (Haykin, 1999; Hassoun, 1995). Each neuron in one layer is directly connected to the neurons of the subsequent layer. A fully connected two-layered MLP architecture is used in this chapter. A NETLAB® toolbox (Nabney, 2001) that runs in MATLAB® is used to implement the MLP neural network. The two-layered MLP architecture, shown in Figure 2.1 is used because of the universal

Figure 2.1. Feed-forward multi-layer perceptron network having two layers of adaptive weights

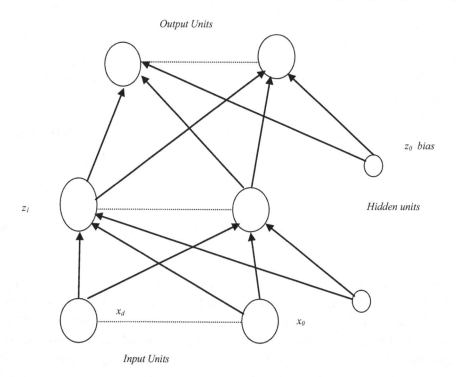

approximation theorem, which states that a two-layered architecture is adequate for MLP and, therefore, it is able to approximate data of arbitrary complexity (Nabney, 2001; Haykin, 1999).

The network can be described as follows (Bishop, 1995):

$$y_k = f_{outer} \left(\sum_{j=1}^{M} w_{kj}^{(2)} f_{inner} \left(\sum_{i=1}^{d} w_{ji}^{(1)} x_i + w_{j0}^{(1)} \right) + w_{k0}^{(2)} \right)$$

(2.1)

In equation 2.1, $w_{ji}^{(1)}$ and $w_{kj}^{(2)}$ indicate weights in the first and second layers, respectively, going from input i to hidden unit j or output unit k, M is the number of hidden units, d is the number of output units while $w_{j0}^{(1)}$ and $w_{k0}^{(2)}$ are the free parameters that indicate the biases for the hidden unit j and the output unit k. These free parameters can be viewed as a mechanism that makes the model actually understand the data. In this chapter, the parameter $f_{outer}(\bullet)$ is chosen as the *logistic function* while f_{inner} is the *hyperbolic tangent function*. The choice of these parameters allows the system to model linear and nonlinear data of any order. The logistic function is defined as follows (Bishop, 1995):

$$f_{outer}(v) = \frac{1}{1 + e^{-v}}$$

(2.2)

The logistic function maps the interval $(-\infty, \infty)$ onto a $(0, 1)$ interval and can be approximated by a linear function provided the magnitude of v is small.

The hyperbolic tangent function is:

$$f_{inner}(v) = \tanh(v)$$

(2.3)

Training the Multi-layer Perceptron

Given the data, the training of the neural network consists of identifying the weights in equations 2.1. This training process results in a model that embodies the rules and inter-relationships that govern the data. As described before, an objective function must be chosen for optimization that represents some distance between the model prediction and the observed target data with the free parameters as unknown variables. Given the training data, a minimizing of this cost function, therefore, identifies the free parameters, known as weights, in equation 2.1. An *objective function* is a mathematical representation of the overall objective of the problem. In this chapter, the main objective, used to construct a cost function, is to identify a set of neural network weights that would map the input variables onto themselves in an auto-associative way. If the training data set $D = \{x_k, y_k\}_{k=1}^{N}$ is used and we assume that the targets y are sampled independently, given the inputs x_k and the weight parameters, w_{kj}, the objective function, E, may therefore be written as follows using the sum-of-squares of errors objective function (Bishop, 1995):

$$E = \sum_{n=1}^{N} \sum_{k=1}^{K} \{t_{nk} - y_{nk}\}^2$$

$$= \sum_{n=1}^{N} \sum_{k=1}^{K} \{t_{nk} - y_{nk}(\{x\}, \{w\})\}^2$$

(2.4)

In equation 2.4, n is the index for the training example, k is the index for the output units, $\{x\}$ is the input vector and $\{w\}$ is the weight vector.

Before neural network training is performed, the network architecture needs to be constructed by choosing the number of hidden units, *M*. If *M* is too small, the neural network will be insufficiently flexible and will give poor generalization of the data because of high bias. However, if *M* is too large, the neural network will be too complex and, therefore, unnecessarily flexible, and will consequently provide poor generalization due to a phenomenon known as *over-fitting* caused by high variance. The process of choosing an appropriate *M* is known as *model selection* and is discussed in detail in this chapter.

To train the MLP network using the maximum-likelihood method, as is done in this chapter, a procedure called *back-propagation*, the subject of the next section, needs to be implemented. Back-propagation is essentially a method for finding the derivatives of the error, shown in equation 2.4, with respect to the network weights. This then allows us to implement a standard gradient-based optimization method to identify the optimal free parameters that are can best describe the observed training data.

Back-Propagation Method

To identify the network weights given the training data, an optimization method can be implemented within the context of the maximum-likelihood framework. In general, the weights can be identified using the following iterative method (Werbos, 1974):

$$\{w\}_{i+1} = \{w\}_i - \eta \frac{\partial E}{\partial \{w\}}(\{w\}_i) \tag{2.5}$$

In equation 2.5, the parameter η is the learning rate while $\{\}$ represents a vector. The minimization of the objective function, *E,* is achieved by calculating the derivative of the errors in equations 2.4, with respect to the network weight. The derivative of the error is calculated with respect to the weight that connects the hidden to the output layer and using the chain rule may be written as follows (Bishop, 1995):

$$\frac{\partial E}{\partial w_{kj}} = \frac{\partial E}{\partial a_k}\frac{\partial a_k}{\partial w_{kj}}$$

$$= \frac{\partial E}{\partial y_k}\frac{\partial y_k}{\partial a_k}\frac{\partial a_k}{\partial w_{kj}}$$

$$= \sum_n f'_{outer}(a_k)\frac{\partial E}{\partial y_{nk}}z_j \tag{2.6}$$

In equation 2.6, $z_j = f_{inner}(a_j)$ and $a_k = \sum_{j=0}^{M} w_{kj}^{(2)} y_j$. Using the chain rule, the derivative of the error with respect to weight that connects the hidden to the output layer may be written as follows (Bishop, 1995):

$$\frac{\partial E}{\partial w_{kj}} = \frac{\partial E}{\partial a_k}\frac{\partial a_k}{\partial w_{kj}}$$

$$= \sum_n f'_{inner}(a_j)\sum_k w_{kj} f'_{outer}(a_k)\frac{\partial E}{\partial y_{nk}} \tag{2.7}$$

In equation 2.7, $a_j = \sum_{i=1}^{d} w_{ji}^{(1)} x_i$. The derivative of the sum of square cost function in equation 2.4 may thus be written as:

$$\frac{\partial E}{\partial y_{nk}} = t_{nk} - y_{nk} \qquad (2.8)$$

while that of the hyperbolic tangent function is:

$$f'_{inner}(a_j) = \sec h^2(a_j) \qquad (2.9)$$

Now that it has been determined how to calculate the gradient of the error with respect to the network weights using back-propagation algorithms, equation 2.5 can be used to update the network weights using an optimization process until some pre-defined stopping condition is achieved. If the learning rate, in equation 2.5, is fixed, then this is known as the *steepest descent optimization method* (Robbins & Monro, 1951). On the other hand, the steepest descent method is not computationally efficient and, therefore, an improved method needs to be found. In this chapter the scaled conjugate gradient method is implemented (Møller, 1993), which is the subject of the next section.

Scaled Conjugate Gradient Method

The mechanism by which the free parameters (network weights) are deduced from the data is through using some nonlinear optimization method (Mordecai, 2003), and in this chapter the scaled conjugate gradient method. Before the scaled conjugate gradient method is described, it is vital to understand how it works. As indicated before, the weight vector, which gives the minimum error, is achieved by taking successive steps through the weight space as shown in equation 2.5 until some stopping criterion is attained. Different algorithms choose this learning rate differently. In this section, the gradient descent method is discussed, followed by how it is extended to the conjugate gradient method (Hestenes & Stiefel, 1952). For the gradient descent method, the step size is defined as $-\eta \partial E / \partial w$, where the parameter η is the learning rate and the gradient of the error is calculated using the back-propagation technique described in the previous section.

If the learning rate is sufficiently small, the value of error decreases at each successive step until a minimum value for the error between the model prediction and training target data is obtained. The disadvantage with this approach is that it is computationally expensive when compared to other techniques. For the conjugate gradient method, the quadratic function of error is minimized at each iteration over a progressively expanding linear vector space that includes the global minimum of the error (Luenberger, 1984; Fletcher, 1987; Bertsekas, 1995).

For the conjugate gradient procedure, the following steps are followed (Haykin, 1999):

1. Choose the initial weight vector $\{w\}_0$.

2. Calculate the gradient vector $\frac{\partial E}{\partial \{w\}}(\{w\}_0)$.

3. At each step n use the line search to find $\eta(n)$ that minimizes $E(\eta)$ representing the cost function expressed in terms of η for fixed values of w and $-\dfrac{\partial E}{\partial\{w\}}(\{w_n\})$.

4. Check that the Euclidean norm of the vector $-\dfrac{\partial E}{\partial w}(\{w_n\})$ is sufficiently less than that of $-\dfrac{\partial E}{\partial w}(\{w_0\})$.

5. Update the weight vector using equation 2.5.

6. For w_{n+1} compute the updated gradient $\dfrac{\partial E}{\partial\{w\}}(\{w\}_{n+1})$.

7. Use Polak-Ribiére method to calculate:
$$\beta(n+1) = \frac{\nabla E(\{w\}_{n+1})^T (\nabla E(\{w\}_{n+1}) - \nabla E(\{w\}_n)))}{\nabla E(\{w\}_n)^T \nabla E(\{w\}_n)}$$

8. Update the direction vector $\dfrac{\partial E}{\partial\{w\}}(\{w\}_{n+2}) = \dfrac{\partial E}{\partial\{w\}}(\{w\}_{n+1}) - \beta(n+1)\dfrac{\partial E}{\partial\{w\}}(\{w\}_n)$.

9. Set $n=n+1$ and go back to step 3.

10. Stop when the following condition is satisfied: $\dfrac{\partial E}{\partial\{w\}}(\{w\}_{n+2}) = \varepsilon\dfrac{\partial E}{\partial\{w\}}(\{w\}_{n+1})$ where ε is a small number.

The scaled conjugate gradient method differs from conjugate gradient method in that it does not involve the line search, described in step 3, in the previous section. The step-size (see step 3) is calculated directly by using the following formula (Møller, 1993):

$$\eta(n) = 2\left(\eta(n) - \left(\frac{\partial E(n)}{\partial\{w\}}(n)\right)^T H(n)\left(\frac{\partial E(n)}{\partial\{w\}}(n)\right) + \eta(n)\left\|\left(\frac{\partial E(n)}{\partial\{w\}}(n)\right)\right\|^2 \middle/ \left\|\left(\frac{\partial E(n)}{\partial\{w\}}(n)\right)\right\| \right)^2 \tag{2.10}$$

where H is the Hessian of the gradient.

The scaled conjugate gradient method is used because it has been found to solve the optimization problems encountered in training an MLP network more computationally efficiently than the gradient descent and conjugate gradient methods (Bishop, 1995).

Radial-Basis Function (RBF)

The other neural network architecture that is used in this chapter is the *Radial Basis Function* (RBF). RBF neural networks are feed-forward networks trained using a supervised training algorithm (Haykin, 1999; Buhmann & Ablowitz, 2003). The RBF is typically configured with a single hidden layer of units whose activation function is selected from a class of functions called *basis functions*. The activation of the hidden units in an RBF neural network is given by a nonlinear function of the distance between the input vector and a prototype vector (Bishop, 1995).

While similar to back-propagation in many aspects, radial basis function networks have several advantages. They usually train much faster than the MLP networks and are less prone to problems with non-stationary inputs due to the behavior of the radial basis function (Bishop, 1995). The RBF network

Figure 2.2. Radial basis function network having two layers of adaptive weights

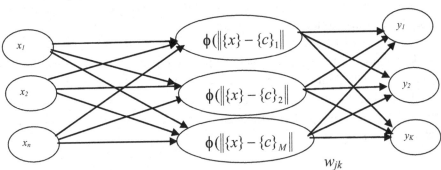

can thus be described as shown in Figure 2.2 and is described mathematically as follows (Buhmann & Ablowitz, 2003):

$$y_k(\{x\}) = \sum_{j=1}^{M} w_{jk} \phi_j \left(\left\| \{x\} - \{c\}_j \right\| \right) \tag{2.11}$$

In equation 2.11, w_{jk} represents the output weights, each corresponding to the connection between a hidden unit and an output unit, M represents the number of hidden units, $\{c\}_j$ is the center for the jth neuron, $\phi_j(\{x\})$ is the jth nonlinear activation function, $\{x\}$ the input vector, and $k = 1,2,3,...,M$ (Bishop, 1995). Again as in the MLP, the selection of the number of hidden nodes M, is part of the model selection process.

In this chapter, the activation in the hidden layers is assumed to be a Gaussian distribution $(\phi(\|\{x\} - \{c\}\|) = \exp(-\beta(\{x\} - \{c\})^2))$ where β is constant. The radial basis function differs from the multilayer perceptron in that it only has weights in the outer layer, while the hidden nodes have what are called the *centers*. Training the radial basis function network entails identifying two sets of parameters, namely the centers and the output weights. Both these can be viewed as free parameters in a regression framework. Even though the centers and network weights can both be determined simultaneously in this chapter, a two stage training process is used to identify the centers. The first stage uses self-organizing maps to determine the centers, and in this chapter the *k-means* clustering method is used (Hartigan, 1975). The step of identifying the centers only considers the input space while the identification of the network weights considers both the input and output space.

The *k-means algorithm* is aimed at clustering objects based on attributes into k partitions. In this chapter, k, is equal to the number of centers M. Its objective is to discover the centers of natural clusters in the data and assumes that the object attributes form a vector space. It achieves this by minimizing the total intra-cluster variance, or the squared error function (Hartigan & Wong, 1979):

$$E = \sum_{i=1}^{C} \sum_{x_j \in S_i} \left(\{x\}_j - \{c\}_i \right)^2 \tag{2.12}$$

In equation 2.12, C is the number of clusters S_i, $i = 1,2,...,M$ and $\{c\}_i$ is the center of all the points $x_j \in S_i$. In this chapter, the Lloyd algorithm is used to identify the cluster centers (Lloyd, 1982). Lloyd's algorithm is initialized by randomly dividing the input space into k initial sets or using heuristic data. Then the mean point is calculated for each set and a new partition is constructed by associating each point with the closest center. The centroids are then re-calculated for the new clusters, and the process is repeated by changing these two steps until convergence. Convergence is achieved when the centroids no longer change or the points no longer switch clusters.

Now that the method in which the centers are identified has been described, the next step is to calculate the network weights in equation 2.11 given the training data. To achieve this, the Moore-Penrose pseudo inverse (Moore, 1920; Penrose, 1955; Golub & Van Loan, 1996) is used. It should be noted that once the centers have been identified, then the estimation of the network weights becomes a linear process (Golub & Van Loan, 1996). Given the training data and the centers identified, equation 2.11 can be rewritten as follows:

$$[y_{ij}] = [\phi_{ik}][w_{kj}] \tag{2.13}$$

Here $[y_{ij}]$ is the output matrix, with i representing the number of training examples, while j represents the number of output; parameter $[\phi_{ik}]$ is the activation function matrix in the hidden layer with i representing the training examples, while k is the number of hidden neurons while $[w_{kj}]$ is the weight matrix, while j is the number of outputs. From equation 2.13 it can be observed that to solve for the weight matrix $[w_{kj}]$, what needs to be done, is to invert the activation function matrix $[\phi_{ik}]$. However, this matrix is not square and, therefore, cannot be inverted using standard tools. This matrix can be inverted using the Moore-Penrose pseudo-inverse which can be written as follows:

$$[\phi_{ik}]^* = \left([\phi_{ik}][\phi_{ik}]^T\right)^{-1}[\phi_{ik}]^T \tag{2.14}$$

This, therefore, implies that the weight matrix may be estimated as follows:

$$[w_{kj}] = [\phi_{ik}]^*[y_{ij}] \tag{2.15}$$

A NETLAB® toolbox (Nabney, 2001) that runs in MATLAB® is used to implement the RBF neural network in this chapter.

Model Selection

After identifying the weights through training the MLP or RBF model, an appropriate model given by the size of the hidden nodes is obtained from the validation data set. As indicated before, to conduct a missing data estimation procedure successfully, it is important to identify the correct model, whether it is the MLP or the RBF. The process of selecting an appropriate model is known as *model selection* (Burnham & Anderson, 2002). The process of deriving a model from data is a non-unique problem. This is because many models can fit the training data and, therefore, it becomes impossible to identify the most appropriate model. The general approach to model selection is based on two principles, namely the goodness of fit and complexity. The *goodness of fit* essentially implies that a good model should be able to predict the validation data, which it has not seen during the training stage. The *complexity of the model* is based on the Occam's principle that states that the best model is the simplest one.

In selecting the best model crucial questions that need to taken into account. Some of these are:

- What is the best balance between the goodness of fit and the complexity of the model?
- How are these attributes actually implemented?

This chapter uses a statistical approach, called *cross-validation,* to select the best model that defines the inter-relationships and the rules in the data. In this chapter, the goodness of fit is measured by the error between the model prediction and the validation set, while the complexity of the model is measured by the number of free parameters in the data. As stated before, free parameters in the MLP model are defined as the network weights and biases, while in the RBF network they are defined as the network centers and weights. In this chapter, model selection is viewed as a mechanism of selecting a model that has a good probability of estimating the validation data that it has not seen during the training stage. The bias and variance are measures of the ability of this model to operate in an acceptable way.

In this chapter, the cross-validation method is used (Devijver, 1982; Chang, Luo, & Su, 1992; Kohavi, 1995). The way in which cross-validation is conducted, is to divide the data into three segments. The first segment is called the training set, which is used to train the MLP and RBF networks. In this chapter, for the MLP scaled conjugate gradient method and the training data are used to estimate the free parameters (weights and biases in equation 2.1) while for the RBF, *k-means* and pseudo-inverse techniques are used to estimate the free parameters (centers and weights in equation 2.11). The way in which this training process is conducted is to train several models (with different numbers of hidden nodes). The second set of data is called *the validation data set,* and is used in this chapter to select the best MLP and RBF models and this constitutes the model selection stage. Finally, the third data set of data called the *testing set* is used to evaluate the effectiveness of the selected models.

MISSING DATA ESTIMATION METHOD

Now that the MLP and RBF have been discussed, the next step is to determine how they can be used for missing data estimation process. The missing data estimation algorithms presented in this chapter entail the use of a neural network model (RBF or MLP) which is trained to recall itself (i.e., predict its input vector) and is, therefore, called the auto-associative neural network. From the explanations of the MLP and RBF given in this chapter, this auto-associative neural network can be written mathematically as follows:

$$\{y\} = f(\{x\}, \{w\}) \tag{2.16}$$

In equation 2.16, $\{y\}$ is the output vector, $\{x\}$ the input vector and $\{w\}$ is the vector of network weights. Note that in equation 2.16, the network weights vector is simplified to include biases for the MLP and centers for the RBF. Since the network is trained to predict its own input vector, the input vector $\{x\}$ is approximately equal to output vector $\{y\}$ and therefore $\{x\} \approx \{y\}$.

In essence, the input vector $\{x\}$ and output vector $\{y\}$ will not always be perfectly the same. Hence, an error function expressed as the difference between the input and output vector can be formulated as follows:

$$\{e\} = \{x\} - \{y\} \tag{2.17}$$

Substituting the value of $\{y\}$ from equation 2.16 into equation 2.17, the following expression is obtained:

$$\{e\} = \{x\} - f(\{x\}, \{w\}) \tag{2.18}$$

It is intended that the error to be minimized must be non-negative and hence, the error function can be re-written as a square of equation 2.18:

$$\{e\} = (\{x\} - f(\{x\}, \{w\}))^2 \tag{2.19}$$

In the case of missing data, some of the values of the input vector $\{x\}$ are not available. Hence, the input vector elements can be categorized into $\{x\}$ known vectors represented by $\{x_k\}$ and $\{x\}$ unknown vectors represented by $\{x_u\}$. Re-writing equation 2.19 in terms of $\{x_k\}$ and $\{x_u\}$ gives:

$$\{e\} = \left(\left\{ \begin{array}{c} \{x_k\} \\ \{x_u\} \end{array} \right\} - f\left(\left\{ \begin{array}{c} \{x_k\} \\ \{x_u\} \end{array} \right\}, \{w\} \right) \right)^2 \tag{2.20}$$

The error vector in equation 2.20 can be reduced into a scalar by integrating over the size of the input vector and the number of training examples as follows:

$$E = \left\| \left(\left\{ \begin{array}{c} \{x_k\} \\ \{x_u\} \end{array} \right\} - f\left(\left\{ \begin{array}{c} \{x_k\} \\ \{x_u\} \end{array} \right\}, \{w\} \right) \right) \right\| \tag{2.21}$$

Here $\| \ \|$ is the Euclidean norm. Equation 2.21 is called the *Missing Data Estimation Error Function* (MDEEF). To approximate the missing input values, equation 2.21 is minimized and in this chapter, the genetic algorithm is used (Holland, 1975; Koza, 1992; Falkenauer, 1997; Goldberg, 2002; Fogel, 2006). However, it must be noted that any optimization procedure or a combination of optimization method can be used to achieve this task, as will be observed in later chapters. The genetic algorithm is chosen because it has a higher probability of finding the global optimum solution than traditional optimization methods such as the scaled conjugate gradient method, which is used in this chapter for training the MLP network (Goldberg, 1989; Michalewicz, 1996; Jones & Konstam, 1999). It should be noted here that for this process to be successful, the identification of a global optimum solution as opposed to local one is crucial because if this is not achieved then a wrong estimation of the missing data will be reached. It is important to note that this is not the case for neural network training because a global optimum solution will over-train the network and hence result in a poor generalization. Consequently, for the neural network training process, the local optimization search called the scaled conjugate gradient method, described earlier, is used as opposed to global optimization method such as the genetic algorithm, which is used to estimate missing data.

For the individual cases of the MLP and RBF, equation 2.21 is used as the fitness function, where f in the equation refers to either the MLP or RBF auto-associative network, accordingly. The missing data process, described in this section, is illustrated in Figure 2.3.

In summary, the fitness function is derived from the MDEEF of the input and the output vector is obtained from the trained auto-associative neural networks. The missing data estimation error function is then minimized using a genetic algorithm to approximate the missing variables, given the observed

Figure 2.3. Schematic representation of the missing data estimation model

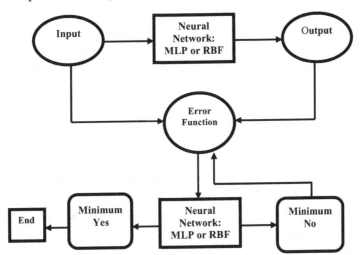

variables $\{x_k\}$ and the model f describing the inter-relationships and the rules describing the data. The genetic algorithm as implemented in this chapter is described in detail in the next section.

GENETIC ALGORITHMS (GAs)

This section describes the use of genetic algorithms to optimize equation 2.21. The GA is an algorithm that is used to find approximate solutions to difficult problems through the application of the principles of evolutionary biology to computer science (Michalewicz, 1996; Mitchell, 1996; Forrest, 1996; Vose, 1999; Tettey & Marwala, 2006). It uses biologically derived techniques such as inheritance, mutation, natural selection and recombination to approximate an optimal solution to difficult problems (Banzhaf et al., 1998). Genetic algorithms view learning as a competition among a population of evolving candidate problem solutions. A fitness function evaluates each solution to decide whether it will contribute to the next generation of solutions. Through operations analogous to gene transfer in sexual reproduction, the algorithm creates a new population of candidate solutions (Banzhaf et al., 1998; Goldberg, 1989).

From (Michalewicz, 1996; Houck, Joines, & Kay, 1995) the three most important aspects of using genetic algorithms are: definition of the objective function, definition and implementation of the genetic representation, and definition as well as implementation of the genetic operators. GAs have been proven to be successful in optimization problems such as wire routing, scheduling, adaptive control, game playing, cognitive modeling, transportation problems, traveling salesman problems, optimal control problems and database query optimization (Michalewicz, 1996; Pendharkar & Rodger, 1999; Marwala et al., 2001; Marwala, 2002; Marwala, 2004; Marwala, 2007b; Marwala & Chakraverty, 2006; Marwala, 2007a; Crossingham & Marwala, 2007; Hulley & Marwala, 2007). The MATLAB® implementation of genetic algorithm described in Houck, Joines, and Kay (1995) is used to implement GA in this chapter. After executing the program with different genetic operators, optimal operators that give the best results are selected and used in conducting the experiment. To implement genetic algorithms, the following steps are followed: initialization, selection, reproduction and termination.

Initialization

In the beginning, a huge number of possible individual solutions are randomly generated to form an initial population. This initial population ought to cover a good representation of the solution space. The size of the population should depend on the nature of the problem, which is determined by the number of variables. For example, if there are two variables missing, the size of the population must be higher than if there is one variable missing.

Selection

For every generation, a selection of the proportion of the existing population is chosen to breed a new population. This selection is conducted using the fitness-based process, where solutions that are fitter, as measured by the MDEEF, in equation 2.21, are given a higher probability of being selected. Some selection methods rank the fitness of each solution and choose the best solutions while for computational efficiency other procedures rank a randomly chosen sample of the population.

Many selection functions tend to be stochastic in nature. Thus, they are designed in such a way that a selection process is conducted on a small proportion of less fit solutions. This ensures that the diversity of the population of possible solutions is maintained at high level and, therefore, avoiding convergence on poor and incorrect solutions. There are many selection methods and these include roulette wheel selection, which is used in this chapter.

Roulette-wheel selection is a genetic operator used to select potentially useful solutions for the genetic algorithm optimization process. In this method, each possible procedure is assigned the fitness function that is used to map the probability of selection with each individual solution. Suppose the fitness f_i is of individual i in the population and the probability that this individual is selected is:

$$p_i = \frac{f_i}{\sum_{j=1}^{N} f_j} \tag{2.22}$$

In equation 2.22, N is the total population size. This process ensures that candidate solutions with a higher fitness have a lower probability that they may be eliminated than those with lower fitness. Similarly, solutions with low fitness have a low probability of surviving the selection process. The advantage of this is that even though a solution may have low fitness, it may still contain some components that may be useful in the future.

Reproduction

Reproduction generates subsequent population of solutions from those selected through the genetic operators, which are the cross-over and mutation operators.

The *cross-over operator* mixes genetic information in the population by cutting pairs of chromosomes at random points along their length and exchanging over the cut sections. This has a potential of joining successful operators together. Cross-over occurs with a certain probability. In many natural systems, the probability of cross-over occurring is higher than the probability of mutation occurring. A simple cross-over technique is used in this chapter (Goldberg, 1989). For the simple cross-over technique, one cross-over point is selected, a binary string from the beginning of a chromosome to the cross-over point is copied from one parent, and the rest is copied from the second parent. For example, when the chromosome **11001**011 undergoes a simple cross-over with the chromosome 11011**111**, it becomes **11001111.**

The *mutation operator* picks a binary digit of the chromosomes at random and inverts it. This has a potential of introducing to the population new information. Mutation occurs with a certain probability. In many natural systems, the probability of mutation is low (i.e., is less than 1%). In this chapter, binary mutation is used (Goldberg, 1989). When binary mutation is used, a digit written in binary form is chosen and its value is inverted. For example: chromosome 11001011 may become chromosome 11000011.

The processes described result in the subsequent generation of population of solutions that are different from the previous generation and that have an average fitness that is higher than the previous generation.

Termination

The process described is repeated until a termination condition has been achieved because either a desired solution is found that satisfies the MDEEF objective, a specified number of generations has been reached or the solution's fitness converged or any combinations of these. The process described above can be written in pseudo algorithmic code as (Goldberg, 1989):

1. Select the initial population.
2. Calculate the fitness of each chromosome in the population using the MDEEF.
3. Repeat:
    a. Choose chromosomes with higher fitness to reproduce.
    b. Generate new population using cross-over and mutation to produce offspring.
    c. Calculate the fitness of each offspring.
    d. Replace low fitness section of population with offspring.
4. Until termination

EXPERIMENTATION, RESULTS AND DISCUSSIONS

The data used in this chapter is information that is used to construct an artificial beer taster. In general, the parameters that human beings are sensitive to when tasting beer are: (1) color of beer; (2) smell; and (3) chemical components. In human beings, the color of objects is captured using the eye; smell is captured using the nose and taste using the tongue. An artificial beer taster, that is under consideration in this chapter, contains the following variables: alcohol, present extract, real extract, present extract minus limit extract (PE-LE); pH; iron; acetaldehyde; dimethyl sulfide (DMS), eythyl acetate, iso-amyl acetate, total higher alcohols, color, bitterness and amount of carbohydrates. These variables capture the color, smell and chemical components of the beer. They are normally used to predict the beer taste score, which is conventionally obtained from a panel of tasters, using some regression analysis such as neural networks.

All these parameters are measured. However, sometimes one or more of these measurements may not be available due to problems such as instrumentation failure. In such a situation, it is important to estimate these missing values because it would be impossible to predict the taste scores. The reason why the data points are missing in this artificial taster case is random and is therefore MCAR. However, the fact that the missing data mechanism in this example is MCAR is good to know, but is not factored into the design of the auto-associative neural networks nor into the implementation of a genetic algorithm based missing data estimator. More information on this artificial taster can be found in Marwala (2005).

The following terms are used to measure the modeling quality: (i) Standard error (S_e) and (ii) Correlation coefficient (r). For a given data $x_1; x_2; ..., x_n$ and corresponding approximated values $\hat{x}_1, \hat{x}_2,..., \hat{x}_n$ the Standard error (S_e) is computed as follows (Draper & Smith, 1998):

$$Se = \sqrt{\frac{\sum_{i=1}^{n}(x_i - \hat{x}_i)^2}{n}} \tag{2.23}$$

and the correlation coefficient (r) is computed as follows (Draper & Smith, 1998):

$$r = \frac{\sum_{i=1}^{n}(x_i - \overline{x}_i)(\hat{x}_i - \overline{\hat{x}}_i)}{\left[\sum_{i=1}^{n}(x_i - \overline{x}_i)^2 \sum_{i=1}^{n}(\hat{x}_i - \overline{\hat{x}}_i)^2\right]^{1/2}} \tag{2.24}$$

where $\overline{x}_i$ and $\overline{\hat{x}}_i$ are the means of actual missing data and approximated values, respectively. The error "S_e" estimates the capability of the model to predict the known data set (Kolarik & Rudorfer, 1994). The higher the value of "S_e", the less reliable are the approximations and vice-versa. The correlation coefficient measures the linear relationship between two variables. The absolute value of "r" provides an indication of the strength of the relationship. The value of "r" varies between negative 1 and positive 1, with -1 or 1 indicating a perfect linear relationship, and $r = 0$ indicating no relationship. The sign of the correlation coefficient indicates whether the two variables are positively or negatively related (Draper & Smith, 1998). Here, "r" measures the degree of relationship between the actual missing data and corresponding approximated values using the model. A positive value indicates a direct relationship between the actual missing data and its approximated value using the model. The closer the value of "r" is to 1, the stronger the relationship. As the strength of the relationship between the predicted values and actual values increases, so does the correlation coefficient.

These beer characteristic variables are then used to train the MLP and RBF neural networks. The neural networks essentially map all the 14 variables onto themselves, in an auto-associative way, as described earlier in the chapter, using the MLP and RBF networks. A data set of 1200 examples was provided; 400 were used for training, a further 400 were used for validation and 400 for testing both the MPL and RBF neural network architectures.

The sizes of the hidden nodes of these networks that are evaluated vary from 7 to 11 and the performances of these networks are shown in Table 2.2. This table shows that the optimal network architectures, which in turn appear in Table 2.3, contain 14 inputs, 10 hidden neurons and 14 outputs for the MLP network as well as 14 inputs, 9 hidden neurons and 14 outputs for the RBF network. The activation functions for the MLP network are the hyperbolic tangent function in the hidden units and linear activation function in the outer units. For the RBF, a Gaussian activation function is used for the hidden nodes and linear activation function for the output layer. An additional set of 100 data points are used to test the ability of the proposed procedure to estimate missing values. The genetic algorithm in the proposed missing data estimation procedure is implemented with 70% probability of reproduction, 3% mutation rate and 60% cross-over rate.

Table 2.2. Standard error

Number of Hidden Nodes	7	8	9	10	11
RBF	11.6	10.8	10.7	11.2	11.5
MLP	11.9	11.4	11.1	10.7	10.9

Table 2.3. Standard error

	Training	Validation	Testing
RBF	9.32	10.67	11.94
MLP	9.29	10.72	11.03

Table 2.4. Correlation coefficient

	Number of Missing Variables				
	1	2	3	4	5
RBF	0.945	0.938	0.937	0.934	0.936
MLP	0.967	0.968	0.966	0.969	0.969

Table 2.5. Standard error

	Number of Missing Variables				
	1	2	3	4	5
RBF	12.26	12.67	12.08	12.13	12.41
MLP	11.99	11.72	11.71	11.90	12.05

Cases of 1, 2, 3, 4, and 5 missing values in a single record were examined to investigate the accuracy of the approximated values as the number of missing cases within a single record increases. To assess the accuracy of the values approximated using the model, the standard error and the correlation coefficient were calculated for each missing case. The resulting standard error and correlation measures obtained from the investigation are given in Tables 2.4 and 2.5.

The results are also depicted in Figures 2.4 and 2.5 for easy comparison between the results found for MLP and RBF. The results show that the models' approximations of missing data are highly accurate. There is no significant difference among the approximations obtained for different number of missing cases within a single record. Approximations obtained using the MLP in all the missing cases are better than corresponding values found using the RBF. This could be because the MLP is more complex than the RBF in terms of the order of nonlinearity. The RBF is found to be less stable than the MLP due to the calculation of the pseudo-inverse, which some times results in singular matrices. However, the RBF is found to be more computationally efficient than the MLP. A sample of the actual missing data and approximated values, using the model for the 14 variables used in the model, is presented in Tables 2.6 and 2.7, and Figures 2.6, 2.7, 2.8, 2.9 and 2.10.

Figure 2.4. Correlation coefficient for MLP and RBF

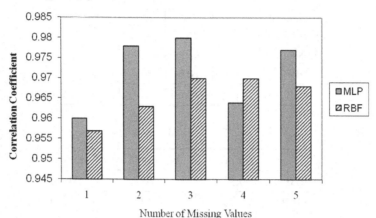

Figure 2.5. Standard error for MLP and RBF

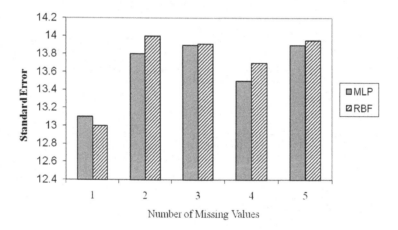

The results show that the model's approximate values of the missing data are similar to the actual missing values. It is also observed that the estimates found for 1, 2, 3, 4 and 5 missing cases are not significantly different from each other.

CONCLUSION

Neural networks and genetic algorithms can be used to predict missing data in a database. Auto-associative neural networks may be trained to predict their own input. The error function was derived as the square of the difference of the output vector from the trained neural network and the input vector. Since some of the input vectors are missing, the error function was expressed in terms of the known and unknown components of the input vector. A genetic algorithm was used to approximate the missing values in the input vector that best minimize the error function. The RBF and MLP neural networks were used to train the neural network. It was found that the models can approximate missing values

Figure 2.6. One missing case: Actual vs. approximated using MLP and RBF

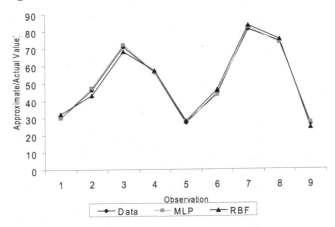

Figure 2.7. Two missing case: Actual vs. approximated using MLP and RBF

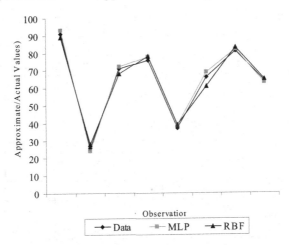

Figure 2.8 Three missing case: Actual vs. approximated using MLP and RBF

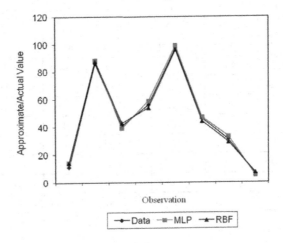

Table 2.6. Actual and approximated values using RBF

Data	Number of missing values in a record				
	1	2	3	4	5
4.28	4.54	4.54	4.53	4.47	4.07
7.5	6.86	6.79	6.41	6.8	6.52
23.8	21.2	20.9	21.3	21	22
1.8	2.48	2.41	1.81	2.54	2.21
0.4	0.1	0.104	0.72	0.22	0.72
0.2	0.58	0.06	0.02	0.11	0.159
40	38.1	37.8	38.4	37.2	38
5.7	6.64	6.66	6.96	5.82	5.67
24	22.1	22.4	22.3	23	23.2
2.9	3.23	3.86	3.74	3.83	3.97
0.4	0.1	0.104	0.72	0.22	0.72
0.2	0.58	0.06	0.02	0.11	0.159
40	38.1	37.8	38.4	37.2	38
5.7	6.64	6.66	6.96	5.82	5.67
24	22.1	22.4	22.3	23	23.2
2.9	3.23	3.86	3.74	3.83	3.97

Table 2.7. Actual and approximated values using MLP

Data	Number of missing values in a record				
	1	2	3	4	5
4.28	4.21	4.2	4.12	4.13	4.25
17.5	17.89	18.79	18.71	18.65	18.21
17	16.96	17.16	16.04	15.95	12.48
75	83.92	74.84	75.96	75.7	78.79
1.8	1	1.14	2.15	2.01	1.73
0.4	0.7	0.71	0.76	0.71	0.55
10.2	10.1	10.1	10.09	10.11	10.16
40	56.45	57.73	61.73	62.65	62.16
5.7	9.79	9.3	43	6.54	9.33
24	22.4	22.52	27.81	34.45	36.79
1.7	1	1.13	2.14	2.02	1.71
10.3	10.7	10.70	10.75	10.72	10.54
1.2	1.2	1.0	1.09	1.11	1.16
40	56.45	57.73	61.73	62.65	62.16
15.7	19.79	19.3	143.4	16.54	19.35
34	32.4	32.52	37.81	34.45	36.79

Figure 2.9. Four missing case: Actual vs. approximated using MLP and RBF

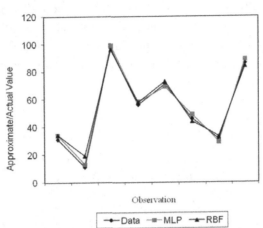

Figure 2.10. Five missing case: Actual vs. approximated using MLP and RBF

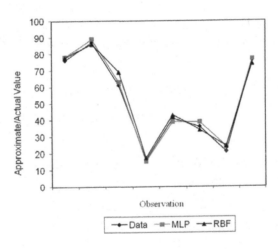

with great accuracy. Though there is a slight decrease in correlation coefficient, there is no significant reduction in accuracy of the results observed as the number of missing cases within a single record gets larger. Results found using the MLP were observed to be better than the RBF. Furthermore, the RBF was found to be unstable because of a pseudo-inverse process, which sometimes results in badly scaled matrices.

FURTHER WORK

This chapter introduced neural networks and genetic algorithms for missing data estimation. Further work can be done on methods to design the entire neural-network-genetic-algorithm-system for maximum efficiency in terms of computational time as well as accuracy of missing data estimation. Another issue that needs to be addressed is how theoretically to determine the relationship between the number of missing variables and the accuracy of missing data estimation given the level of errors in the measured variables and the data used for training. The conclusions reached in this chapter are highly dependent on the nature of the data used in the analysis. Therefore, further statistical tests should be conducted to ensure that the conclusions reached are not dependant on the data used. In particular, a relationship must be identified between the nature of data and the auto-associative neural network used.

REFERENCES

Abdella, M. (2005). *The use of genetic algorithms and neural networks to approximate missing data in database.* Unpublished master's thesis, University of the Witwatersrand, Johannesburg.

Abdella, M., & Marwala, T. (2005). Treatment of missing data using neural networks. In *Proceedings of the IEEE International Joint Conference on Neural Networks* (pp. 598-603).

Allison, P. (2000). Multiple imputation for missing data: A cautionary tale. *Sociological Methods and Research, 28*, 301-309.

Banzhaf, W., Nordin, P., Keller, R., & Francone, F. (1998). *Genetic programming-An introduction: On the automatic evolution of computer programs and its applications.* 5th Edition, California: Morgan Kaufmann Publishers.

Bertsekas, D. P. (1995). *Nonlinear programming.* Belmont, Massachusetts: Athenas Scientific.

Bishop, C. M. (1995). *Neural networks for pattern recognition.* Oxford, UK: Oxford University Press.

Buhmann, M. D., & Ablowitz, M. J. (2003). *Radial basis functions: Theory and implementations.* Cambridge, UK: Cambridge University Press.

Burnham, K. P., & Anderson, D. R. (2002). *Model selection and multi-model inference: A practical-theoretic approach.* Heidelberg: Springer-Verlag.

Chang, J., Luo, Y., & Su, K. (1992). GPSM: A generalized probabilistic semantic model for ambiguity resolution. In *Proceedings of the 30th Annual Meeting on Association For Computational Linguistics* (pp. 177-184).

Crossingham, B., & Marwala, T. (2007). Using genetic algorithms to optimise rough set partition sizes for HIV data analysis. *Studies in Computational Intelligence, 78*, 245-250.

Devijver, P. A., & Kittler, J. (1982). *Pattern recognition: A statistical approach.* London, UK: Prentice-Hall.

Draper, N., & Smith, H. (1998). *Applied regression analysis.* New York: John Wiley.

Falkenauer, E. (1997). *Genetic algorithms and grouping problems.* Chichester, England: John Wiley & Sons.

Fletcher, R., (1987). *Practical methods of optimization.* New York: John Wiley.

Fogel, D. B. (2006). *Evolutionary computation: Toward a new philosophy of machine intelligence.* Piscataway, NJ: IEEE Press.

Forrest, S. (1996). Genetic algorithms. *ACM Computing Survey, 28*, 77-80.

Freeman, J., & Skapura, D. (1991). *Neural networks: Algorithms, applications and programming Techniques.* Reading, MA: Addison-Wesley.

Goldberg, D. E. (1989). *Genetic algorithms in search, optimization, and machine learning.* Reading, MA: Addison-Wesley.

Goldberg, D. E. (2002). *The design of innovation: Lessons from and for competent genetic algorithms.* Reading, MA: Addison-Wesley.

Golub, G. H., & Van Loan, C. F. (1996). *Matrix computation.* Baltimore, Maryland: Johns Hopkins University Press.

Hartigan, J. A. (1975). *Clustering algorithms.* New York: Wiley & Sons.

Hartigan, J. A., & Wong, M. A. (1979). A K-means clustering algorithm. *Applied Statistics, 28*, 100-108.

Hassoun, M. H. (1995). *Fundamentals of artificial neural networks.* Cambridge, Massachusetts: MIT Press.

Haykin, S. (1999). *Neural networks.* New Jersey: Prentice-Hall.

Hestenes, M. R., & Stiefel, E. (1952), Methods of conjugate gradients for solving linear systems. *Journal of Research of the National Bureau of Standards, 6*, 409-436.

Holland, J. H. (1975). *Adaptation in natural and artificial systems.* Ann Arbor, MI: University of Michigan Press.

Houck, C. R., Joines, J. A., & Kay, M. G. (1995). *A genetic algorithm for function optimisation: A MATLAB implementation* (Tech. Rep. NCSU-IE TR 95-09). Chapel Hill, North Carolina: North Carolina State University.

Hu, M., Savucci, S., & Choen, M. (1998). Evaluation of some popular imputation algorithms. In *Proceedings of the Survey Research Methods Section of the American Statistical Association* (pp. 308-313).

Hulley, G., & Marwala, T. (2007). Genetic algorithm based incremental learning for optimal weight and classifier selection. *Computational Models for Life Sciences, 952*, 258-268.

Jones, M., & Konstam, A. (1999). The use of genetic algorithms and neural networks to investigate the Baldwin effect. In *Proceedings of the 1999 ACM Symposium on Applied computing* (pp. 275-279).

Kohavi, R. (1995). A study of cross-validation and bootstrap for accuracy estimation and model selection. In *Proceedings of the Fourteenth International Joint Conference on Artificial Intelligence, 2*(12), 1137–1143.

Kolarik, T., & Rudorfer, G. (1994). Time series forecasting using neural networks. In *Proceedings of the International Conference on APL: The Language and its Applications* (pp. 86-94).

Koza, J. (1992). *Genetic programming: On the programming of computers by means of natural selection.* Cambridge, MA: MIT Press.

Lagazio, M., & Marwala, T. (2005). Assessing different Bayesian neural network models for militarized interstate dispute. *Social Science Computer Review, 24*(1), 1-12.

Leke, B. B., & Marwala, T. (2006). Ant colony optimization for missing data estimation. In *Proceeding of the Pattern Recognition of South Africa* (pp. 183-188).

Leke, B., & Marwala, T and Tettey, T. (2007). Using inverse neural network for HIV adaptive control. *International Journal of Computational Intelligence Research, 3*(1), 11-15.

Little, R., & Rubin, D. (1987). *Statistical analysis with missing data.* New York: John Wiley and Sons.

Lloyd, S. O. (1982). Least squares quantization in PCM. *IEEE Transactions on Information Theory, 28*, 129-137.

Luenberger, D. G. (1984). *Linear and nonlinear programming.* Reading, Massachusetts, USA: Addison-Wesley.

Lunga, D., & Marwala, T. (2006). Online forecasting of stock market movement direction using the improved incremental algorithm. *Lecture Notes in Computer Science, 4234*, 440-449.

Marwala, T. (2000). On damage identification using a committee of neural networks. *American Society of Civil Engineers, Journal of Engineering Mechanics, 126*, 43-50.

Marwala, T. (2001a). Probabilistic fault identification using a committee of neural networks and vibration data. *American Institute of Aeronautics and Astronautics, Journal of Aircraft, 38,* 138-146.

Marwala, T. (2001b) Scaled conjugate gradient and Bayesian training of neural networks for fault identification in cylinders. *Computers and Structures, 79/32*, 2793-2803.

Marwala, T. (2001c). *Fault identification using neural networks and vibration data.* Unpublished doctoral dissertation, University of Cambridge, Cambridge, UK.

Marwala, T. (2002). Finite element updating using wavelet data and genetic algorithm. *American Institute of Aeronautics and Astronautics, Journal of Aircraft, 39,* 709-711.

Marwala, T. (2003). Fault classification using pseudo modal energies and neural networks. *American Institute of Aeronautics and Astronautics Journal, 41,* 82-89.

Marwala, T. (2004). Control of complex systems using Bayesian neural networks and genetic algorithm. *International Journal of Engineering Simulation, 5*, 28-37.

Marwala, T. (2005). The artificial beer taster. *Electricity+Control* (pp. 22-23).

Marwala, T. (2007a). Bayesian training of neural network using genetic programming. *Pattern Recognition Letters, 28*, 1452–1458

Marwala, T. (2007b). *Computational intelligence for modelling complex systems.* New Delhi, India: Research India Publications.

Marwala, T., & Chakraverty, S. (2006). Fault classification in structures with incomplete measured data using auto-associative neural networks and genetic algorithm. *Current Science, 90*, 542-548.

Marwala, T., de Wilde, P., Correia, L., Mariano, P., Ribeiro, R., Abramov, V., Szirbik, N., and Goossenaerts, J. (2001). Scalability and optimisation of a committee of agents using genetic algorithm. In *Proceedings of the International Symposia on Soft Computing and Intelligent Systems for Industry* (pp. 1-7).

Marwala, T., & Hunt, H.E.M. (1999). Fault identification using finite element models and neural networks. *Mechanical Systems and Signal Processing, 13*, 475-490.

Michalewicz, Z. (1996). *Genetic algorithms + Data structures = Evolution programs.* Heidelberg: Springer-Verlag.

Mitchell, M. (1996). *An introduction to genetic algorithms.* Cambridge, MA: MIT Press.

Mohamed, S., & Marwala, T. (2005). Neural network based techniques for estimating missing data in databases. In *Proceedings of the 16th Annual Symposium of the Pattern Recognition Society of South Africa* (pp. 27-32).

Mohamed, A.K., Nelwamondo, F.V., & Marwala, T. (2007). Estimation of missing data: Neural networks, principal component analysis and genetic algorithms. In *Proceedings of the 18th Annual Pattern Recognition Association of South Africa* (pp. 656-2), CD-Rom ISBN: 978-86840.

Mohamed, N, Rubin, D., & Marwala, T. (2006). Detection of epileptiform activity in human EEG signals using Bayesian neural networks. *Neural Information Processing – Letters and Reviews, 10*, 1-10.

Mohamed, S., Tettey, T., & Marwala, T. (2006). An extension neural network and genetic algorithm for bearing fault classification. In *Proceedings of the IEEE International Joint Conference on Neural Networks* (pp. 7673-7679).

Møller, A. F. (1993). A scaled conjugate gradient algorithm for fast supervised learning. *Neural Networks, 6*, 525-533.

Moore, E. H. (1920). On the reciprocal of the general algebraic matrix. *Bulletin of the American Mathematical Society, 26*, 394-395.

Mordecai, A. (2003). *Nonlinear programming: Analysis and methods.* New York: Dover Publishing.

Nabney, I. T. (2001). *Netlab: Algorithms for pattern recognition.* Heidelberg: Springer-Verlag.

Nelwamondo, F. V. 2008, *Computational intelligence techniques for missing data imputation.* Unpublished doctoral dissertation, University of the Witwatersrand, Johannesburg.

Nelwamondo, F. V., & Marwala, T. (2007a). Handling missing data from heteroskedastic and nonstationary data. *Lecture Notes in Computer Science, 4491*, 1297-1306.

Nelwamondo, F. V., & Marwala, T. (2007b) Fuzzy ARTMAP and neural network approach to online processing of inputs with missing values. *SAIEE Africa Research Journal, 98*, 45-51.

Nelwamondo, F. V., & Marwala, T. (2007c) Rough set theory for the treatment of incomplete data. In *Proceedings of the IEEE Conference on Fuzzy Systems* (pp. 338-343).

Nelwamondo, F. V., & Marwala, T. (2008). Techniques for handling missing data: applications to on-line condition monitoring. *International Journal of Innovative Computing, Information and Control, 4,* 1507-1526.

Nelwamondo, F. V., Mohamed, S., & Marwala, T. (2007). Missing data: A comparison of neural network and expectation maximisation techniques. *Current Science, 93,* 1514-1521.

Patel, P., & Marwala, T. (2006). Neural networks, fuzzy inference systems and adaptive-neuro fuzzy inference systems for financial decision making. *Lecture Notes in Computer Science, 4234,* 430-439.

Pendharkar, P. C., & Rodger, J. A. (1999). An empirical study of non-binary genetic algorithm-based neural approaches for classification. In *Proceedings of the 20th International Conference on Information Systems,* 155-165.

Penrose, R. (1955). A generalized inverse for matrices. In *Proceedings of the Cambridge Philosophical Society, 51,* 406-413.

Robbins, H., & Monro, S. (1951). A stochastic approximation method. *Annals of Mathematical Statistics, 22,* 400-407.

Roth, P. (1994). Missing data: A conceptual overview for applied psychologists. *Personnel Psychology, 47,* 537-560.

Rubin, D. B. (1978). Multiple imputations in sample surveys - A phenomenological Bayesian approach to non-response. In *Proceedings of the Survey Research Methods Section of the American Statistical Association* (pp. 20-34).

Tettey, T., & Marwala, T. (2006). Controlling interstate conflict using neuro-fuzzy modeling and genetic algorithms. In *Proceedings of the 10th IEEE International Conference on Intelligent Engineering Systems* (pp. 30-44).

Vilakazi, B. C., & Marwala, T. (2007). Condition monitoring using computational intelligence. In D. Laha and P. Mandal (Eds.), *Handbook on Computational Intelligence in Manufacturing and Production Management* (pp. 106-143). New York: Idea Group Inc (IGI).

Vose, M. D. (1999). *The simple genetic algorithm: Foundations and theory.* Cambridge, MA: MIT Press.

Werbos, P. J. (1974). *Beyond regression: New tool for prediction and analysis in the behavioral sciences.* Unpublished doctoral dissertation, Harvard University, Cambridge.

Yoon, Y., & Peterson, L. L. (1990). Artificial neural networks: an emerging new technique. In *Proceedings of the 1990 ACM SIGBDP Conference on Trends and Directions in Expert Systems* (pp. 417-422).

Yuan, Y. (2000). Multiple imputation for missing data: Concepts and new development. *In SUGI Paper,* 267-325.

Chapter III
A Hybrid Approach to Missing Data:
Bayesian Neural Networks, Principal Component Analysis and Genetic Algorithms

ABSTRACT

The problem of missing data in databases has recently been dealt with through the use computational intelligence. The hybrid of auto-associative neural networks and genetic algorithms has proven to be a successful approach to missing data imputation. Similarly, two auto-associative neural networks are developed to be used in conjunction with genetic algorithm to estimate missing data, and these approaches are compared to a Bayesian auto-associative neural network and genetic algorithm approach. One technique combines three neural networks to form a hybrid auto-associative network, while the other merges principal component analysis and neural networks. The hybrid of the neural network and genetic algorithm approach proves to be the most accurate when estimating one missing value, while a hybrid of principal component and neural networks is more consistent and captures patterns in the data more efficiently.

INTRODUCTION

As explained in earlier chapters, the occurrence of missing data in databases, such as those that house medical data, for example ante-natal HIV data, is a common problem (Schafer and Graham, 2002; Nelwamondo & Marwala, 2007). Missing data may come about as a consequence of inefficiencies in the data acquisition or data storage processes. Non-responses to various fields in a questionnaire, a break in the transmission line or failure of hardware are some of the general causes of missing data (Abdella & Marwala, 2006). Many knowledge discovery and data analysis techniques for databases depend heavily on complete data (Fujikawa, 2001; Yim and Mitchell, 2005). For that reason, an effective method for tackling the problem of missing data is required (Abdella & Marwala, 2006).

In many applications, cases with missing data are simply deleted. However, this approach may result in biased or erroneous results and, thereby, wrong conclusions. Furthermore, it is not always an option in engineering applications where decisions have to be made, despite the fact that some aspect of the data is missing (Yuan & Bentler, 2000; Wayman, 2003; Markey et al., 2006). If a sensor fails, the value for that sensor will need to be estimated quickly and accurately, based on the values of the other sensors (Abdella & Marwala, 2006). This scenario illustrates the idea of data imputation, where missing data are predicted based on existing data (Wayman, 2003).

In recent times, there has been an increased interest in dealing with missing data problem by estimation or imputation (Nelwamondo & Marwala, 2007; Abdella & Marwala, 2006). The auto-associative neural network (AANN) coupled with the genetic algorithm (GA) has been shown to be a successful approach to missing data estimation (Nelwamondo, 2008; Abdella & Marwala, 2006) and this is described in detail in Chapter II. The efficient and effective estimation of missing data relies on the extraction and storage of the inter-relationships and rules that govern the data (Nelwamondo, 2008). Auto-associative neural networks facilitate the effective identification of these inter-relationships and rules (Kramer, 1991). However, other simple techniques such as standard the principal component analysis (PCA) can also be used (Jollife, 1986). The PCA is therefore used in this chapter, as it was found to be successful in complex applications such as in power plant monitoring (Pan, Flynn, & Cregan, 2007).

In this chapter, a hybrid auto-associative network is developed and its performance in conjunction with a genetic algorithm is compared to that of the ordinary auto-associative neural network. Hybrid neural network approaches have been successful in many applications such as handwriting recognition (Chiang, 1998) and temperature control (Ng & Hussain, 2004). A combined principal component analysis and neural network missing data estimation system is also developed and compared to the other two systems. A description of this method is presented, followed by the experimental implementation using the HIV ante-natal data. Results from this implementation are presented and conclusions are then drawn.

BACKGROUND

Traditional methods of data imputation, such as mean substitution, regression-based methods and resemblance-based or 'hot deck imputation' may produce biased results (Gold & Bentler, 2000). Regression-based methods predict missing values while resemblance-based methods impute new values based on similar cases (Yuan & Bentler, 2000) and other imputation methods include multiple imputation and Expectation Maximisation (Little & Rubin, 1987; Yuan & Bentler, 2000) and are dealt with by (Wayman, 2003).

Multi-layer perceptron (MLP) and radial basis functions (RBF) auto-associative neural networks combined with optimization algorithms have also been used to successfully estimate missing data (Fariñas & Pedreira, 2002). In this chapter it is assumed that all data are missing via the missing at random (MAR), missing completely at random (MCAR) or missing not at random (MNAR) mechanisms. Furthermore, it is postulated that it is possible to estimate missing entries based on the other data that are measured, regardless of the existence of these mechanisms. This is done using a number of hybrid methods, which are explained in this chapter (Nelwamondo & Marwala, 2007).

As explained earlier, the MCAR describes the case when the probability of a value of a variable missing does not depend on itself or on any of the other variables in the dataset (Abdella & Marwala, 2006). Effectively, the cases with missing entries are the same as the complete cases (Wayman, 2003). The second case, MAR describes the situation where missing data can be described fully by the remaining variables in the dataset (Wayman, 2003). The missing datum depends on the remaining data, but not on itself or on any other missing data (Nelwamondo, 2008; Dhlamini, Nelwamondo, & Marwala, 2006; Dlhamini, 2007). In the last case, MNAR occurs when the missing datum depends on itself or on other missing data (Nelwamondo, 2008). As a consequence, it cannot be deduced from the existing data set and is hence termed non-ignorable case (Wayman, 2003).

In Chapter II, it was shown that the MLP neural network performs better than the RBF network, but the difference in performances is not substantial. For this reason the hybrid approach presented in this chapter uses both the RBF and the MLP networks. In Chapter II, the MLP neural network was formulated in a maximum likelihood framework and trained using the scaled conjugate gradient method. In this chapter some of the MLP networks used are formulated in the Bayesian framework (Nakada et al., 2005) and trained using the hybrid Monte Carlo method (Ökten, Tuffin, & Burago, 2006).

Bayesian Auto-Associative Neural Networks

As described in Chapter II, an *auto-associative neural network*, also referred to as *auto-encoder neural network*, is a specific type of neural network, which is trained to recall its own inputs (Leke, Marwala, & Tettey, 2006). Given a set of inputs, the network predicts these inputs as outputs and thus has the same number of output nodes as there are inputs (Leke et al., 2006). Nevertheless, the hidden layer is characterized by a bottleneck, with fewer hidden nodes than the output nodes (Mohammed, Nelwamondo & Marwala, 2007; Leke et al., 2006). This gives it a butterfly-like structure (Nelwamondo, 2008). The smaller hidden layer projects the inputs onto a smaller space, extracting linear and nonlinear inter-relationships, such as covariance and correlation from the input space and also removing redundant information (Leke et al., 2006). These networks can be used for data reduction and in missing data estimation applications, as is described in Chapter II.

In this chapter, neural networks are viewed as parameterized regression models that make probabilistic assumptions about the data. The probabilistic outlook of these models is facilitated by the use of the Bayesian framework. Learning algorithms are viewed as methods for finding parameter values that look probable in the light of the data. In Chapter II, the learning process was conducted by dividing the data into training, validation and testing sets. This was done for model selection and also for ensuring that the trained network was not biased towards the training data it had seen. Another way of achieving this is by the use of a regularization framework, which comes naturally from the Bayesian formulation and is discussed in detail in this chapter.

There are several types of neural network procedures, two of which have been described in Chapter II. These are multi-layer perceptron and radial basis function (Bishop, 1995). In this chapter, the auto-associative MLP is used for missing data estimation because it provides a distributed representation with respect to the input space due to cross-coupling between hidden units. In this chapter, the auto-associative MLP architecture contains a tangent basis function in the hidden units and linear functions in the output units (Bishop, 1995). This network architecture contains hidden units and output units and has one hidden layer. The relationship between the output y and input x may be may be written as follows (Bishop, 1995), as was introduced in Chapter II:

$$y_k = f_{outer}\left(\sum_{j=1}^{M} w_{kj}^{(2)} f_{inner}\left(\sum_{i=1}^{d} w_{ji}^{(1)} x_i + w_{j0}^{(1)}\right) + w_{k0}^{(2)}\right)$$

(3.1)

As introduced in Chapter II, in equation 3.1, $w_{ji}^{(1)}$ and $w_{kj}^{(2)}$ indicate weights in the first and second layer, respectively, going from input i to hidden unit j or the output unit k, M is the number of hidden units, d is the number of output units while $w_{j0}^{(1)}$ indicates the bias for the hidden unit j, and $w_{k0}^{(2)}$ indicates the biases for the output unit k. In this chapter, as in Chapter II, the function $f_{outer}(\bullet)$ is again logistic and f_{inner} is again a hyperbolic tangent function. The bias parameters in the first layer are now shown as weights from an extra input having a fixed value of x_0=1. The bias parameters in the second layer are weights from an extra hidden unit, with the activation fixed at z_0=1.

The training of the neural network consists of identifying the weights in equation 3.1. An objective function must be chosen to identify these weights. If the training set $D = \{\{x_k\}, \{y_k\}\}_{k=1}^{N}$ is used and assuming that the target vector $\{y\}$ is sampled independently given the input vector $\{x_k\}$ and the weight parameters vector, $\{w_{kj}\}$ the objective function, E, may thus be written as follows by using the sum of square errors (Bishop, 1995):

$$E = \beta \sum_{n=1}^{N} \sum_{k=1}^{K} \{y_{nk} - t_{nk}\}^2 + \frac{\alpha}{2} \sum_{j=1}^{W} w_j^2$$

(3.2)

The sum of squares of the error function is chosen because it has been established to be suited to regression problems (Bishop, 1995). In equation 3.2, n is the index for the training pattern, β is the data contribution to the error and k is the index for the output units. The second term is the regularization parameter and it penalises weights of large magnitudes (Bishop, 1995). This regularization parameter is called the *weight decay* and its coefficient α determines the relative contribution of the regularization term on the training error. This regularization parameter ensures that the mapping function is smooth and, therefore, the training process does not over-train and, thereby, be biased towards the training data set. Including the regularization parameter has been found to give significant improvements in network generalization. The regularization component, in equation 3.2, may be viewed as prior information within the Bayesian framework and this will be elaborated further in this chapter. If α is too high then the regularization parameter over-smoothes the network weights, thereby, giving inaccurate results. On the other hand, if α is too small, then the effect of the regularization parameter is negligible and unless other measures are implemented that control the complexity of the model, such as the early stopping method or cross-validation method, described in Chapter II, (Bishop, 1995), then the trained network becomes too complex and thus performs poorly on the validation set.

Before network training is performed, the network architecture needs to be constructed by choosing the number of hidden units, M. If M is too small, the neural network will be insufficiently flexible and will give poor generalization of the data because of high bias. However, if M is too large, then the neural network will be unnecessarily flexible and will give poor generalization due to the phenomenon mentioned in Chapter II and known as *over-fitting caused by high variance*. In this chapter, the number of hidden nodes is chosen through trial and error. The problem of identifying the weights vector $\{w\}$ may be posed in Bayesian form as follows (Bishop, 1995; Marwala & Sibisi, 2004):

$$P(\{w\} \mid D) = \frac{P(D \mid \{w\}) P(\{w\})}{P(D)}$$

(3.3)

In equation 3.3, the parameter $P(w)$ is the probability distribution function of the weight-space in the absence of any data, is also known as the prior distribution. The matrix $[D] \equiv (x_1,...,x_N,y_1,...,y_N)$ is a matrix containing the input and output data. In this chapter, the input vector $\{x\}$ is the same as the output vector $\{y\}$ for the reason that the model being built is the auto-associative neural network. The quantity $P(\{w\}|[D])$ is the posterior probability distribution after the data have been seen and $P([D]|\{w\})$ is the likelihood function. Equation 3.3 may be expanded using equation 3.2 to give (Marwala, 2004):

$$P(\{w\}|[D]) = \frac{1}{Z_s} \exp\left(-\beta \sum_{n}^{N} \sum_{k}^{K} \{y_{nk} - t_{nk}\}^2 - \frac{\alpha}{2} \sum_{j}^{W} w_j^2 \right) \qquad (3.4)$$

In equation 3.4:

$$Z_S(\alpha, \beta) = \int \exp\left(-\beta E_D - \alpha E_W \right) dw$$
$$= \left(\frac{2\pi}{\beta} \right)^{N/2} + \left(\frac{2\pi}{\alpha} \right)^{W/2} \qquad (3.5)$$

In equation 3.4, the optimal weight vector corresponds to the maximum of the posterior distribution which is identified using the scaled conjugate gradient method in Chapter II. The distribution, in equation 3.4, is a canonical distribution (Haykin, 1995). Training the network using a Bayesian approach automatically penalises highly complex models and one is, therefore, able to select an optimal model without applying independent methods such as cross-validation (Bishop, 1995). It also gives a probability distribution of the output of the networks which can be used to assess the reliability of the estimated missing values. There are many methods that have been applied to solve equation 3.4, including most recently a method proposed by Marwala (2007) which samples the weight space using genetic programming and many others (Vivarelli & Williams, 2001). Further details on Bayesian neural network training and the application of this to condition monitoring in a mechanical system may be found in Neal (1992-1994) and Marwala (2001a).

In this chapter, the method of sampling through a posterior distribution of weights, described in equation 3.4, called the *hybrid Monte Carlo method* (Neal, 1994) is reviewed and then applied. Distributions of this nature (equation 3.4) have been studied extensively in statistical mechanics. In statistical mechanics, macroscopic thermodynamic properties are derived from the state space, i.e., from the position and momentum of microscopic objects such as molecules. The number of degrees of freedom that these microscopic objects have is enormous, so the only way to solve this problem is to formulate it in a probabilistic framework.

Furthermore, in this chapter, the hybrid Monte Carlo method, which uses the gradient of the error, calculated using back-propagation as described in Chapter II, is used to identify the posterior probability of the weight vectors given the training data. The use of the gradient ensures that the simulation samples through the regions of higher probabilities and thus increases the time it takes to converge on a stationary probability distribution function. This technique is viewed as a form of a Markov chain with transition between states achieved by alternating the 'stochastic' and 'dynamic' moves. The 'stochastic' moves allow the algorithm to explore states with different total energy while the 'dynamics' moves are achieved by using the Hamiltonian dynamics and allows the algorithm to explore states with the total energy approximately constant. In simple form, the hybrid Monte Carlo method can be viewed as a combination of Monte Carlo sampling method, which is guided by the gradient of the probability distribution function at each state.

The Stochastic Dynamics Model

As mentioned before, in statistical mechanics the positions and the momentum of all molecules at a given time in a physical system define the state space of the system at that time. The positions of the molecules define the potential energy of a system and the momentum defines the kinetic energy of the system. What is referred to in statistical mechanics as the canonical distribution of the 'potential energy' in this chapter is the posterior distribution in equation 3.4. The canonical distribution of the system's kinetic energy is:

$$P(\{p\}) = \frac{1}{Z_K} \exp(-K(\{p\}))$$

$$= (2\pi)^{-n/2} \exp(-\frac{1}{2}\sum_i p_i^2) \tag{3.6}$$

In molecular dynamics p_i is the momentum of the ith molecule. Here p is not to be mistaken with, P, which indicates probability. In a neural network, p_i is a fictitious parameter that is used to give the procedure a molecular dynamics structure. It should be noted that the weight vector, $\{w\}$, and momentum vector, $\{p\}$, are of the same size. For that reason, the superscript W is used in equation 3.4. The combined kinetic and potential energy is called the *Hamiltonian of the system* and can be written as follows (Neal, 1994; Bishop 1995; Marwala, 2001a):

$$H(w,p) = \beta \sum_{}^{N} \sum_{k}^{K} \{y_{nk} - t_{nk}\}^2 + \frac{\alpha}{2}\sum_{j=1}^{W} w_j^2 + \frac{1}{2}\sum_i^{W} p_i^2 \tag{3.7}$$

In equation 3.7, the first two terms are the potential energy of the system, which is the exponent of the posterior distribution of equation 3.7, and the last term is the kinetic energy. The canonical distribution over the phase space, i.e., position and momentum, can be written as follows (Neal, 1994; Bishop 1995; Marwala, 2001a):

$$P(w,p) = \frac{1}{Z}\exp(-H(w,p)) = P(w|D)P(p) \tag{3.8}$$

By sampling through the distribution in equation 3.8, the posterior distribution of weight is obtained by ignoring the distribution of the momentum vector, p. The dynamics in the phase space may be specified in terms of the Hamiltonian dynamics by expressing the derivative of the 'position' and 'momentum' in terms of fictitious time τ. It should be recalled here that the word 'position' used here is synonymous to network weights. The dynamics of the system may thus be written by using the Hamiltonian dynamics as follows (Neal, 1994; Bishop 1995; Marwala, 2001a):

$$\frac{dw_i}{d\tau} = +\frac{\partial H}{\partial p_i} = p_i \tag{3.9}$$

$$\frac{dp_i}{d\tau} = +\frac{\partial H}{\partial w_i} = -\frac{\partial E}{\partial p_i} \tag{3.10}$$

The dynamics, specified in equations 3.9 and 3.10, cannot be followed exactly. As a result these equations are discretized using a 'leapfrog' method. The leapfrog discretization of equations 3.9 to 3.10 may be written as follows (Neal, 1994; Bishop 1995; Marwala, 2001b):

$$\hat{p}_i(\tau + \frac{\varepsilon}{2}) = \hat{p}_i(\tau) - \frac{\varepsilon}{2}\frac{\partial E}{\partial w_i}(\hat{w}(\tau)) \tag{3.11}$$

$$\hat{w}_i(\tau + \varepsilon) = \hat{w}_i(\tau) + \varepsilon\hat{p}_i(\tau + \frac{\varepsilon}{2}) \tag{3.12}$$

$$\hat{p}_i(\tau + \varepsilon) = \hat{p}_i(\tau + \frac{\varepsilon}{2}) - \frac{\varepsilon}{2}\frac{\partial E}{\partial w_i}(\hat{w}(\tau + \varepsilon)) \tag{3.13}$$

Using equation 3.11, the leapfrog takes a little half step for the momentum vector, $\{p\}$, and using equation 3.12, takes a full step for the 'position', $\{w\}$. Then, using equation 3.13, takes a half step for the momentum vector, $\{p\}$. The combination of these three steps form a single leapfrog iteration that calculates the 'position' and 'momentum' of a system at time $\tau+\varepsilon$ from the network weight vector and 'momentum' at time τ. The above discretization is reversible in time. It almost conserves the Hamiltonian, representing the total energy, and preserves the volume in the phase space, as required by Liouville's theorem (Neal, 1993). The volume preservation is achieved because the moves the leapfrog steps take are shear transformations.

One issue that should be noted is that following Hamiltonian dynamics does not ergodically sample through the canonical distribution, as represented by equation 3.4, because the total energy remains constant, but rather at most samples through the micro-canonical distribution for a given energy. One way used to ensure that the simulation is ergodic, is by introducing 'stochastic' moves by changing the Hamiltonian, H, during the simulation. This is achieved by replacing the 'momentum' vector, $\{p\}$, before the next leapfrog iteration is performed. In this chapter, a normally distributed vector with a zero-mean replaces the 'momentum' vector.

The dynamic steps introduced in this section make use of the gradient of the error with respect to the 'position' (network weights) as shown in equation 3.11. In this section, a procedure on how to move from one state to another is described. This procedure uses Hamiltonian dynamics to achieve dynamic moves and randomly changes the 'momentum' vector to achieve stochastic moves. The next section describes how the states visited are either accepted or rejected.

The Metropolis Algorithm

An algorithm due to Metropolis et al. (1953) has been used extensively to solve problems of statistical mechanics. In the Metropolis algorithm, on sampling a stochastic process $\{X_1, X_2, ..., X_n\}$ consisting of random variables, random changes to X are considered and are either accepted or rejected according to the following criterion:

if $H_{new} < H_{old}$ accept state (w_{new}, p_{new})

else

accept (w_{new}, p_{new}) with probability

$\exp\{-(H_{new} - H_{old})\}$ (3.14)

In this chapter, this procedure is viewed as a way of generating a Markov chain with the transition from one state to another conducted using the criterion in equation 3.14. By investigating carefully equation 3.14, it may be observed that states with high probability form the majority of the Markov chain, and those with low probability form the minority of the Markov chain. However, simulating a distribution by perturbing a single vector, $\{w\}$ is infeasible due to high dimensional nature of the state space and the variation of the posterior probability of weight vector. A technique that exploits the gradient of the Hamiltonian with respect to the weight vector, $\{w\}$, is used to improve the Metropolis algorithm described in this section, and is the subject of the next section.

Hybrid Monte Carlo Method

A Hybrid Monte Carlo method combines the stochastic dynamics model with the Metropolis algorithm and by so doing eliminates the bias introduced by using a non-zero step size, as shown in equations 3.11 to 3.13. The Hybrid Monte Carlo method works by taking a series of trajectories from an initial state, i.e., 'positions' and 'momentum', and moves in some direction in the state space for a given length of time and accepts the final state using the Metropolis algorithm. The validity of the hybrid Monte Carlo rests on three properties of the Hamiltonian dynamics and these are:

1. Time reversibility: it is invariant under t→-t, p→-p.
2. Conservation of energy: the $H(w,p)$ is the same at all times.
3. Conservation of state space volumes apply from Liouville's theorem (Neal, 1993).

For a given leapfrog step size, ε_0, and the number of leapfrog steps, L, the dynamic transition of the hybrid Monte Carlo procedure is conducted as follows:

1. Randomly choose the direction of the trajectory, λ, to be either −1 for a backward trajectory or +1 for a forward trajectory.
2. Starting from the initial state, $(\{w\}, \{p\})$, perform L leapfrog steps (equations 3.11 to 3.13) with the step size $\varepsilon = \varepsilon_0(1 + 0.1k)$ resulting in state $(\{w\}^*, \{p\}^*)$. Here ε_0 is a chosen fixed step size, and k is the number chosen from a uniform distribution and lies between 0 and 1. The reason why this step size is used is explained later in this chapter.
3. Reject or accept $(\{w\}^*, \{p\}^*)$ using the Metropolis criterion. If the state is accepted then the new state becomes $(\{w\}^*, \{p\}^*)$. If rejected, the old state, $(\{w\}, \{p\})$, is retained as a new state.

After implementing step (3), the momentum vector is re-initialized before moving on to generate the subsequent state. In this chapter, the momentum vector is sampled from a Gaussian distribution before starting to generate the subsequent state. This ensures that the stochastic dynamics model samples are not restricted to the micro-canonical ensemble. By replacing the momentums, the total energy is allowed to vary because the momentums of particles are refreshed. This idea of replacing the momentum was introduced by Anderson (1980).

One remark that should be made about the hybrid Monte Carlo method is that it makes use of the gradient information in step (2) above, using the leapfrog steps in equation 3.13. The advantages of using this gradient information is that the hybrid Monte Carlo trajectories move in the direction of high probabilities, resulting in the improved probability that the resulting state is accepted and that the accepted states are not highly correlated. In neural networks, the gradient is calculated using back-propagation (Bishop, 1995), which was explained in Chapter II. The number of leapfrog steps, L, must be significantly higher than one to allow a faster exploration of the state space. The choice of ε_0 and L affects the speed at which the simulation converges to a stationary distribution and the correlation between the states accepted. The leapfrog discretization does not introduce systematic errors due to occasional rejection of states, which result with the increase of the Hamiltonian.

In step (2) of the implementation of the hybrid Monte Carlo method, the step size $\varepsilon = \varepsilon_0(1+0.1k)$ where k is uniformly distributed between 0 and 1 and is not fixed. In effect this ensures that the actual step size for each trajectory is varied so that the accepted states do not have a high correlation (Mackenzie, 1989). The same effect can be achieved by varying the leapfrog steps. In this chapter only the step size is varied.

The application of the Bayesian approach to neural networks results in weight vectors that have a mean and standard deviation and thus have a probability distribution. As a result, the output parameters have a probability distribution. Following the rules of probability theory, the distribution of the output vector $\{y\}$ for a given input vector $\{x\}$ may be written in the following form:

$$p(\{y\}|\{x\},D) = \int p(\{y\}|\{x\},\{w\})p(\{w\}|D)d\{w\} \tag{3.15}$$

In this chapter, the hybrid Monte Carlo method is employed to determine the distribution of the weight vectors, and subsequently, of the output parameters. The integral in equation 3.15 may be approximated as follows:

$$I \equiv \frac{1}{L}\sum_{i=1}^{L} f(\{w\}_i) \tag{3.16}$$

In equation 3.16, L is the number of retained states and f is the MLP network.

The application of Bayesian framework to auto-associative neural network results in a mapping weight vector between the input and output with a probability distribution. This is different from the auto-associative network implemented in Chapter II, where the mapping weight vector does not have a probability distribution. This gives the opportunity of giving confidence intervals to the estimated missing data, a situation that is not possible in Chapter II. The resulting network set-up of a Bayesian trained network can thus be illustrated as shown in Figure 3.1.

In Figure 3.1, the MLP mapping function has weight vector that forms a probability distribution function. These weight vectors are L in number, from the retained states resulting from the hybrid Monte Carlo sampling process.

Figure 3.1. Auto-associative network trained using Bayesian framework

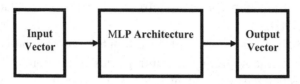

Genetic Algorithm (GA)

The missing data presented in this chapter also uses genetic algorithm as was the case in Chapter II. Consequently only a brief overview of GA will be conducted in this chapter. As explained in Chapter II, the genetic algorithm is a population-based, probabilistic technique that operates to find a solution to a problem from a population of possible solutions (Kubalik & Lazansky, 1999). It is based on Darwin's theory of evolution where members of the population compete to survive and reproduce while the weaker ones die out. Each individual is assigned a fitness value according to how well it meets the objective of solving the problem. New and more evolutionary fit individual solutions are produced during a cycle of generations, wherein selection and re-combination operations, analogous to gene transfer are applied to the current individuals. This continues until a termination condition is met, after which the best individual thus far is considered to be the solution to the problem.

Unlike many optimization algorithms, the genetic algorithm converges to a global optimal solution. In addition, the GA has also been proven to be very successful in many applications including finite element analysis (Marwala, 2003), selecting optimal neural network architecture (Arifovic & Gençay, 2001), training hybrid fuzzy neural networks (Oh & Pedrycz, 2006), solving job scheduling problems (Park, Choi, & Kim, 2003), remote sensing (Stern, Chassidim, & Zofi, 2006) and combinatorial optimization (Zhang & Ishikawa, 2004). Other optimization techniques can also be used in place of genetic algorithm. These include particle swarm optimization as well as simulated annealing (Marwala, 2005). These techniques are the subjects of discussion in later chapters.

For a successful implementation of the GA, the cross-over and mutation should be implemented. So in this chapter, a simple cross-over and non-uniform mutation are used. A simple one-point-cross-over works through cutting a gene, which represents a potential solution, at a cross-over point on both parents' strings and all data after that point in either organism string are exchanged between the two parents and the resulting strings are then called children (Gwiazda, 2006). Non-uniform mutation operates by increasing the probability in such a way that it will be close to 0 as the generation number increases sufficiently. It prevents the population from stagnating in the initial stages of the evolution process, and then permits the algorithm to refine the solution in the end stages of evolution (Michalewicz & Dasgupta, 1997). Normalized geometric selection is a technique where the probability of selecting the ith solution is $\upsilon(\upsilon-1)^{r-1}/(1-(1-\upsilon)^{P})$. Here v is the probability of selecting the best solution, r is the rank of the solution where 1 is the best, and superscript P is the population size (Eiben & Smith, 2003).

Principal Component Analysis

Principal Component Analysis (PCA) is a technique of identifying patterns in the data and displaying those patterns through emphasizing the similarities and differences amongst the data (Jollife, 1986). It is used in data analysis to identify and extract the principal correlation variables amongst data (Kramer, 1991). This permits the dimensionality of the data to be reduced, without the loss of vital information. Hence, the data is effectively compressed. Consequently, principal component analysis has been applied to areas such as image compression, seismic damage detection (Ko, Zhou, & Ni, 2002), environmental forensics (Johnson, Ehrlich, & Full, 2002) and pharmaceutical problems (Adams et al., 2002). PCA has been described as the optimum linear, information-preserving transformation (Fukunaga, & Koontz, 1970) and has been shown to facilitate many types of multivariate analysis, including fault detection and data validation (Marwala, 2001c; Lee, Yoo, & Lee, 2004).

In this chapter, the principal component analysis is used to reduce the input data into independent input data. A principal component analysis orthogonalizes the components of the input vector so that they are uncorrelated with each other. When implementing PCA for data reduction, correlations and interactions among variables in the data are summarized in terms of a small number of underlying factors. Mathematically, a principal component analysis is calculated by first decomposing the input space matrix XT, which is assumed to have zero empirical mean, using singular value decomposition into $V \Sigma W^T$ (Demmel & Kahan, 1990) giving:

$$Y^T = X^T W$$
$$= \Sigma V^T \tag{3.17}$$

In equation 3.17, Y is the data in the reduced space, V is the matrix that consists of eigenvectors and W is the basis vector. In this chapter, the variant of the PCA implemented finds the directions in which the data points have the most variance. These directions are called principal directions. The data are then projected onto these principal directions without the loss of significant information of the data. Here a brief outline of the implementation of principal component analysis adopted in this chapter is described. The first step in the implementation of principal component analysis is to construct a covariance matrix defined as follows (Jollife, 1986):

$$\wedge = \sum_{p=1}^{P} (x^p - \mu)(x^p - \mu)^T \tag{3.18}$$

Here $\wedge$ is the covariance matrix and the superscript P is the number of vectors in the training set, μ is the mean vector of the data set taken over the number of the data set, T is the transpose and x is the input vector. The second step is to calculate the eigenvalues and eigenvectors of the covariance matrix and arrange them from the biggest eigenvalue to the smallest. The first N biggest eigenvalues are chosen. The data are then projected onto the eigenvectors corresponding to the N most dominant eigenvalues.

MISSING DATA ESTIMATION METHOD

For missing data estimation using Bayesian auto-associative neural network model in conjunction with an optimization technique, it is imperative that the Bayesian auto-associative model be as accurate as possible (Leke et al., 2006). For this reason, this chapter attempts to find new and improved ways to capture and model inter-relationships between variables in a dataset and use these inter-relationships all with an optimization technique, the genetic algorithm, to predict missing values.

The approach adopted in this chapter is summarized in Figure 3.2 where $\{x_k\}$ and $\{x_u\}$ are the known and unknown vectors of variables, and when combined they constitute the input space $\{x\}$ (Nelwamondo, 2008).

A genetic algorithm is then used to estimate values for unknown variables and these are input into the auto-associative model along with the known inputs from the measurements (Nelwamondo, 2008). The auto-associative model is trained, using complete data, to extract and store correlations and inter-relationships between variables in the dataset and develop a mapping function F. As a consequence, an output vector $\{y\}$ can, be determined from the trained neural network model with network weights vector $\{w\}$ and the input variables $\{x_k\}$ and $\{x_u\}$, to be written as follows:

$$\{y\} = F\left(\left\{\begin{array}{c}\{x_k\}\\\{x_u\}\end{array}\right\}, \{w\}\right)$$

(3.19)

Because the auto-associative model attempts to replicate the inputs as the outputs, the output is approximately equal to the input. It should be noted that vector $\{w\}$ is probabilistically distributed because a Bayesian based training approach is adopted. The posterior probability distribution, in equation 3.4, is sampled using the hybrid Monte Carlo method. A small error exists between the input and output, the size of which depends on how accurate the auto-associative model is. This error is written as:

$$e = \left\|\left(\left\{\begin{array}{c}\{x_k\}\\\{x_u\}\end{array}\right\} - F\left(\left\{\begin{array}{c}\{x_k\}\\\{x_u\}\end{array}\right\}, \{w\}\right)\right)\right\|$$

(3.20)

Here $\|\ \|$ is Euclidean norm. The error is at a minimum when the output comes closest to matching the input. This occurs only when the data input using the Bayesian auto-associative model carries the same

Figure 3.2. Structure of missing data estimator using an auto-associative model and the genetic algorithm

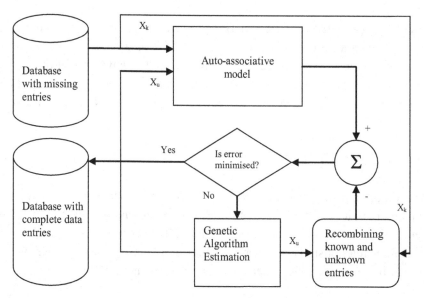

correlations and inter-relationships as those captured during training. Three designs of auto-associative models are presented, i.e. the Bayesian auto-associative neural network, hybrid auto-associative network, which combines multi-layer perceptron network and radial basis function, as well as the combination of principal component analysis and neural networks to form an auto-associative function.

The Bayesian Auto-Associative Neural Network

The regular auto-associative network has a structure similar to that which is described in Chapter II. A Multi-Layer Perceptron architecture is used because preliminary results have shown that it possesses over-all superior performance over radial basis function architecture in the HIV modelling problem studied in this chapter, and also for its superior performance as observed in Chapter II. The MLP is trained using hybrid Monte Carlo method, as described earlier within the context of Bayesian formulation. Because of the Bayesian formulation of the MLP system, it is consequently named the Bayesian MLP. A genetic algorithm is used to optimize the number of hidden nodes and training cycles for the network to make the network as accurate as possible (Leke, et al., 2006). Then again, GA is used to minimize the error in equation 3.20 (Zhong, Lingras, & Sharma, 2004).

The Hybrid Auto-Associative Network

Even though a multi-layer perceptron auto-associative neural network outperforms a radial basis function auto-associative neural network with the data that are used in this chapter, the latter shows superiority in predicting some variables. The design of a hybrid auto-associative network has three objectives:

- To combine the best from both the MLP and RBF architectures.
- To correct some of the distortion introduced by a single auto-associative network.
- To capture complex nonlinear relationships amongst variables more efficiently.

The hybrid approach uses the MLP and a RBF, which is trained using the maximum-likelihood approach, as discussed in Chapter II, as opposed to using Bayesian approach, as discussed in Section 3.3.1. There are a number of hybrid approaches (Taskaya-Temizel & Casey, 2005) that have been implemented in the literature, such as for voltage stability analysis (Wan & Song, 1998) and modeling of polymerization reactors (Chang, Lu, & Chiu, 2007). The next few sections summarize the MLP and RBF networks, as implemented in this chapter in creating the Hybrid auto-associative network.

The Multi-Layer Perceptrons (MLP)

The MLP neural networks consist of multiple layers of computational units, usually interconnected in a feed-forward way. Each neuron in one layer is directly connected to neurons of the subsequent layer. A fully connected two layered MLP architecture is used in this chapter.

The linear activation function was used for the output units and the hyperbolic tangent function is used in the hidden layers. The Scaled Conjugate Gradient (SCG) method was used for training the MLP network (Bishop, 1995). The SCG method was used because it has been found to solve the optimization problems encountered when training an MLP network more efficiently than gradient descent and conjugate gradient methods (Bishop, 1995).

The Radial-Basis Function (RBF)

Radial basis function networks are feed-forward networks trained using a supervised training algorithm (Haykin, 1999). They are typically configured with a single hidden layer of units whose activation function is selected from a class of functions called basis functions. The activation of hidden units in the RBF network is given by a nonlinear function describing a distance between the input vector and a prototype vector (Bishop, 1995).

While similar to MLP in many respects, radial basis function networks have several advantages. They usually train much faster than back-propagation networks and are less prone to problems with non-stationary inputs due to the behavior of the radial basis function (Bishop, 1995). The RBF network can be described as:

$$y_k(\{x\}) = \sum_{j=1}^{M} w_{jk}\phi_j(\{x\})$$

(3.21)

In equation 3.21, the parameters w_{jk} are the output weights, each corresponding to a connection between a hidden unit and an output unit, M represents the number of output units, $\phi_j(\{x\})$ is the jth nonlinear activation function, $\{x\}$ is the input vector, and $k = 1, 2, 3,..., M$ (Bishop, 1995).

The structure of the Hybrid auto-associative network is shown in Figure 3.3, where $\{x\}$ is a set of inputs and $\{y\}$ is the predicted output. The second layer MLP network is trained with a different part of the dataset to that used to train the MLP and RBF auto-associative network. This assists in the corrective ability of the network. The number of hidden nodes in each network is optimized using the GA.

The PCA and Neural Network Approach

To make the auto-associative network more efficient, when working with dimensionally complex and highly nonlinearly related data, a principal component analysis is performed on the data before propaga-

Figure 3.3. Structure of a 3-input, 3-output hybrid auto-associative network

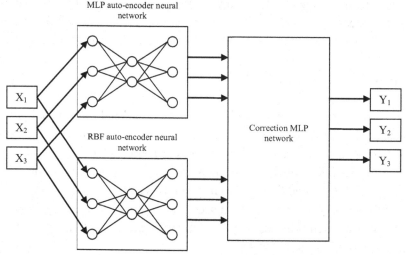

tion through the auto-associative network. This performs much of the linear correlation extraction and reduces the dimensionality of the data, thus reduces the burden on the auto-associative network.

The auto-associative network was trained to recall the reduced data and the inverse PCA is then applied to restore the original data. The MLP neural networks were trained to mimic the principal component extraction and original data reconstruction functions i.e., the PCA and inverse PCA functions. All architectures were optimized using GA. Figure 3.4 depicts the arrangement of the networks, where parameters R and P are the original and predicted dimensionally-reduced data.

Equation 3.21 summarises the auto-associative function, where f_{FWD} is the PCA forward function, f_{AA} is the function of the auto-associative neural network and f_{INV} is the inverse PCA function:

$$\{y\} = f_{INV}\left(\{W\}_3, f_{AA}\left(\{W\}_2, f_{FWD}\left(\{W\}_1, \{x\}\right)\right)\right) \qquad (3.22)$$

In equation 3.22 $\{W\}_1$, $\{W\}_2$ and $\{W\}_3$ are weights of the auto-associative and reconstruction neural networks, respectively.

EXPERIMENTAL IMPLEMENTATION

The dataset used was obtained from South Africa's Antenatal Sero-prevalence Survey of 2001 and consisted of information concerning pregnant women who have visited selected public clinics in South Africa (Leke et al., 2006).

HIV Data Analysis

Only women participating for the first time in this national survey were eligible (Tim & Marwala, 2007). A set of eleven attributes, from the dataset, as successfully found by Leke, Marwala, and Tettey, (2006) are reused here. The set is summarized in Table 3.1.

The HIV status is represented in binary form i.e. a 1 indicating a positive status, while a 0 indicates a negative status. Gravidity refers to the number of combined complete and incomplete pregnancies that

Figure 3.4. Structure of a 3-input, 3 output PCA and neural network auto-associative model

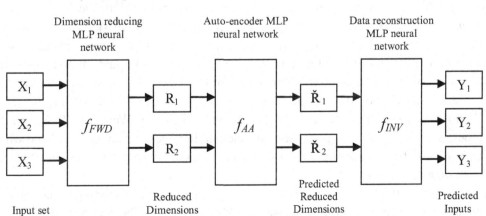

Table 3.1. Summary of variables used from HIV data

Variable number	Variable	Type	Range
1	Age	Integer	14-50
2	Age Gap	Integer	1-7
3	Education	Integer	0-13
4	Gravidity	Integer	0-13
5	Parity	Integer	0-14
6	RPR	Integer	0-2
7	WTREV	Continuous	0.64-1.27
8	Region	Integer	1-77
9	Province	Integer	1-9
10	Race	Integer	1-4
11	HIV status	Binary	[0, 1]

the woman has experienced, while parity refers to the number of occasions on which the woman has given birth. Age gap refers to the difference in age between the pregnant woman and the prospective father of a child. Rapid Plasma Reagin (RPR) refers to a screening test for syphilis for which HIV may cause a false positive test result while WTREV is the income to expenditure ratio. Qualitative variables, stored as text, such as the race of the mother and her province and region of origin, are encoded as integers. The education level of the pregnant woman is represented as an integer corresponding to the highest grade completed, with 13 corresponding to tertiary education.

To deal with the missing data problem effectively, it is important to understand ways data becomes missing so as to identify a possible cause of missing data (Little & Rubin, 1987; Teegavarapu & Chandramouli, 2005). Causes for missing data are commonly described as belonging to three categories, as discussed in detail in Chapter 1 (Wayman, 2003). These are MCAR, MAR and MNAR. In the present dataset, missing data occur following all these three mechanisms. For example, a variable called age gap is missing, which indicates a MNAR mechanism, particularly when the age gap between the couples is large. This is because there is a general stigma in South Africa associated with couples having a large age gap and, therefore, this variable can be missing due to the fact that it is large.

Another variable such as parity goes missing because some people feel that because they have given the variable gravidity, it is not necessary to give this variable, and this missing data mechanism indicates a MAR mechanism. Other missing variables are from a MCAR mechanism, because the reason they become missing depends on another unknown variable. The MAR and MCAR mechanisms are referred to as accessible mechanisms, where the cause of the data going missing can be accounted for (Wayman, 2003). In the case of MNAR, the mechanism is described as an inaccessible mechanism due to lack of knowledge concerning the cause of the data going missing (Wayman, 2003), and there is a belief that there is no choice but to apply list-wise deletion, in which variables are deleted where there are cases of missing data (Yuan & Bentler, 2000). In this chapter, missing data are estimated regardless of the missing data mechanism.

The data consists of 16608 instances. However, many of them contain missing variables or outliers. Such instances were removed, leaving a dataset of 10829 instances. These were normalized and randomized to improve the neural network performance and, thereafter, were split equally into training, validation and testing datasets. Hence 3609 training examples are used.

The neural network architectures were optimized using the validation dataset, while their correct performances was verified and tested using the testing dataset. A total of 100 instances were randomly extracted from the test dataset and the variables, therein, were removed. The resulting dataset, with missing values, was used to test the performance of the three missing data estimation schemes.

Neural Network Optimization

MATLAB® and NETLAB® toolbox (Nabney, 2007) were used to implement all the neural networks. The Bayesian auto-associative network had 11 input nodes, 9 hidden units and 11 output nodes (a 11-9-11) structure and was trained using a hybrid Monte Carlo simulation by returning 1000 samples to form the posterior probability. The MLP auto-associative of the Hybrid auto-associative network also had a similar structure, while its RBF counterpart had one less hidden node in its architecture. The correction MLP was optimized to a 22-19-11 structure.

The optimum number of principal components was found to be 9. These accounted for 99% of the correlation variances in the input space. Fewer principal components resulted in an ineffective reconstruction function. The dimensionality reduction, principal component auto-associative and data reconstruction neural networks were optimized to 11-20-9, 9-8-9 and 9-13-11 node structures, respectively.

GA Implementation

The Genetic Algorithm Optimization Toolbox (GAOT®) was used to implement the GA (Houck, Joines, & Kay, 1995). The initial population size was set to 100, and the process was run for 40 generations. Simple cross-over, non-uniform mutation and normalized geometric selection were used as these were found to produce satisfactory results, not surpassed by using other combinations of GA parameters.

The Performance Evaluation

The effectiveness of the missing data estimation system was evaluated using Standard Error (SE), Correlation Coefficient (r) and the relative prediction accuracy (A). The Standard Error measures the error between the actual values and the predicted values and gives an indication of capability of prediction (Abdella & Marwala, 2006). It is given by equation 3.23, where x_i is the actual value, $\hat{x}_i$ is the predicted value and n is the number of missing values.

$$SE = \sqrt{\frac{\sum_{i=1}^{n}(x_i - \hat{x}_i)^2}{n}}$$

$$(3.23)$$

The Correlation Coefficient measures the linear similarity between the predicted and actual data. The parameter r ranges between -1 and 1, where its absolute value relates the strength of the relationship and the sign of r indicates the direction of relationship. Hence, a value close to 1 indicates a strong predictive capability. The formula where $\bar{x}$ is the mean of the data (Abdella & Marwala, 2006) can be written as:

$$r = \frac{\sum_{i=1}^{n}\left(x_i - \overline{x}_i\right)\left(\hat{x}_i - \overline{\hat{x}}_i\right)}{\left[\sum_{i=1}^{n}\left(x_i - \overline{x}_i\right)^2 \sum_{i=1}^{n}\left(\hat{x} - \overline{\hat{x}}_i\right)^2\right]^{1/2}}$$

(3.24)

The relative prediction accuracy is a measure of how many predictions are made within a certain tolerance, where the tolerance can be set based on the sensitivity demanded by the application (Nelwamondo, 2008). When applied to HIV data, the tolerance was set to 10% as is done by Nelwamondo (2008), since it seems suitable to this application. Using the above mentioned performance parameters, the performance of the three missing data estimation systems were evaluated by estimating each of the 11 attributes individually. Their abilities to estimate two, three and four missing attributes simultaneously were examined through the estimation of the Age, Age Gap, Parity and Gravidity variables.

EXPERIMENTAL RESULTS AND DISCUSSION

Comparisons of the Standard Errors and Correlation Coefficients and relative prediction accuracies for the three systems, when estimating a single missing value, are shown in Figures 3.5 and 3.6. Figure 3.7, 3.8, 3.9 and 3.10 depict the respective estimations of a few instances of the age attribute, when one, two, three and four values are missing from an instance. The average Standard Errors, Correlation Coefficients and relative prediction accuracies for the estimation of 1–4 simultaneous missing values are summarized in Tables 3.2 –3.4.

Table 3.2 summarizes the mean performance of the three systems when estimating a single variable. All the models show effectiveness in estimating certain variables, such as age, age gap, gravidity and parity. This is indicated by the high correlation coefficients, and low errors associated with these variables, as shown in Figures 3.5 and 3.6. Conversely, all estimations of some variables such as HIV status and RPR are poor. This may be due to weak or highly nonlinear inter-relationships between these variables and the rest of the variables in the dataset.

Figure 3.5. Standard errors for all variables when one missing value is estimated by the three estimator systems

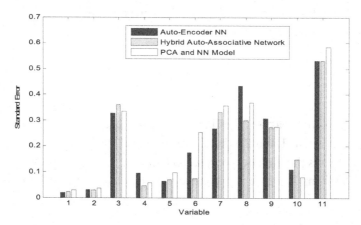

Figure 3.6. Correlation Coefficients for all variables when one missing value is estimated by the three estimator systems

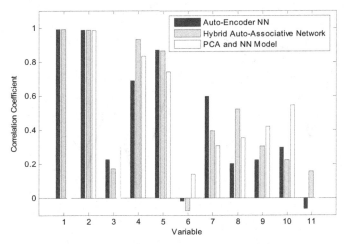

Table 3.2. Overall performance summary for 1 missing value

Method	Standard Error	Correlation Coefficient	Relative Accuracy
Bayesian MLP	0.2248	0.4608	78.80
Hybrid Approach	0.1971	0.4981	82.55
PCA-Neural Network	0.2210	0.5230	72.53

Figure 3.7. Comparing the estimation of the normalized age variable for 1 missing value

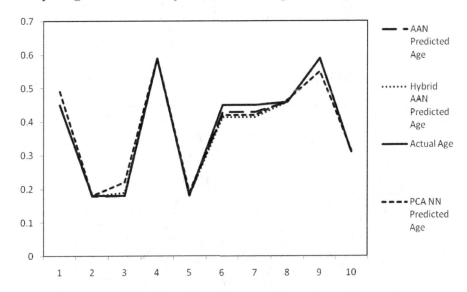

Figure 3.8. Comparing the estimation of the normalized age variable for 2 missing values

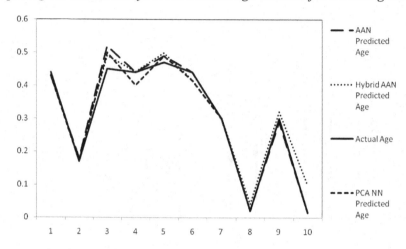

Figure 3.9. Comparison of the estimation of the normalized age variable for 3 missing values

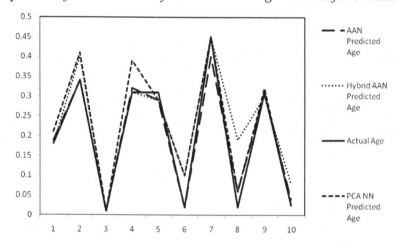

Figure 3.10. Comparison the estimation of the normalized age variable for 4 missing values

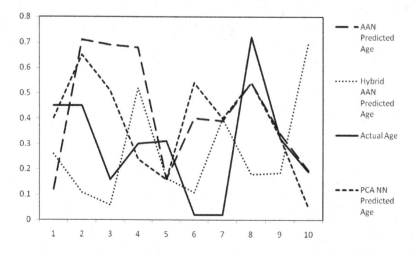

Figures 3.5 to 3.6 show that the Hybrid auto-associative network, coupled with GA is the most accurate when predicting a single missing value. This is supported by Table 3.2, which shows that hybrid auto-associative system produces the smallest mean Standard Error and has the highest overall correlation coefficient and relative prediction accuracy. Overall, the hybrid network performs better than a regular auto-associative neural network. The PCA and neural network auto-associative model demonstrates an increased ability to capture correlations within the data. This is shown in Figure 3.6, where the model produces the highest correlations for most of the poorly estimated variables. Tables 3.3 to 3.4 show that the PCA and neural network auto-associative model coupled with the GA performs best for multiple simultaneous estimations. The considerably higher correlation coefficients shown in Table 3.4, for cases requiring multiple estimations, shows the superiority over the other two models when it comes to capturing patterns and correlations within the data.

For this reason PCA and neural network system would best be suited for implementation in a real system since more than one missing value within an instance is likely, as is the case with the HIV data used.

Figures 3.7 to 3.9 show that all the models are satisfactory for estimating of up to three missing values and that accuracy decreases for increasing simultaneous estimations. However, Figure 3.10 shows the poor estimation capabilities of all the models for the case of 4 missing values. These results are achieved despite the fact that a missing data mechanism is not assumed in designing the missing data estimation method.

CONCLUSION

This chapter investigated the estimation of missing data through novel hybrid techniques. The estimation system involved an auto-associative model to predict the input data, coupled with the genetic algorithm to approximate the missing data. Three auto-associative models were investigated i.e., a Bayesian auto-associative neural network, a hybrid auto-associative network consisting of three neural networks, and a series combination of three neural networks to incorporate a principal component analysis into an

Table 3.3. Mean standard error for variables 1, 2, 4 and 5

Number of missing values	1	2	3	4
Bayesian MLP	0.0531	0.1443	0.1909	0.3154
Hybrid Network	0.0422	0.1446	0.1874	0.2739
PCA-Neural network	0.0542	0.1404	0.1286	0.2621

Table 3.4. Mean correlation coefficient for variables 1, 2, 4 and 5

Number of missing values	1	2	3	4
Bayesian MLP	0.8855	0.7526	0.4465	0.0887
Hybrid Network	0.9458	0.7925	0.4372	0.1191
PCA-Neural network	0.8887	0.8219	0.7133	0.1815

auto-associative function. The performance of each model in conjunction with genetic algorithm was investigated. Results show that the hybrid network is most accurate with single missing value estimation, while a principal component analysis and neural network model provides more consistency for multiple estimations. The latter also appears to perform better than the other two models when dealing with data exhibiting very little inter-dependencies.

FURTHER WORK

This chapter introduced Bayesian neural network for missing data estimation. Hybrid Monte Carlo method was implemented to train the neural network. An alternative to the hybrid Monte Carlo method is a genetic-programming-based Bayesian training of neural networks. For further studies, a comparison between the Hybrid Monte Carlo and genetic programming for Bayesian network training should be made. Secondly, a hybrid network was introduced for missing data estimation. For future work, the best way in which this hybrid network should be constructed should be identified. The methods proposed were implemented on HIV data and the conclusions that are drawn are strictly within the context of such data. In future studies, more databases should be evaluated to study whether the conclusions reached in this chapter are still valid.

REFERENCES

Adams, E., Walczak, B., Vervaet, C., Risha, P. G., & Massart, D. L. (2002). Principal component analysis of dissolution data with missing elements. *International Journal of Pharmaceutics, 234* (1-2), 169-178.

Abdella, M., & Marwala, T. (2006). The use of genetic algorithms and neural networks to approximate missing data in database. *Computing and Informatics, 24,* 1001-1013.

Anderson, H. C. (1980). Molecular dynamics simulations at constant pressure and/or temperature. *Journal of Chemical Physics, 72,* 2384-2393.

Arifovic, J., & Gençay, R. (2001). Using genetic algorithms to select architecture of a feedforward artificial neural network. *Physica A: Statistical Mechanics and its Applications, 289,* 574-594.

Bishop, C. M. (1995). *Neural networks for pattern recognition.* Oxford: Oxford University Press.

Chang, J., Lu, S., & Chiu, Y. (2007). Dynamic modeling of batch polymerization reactors via the hybrid neural-network rate-function approach. *Chemical Engineering Journal, 130,* 19-28.

Chiang, J. (1998). A hybrid neural network model in handwritten word recognition. *Neural Networks, 11,* 337-346.

Demmel, J., & Kahan, W. (1990). Computing small singular values of bi-diagonal matrices with guaranteed high relative accuracy. *SIAM Journal of Scientific and Statistical Computing, 11,* 873-912.

Dhlamini, S. M., Nelwamondo, F. V., & Marwala, T. (2006). Condition monitoring of HV bushings in the presence of missing data using evolutionary computing. *WSEAS Transactions on Power Systems, 1,* 296-302.

Dhlamini, S. M. (2007). *Bushing diagnosis using artificial intelligence and dissolved gas analysis.* Unpublished doctoral dissertation, University of the Witwatersrand, Johannesburg.

Eiben, A. E., & Smith, J. E. (2003). *Introduction to evolutionary computing.* Berlin: Springer.

Fariñas, M., & Pedreira, C. E. (2002). Missing data interpolation by using local-global neural networks. *Engineering Intelligent Systems for Electrical Engineering and Communications, 10,* 85-92.

Fujikawa, Y., (2001). *Efficient algorithms for dealing with missing values in knowledge discovery.* Unpublished master's thesis, Japan Advanced Institute of Science and Technology, Japan.

Fukunaga, K., & Koontz, W. (1970). Application of Karhunen-Loeve expansion to feature selection and ordering. *IEEE Transactions on Computers, 19,* 311-318.

Gold, M. S., & Bentler, P. M. (2000). Treatments of missing data: A Monte Carlo comparison of RB-HDI, iterative stochastic regression imputation, and expectation-maximization. *Structural Equation Modelling, 7,* 319-355.

Gwiazda, T. D. (2006). *Genetic algorithms reference: Crossover for single-objective numerical optimization problems.* Tomasz Gwiazda Publisher.

Haykin, S. (1999). *Neural networks.* New Jersey: Prentice-Hall.

Houck, C. R., Joines, J. A., & Kay, M. G. (1995). *A genetic algorithm for function optimization: A MATLAB implementation* (North Carolina State University-IE Tech. Rep. 95-09). Chapel Hill, North Carolina.

Johnson, G. W., Ehrlich, R., & Full, W. (2002). Principal components analysis and receptor models in environmental forensics. In B. Murphy and R. Morrison (Eds.), *An introduction to environmental forensics* (pp. 461-515). San Diego: Academic Press.

Jollife, I. T. (1986). *Principal component analysis.* New York: Springer-Verlag.

Ko, J. M., Zhou, X. T., & Ni, Y. Q. (2002). Seismic damage evaluation of a 38-storey building model using measured FRF data reduced via principal component analysis. In *Advances in Building Technology* (pp. 953-960).

Kramer, F. A. (1991). Nonlinear principal component analysis using auto-associative neural networks. *AIChE Journal, 3,* 233-243.

Kubalík, J., & Lazanský, J. (1999). Genetic algorithms and their testing. *American Institute of Physics Proceedings, 465,* 217-229.

Lee, J-M., Yoo, C-K, & Lee, I-B. (2004). Fault detection of batch processes using multiway kernel principal component analysis. *Computers & Chemical Engineering, 28,* 1837-1847.

Leke, B. B., Marwala, T., & Tettey, T. (2006). Autoencoder networks for HIV classification. *Current Science, 91,* 1467-1473.

Leke, B. B., Marwala, T., Tim, T., & Lagazio, M. (2006). Prediction of HIV status from demographic data using neural networks. In *Proceedings of the IEEE International Conference on Systems, Man and Cybernetics* (pp. 2339-2344).

Little, R. J. A., & Rubin, D. B. (1987). *Statistical analysis with missing data.* New York: Wiley.

Mackenzie, P. B. (1989). An improved hybrid Monte Carlo method. *Physics Letters B, 226,* 369-371.

Markey, M. K., Tourassi, G. D., Margolis, M., & DeLong, D. M. (2006). Impact of missing data in evaluating artificial neural networks trained on complete data. *Computers in Biology and Medicine, 36,* 516-525.

Marwala, T. (2001a). *Fault identification using neural networks and vibration data.* Unpublished doctoral dissertation, University of Cambridge, Cambridge.

Marwala, T. (2001b). Probabilistic fault identification using a committee of neural networks and vibration data. *American Institute of Aeronautics and Astronautics, Journal of Aircraft, 38,* 138-146.

Marwala, T. (2001c). Scaled conjugate gradient and Bayesian training of neural networks for fault identification in cylinders. *Computers and Structures, 79/32,* 2793-2803.

Marwala, T. (2003). Control of fermentation process using Bayesian neural networks and genetic algorithm. In *Proceedings of the 1st African Control Conference,* Cape Town (pp. 449-454).

Marwala, T. (2004). Fault classification using pseudo modal energies and probabilistic neural networks. *American Society of Civil Engineers, Journal of Engineering Mechanics, 130,* 1346-1355.

Marwala, T. (2005) Finite element model updating using particle swarm optimization. *International Journal of Engineering Simulation, 6,* 25-30.

Marwala, T. (2007). Bayesian training of neural network using genetic programming. *Pattern Recognition Letters, 28,* 1452–1458

Marwala, T., & Sibisi, S. (2005). Finite element updating using Bayesian framework and modal properties. *American Institute of Aeronautics and Astronautics, Journal of Aircraft, 42,* 275-278.

Metropolis, N., Rosenbluth, A. W., Rosenbluth, M. N., Teller, A. H., and Teller, E. (1953). Equations of state calculations by fast computing machines. *Journal of Chemical Physics, 21,* 1087-1092.

Michalewicz, Z., & Dasgupta, D. (1997). *Evolutionary algorithms in engineering applications.* New York: Springer-Verlag.

Mohammed, A. K., Nelwamondo, F. V., & Marwala, T. (2007). Estimation of missing data: neural networks, principal component analysis and genetic algorithms. In *Proceedings of the 18th Annual Pattern Recognition Association of South Africa,* CD-Rom, ISBN: 978-86840-656-2.

Nabney, I., (2007). *Netlab neural network software.* Retrieved May 17, 2008, http://www.ncrg.aston.ac.uk/netlab/.

Nakada, Y., Matsumoto, T., Kurihara, T., & Yosui, K. (2005). Bayesian reconstructions and predictions of nonlinear dynamical systems via the hybrid Monte Carlo scheme. *Signal Processing, 85,* 129-145.

Neal, R. M. (1992). *Bayesian training of backpropagation networks by hybrid Monte Carlo method* (University of Toronto Tech. Rep. CRG-TR-92-1). Toronto, Canada: Department of Computer Science.

Neal, R. M. (1993). *Probabilistic inference using Markov chain Monte Carlo methods* (University of Toronto Tech. Rep. CRG-TR-93-1). Toronto, Canada: Department of Computer Science.

Neal, R. M. (1994). *Bayesian learning for neural networks.* Unpublished doctoral thesis, University of Toronto, Toronto.

Nelwamondo, F. V. (2008). *Computational intelligence techniques for missing data imputation. Unpublished doctoral dissertation.* University of the Witwatersrand, Johannesburg.

Nelwamondo, F. V., & Marwala, T. (2007). Handling missing data from heteroskedastic and nonstationary data. *Lecture Notes in Computer Science, 449,* 1297-1306.

Ng, C. W., & Hussain, M. A. (2004). Hybrid neural network—prior knowledge model in temperature control of a semi-batch polymerization process. *Chemical Engineering and Processing, 43,* 559-570.

Oh, S., & Pedrycz, W. (2006). Genetic optimization-driven multi-layer hybrid fuzzy neural networks. *Simulation Modelling Practice and Theory, 14,* 597-613.

Ökten, G., Tuffin, B., & Burago, V. (2006). A central limit theorem and improved error bounds for a hybrid-Monte Carlo sequence with applications in computational finance. *Journal of Complexity, 22,* 435-458.

Pan, L., Flynn, D., & Cregan, M. (2007). Sub-space principal component analysis for power plant monitoring. *Power Plants and Power Systems Control, 2006,* 243-248.

Park, B. J., Choi, H. R., & Kim, H. S. (2003). A hybrid genetic algorithm for the job shop scheduling problems. *Computers & Industrial Engineering, 45,* 597-613.

Schafer, J. L., & Graham, J. W. (2002). Missing data: Our view of the state of the art. *Psychological Methods, 7,* 147–177.

Stern, H., Chassidim, Y., & Zofi, M. (2006). Multiagent visual area coverage using a new genetic algorithm selection scheme. *European Journal of Operational Research, 175,* 1890-1907.

Taskaya-Temizel, T., & Casey, M. C. (2005). A comparative study of autoregressive neural network hybrids. *Neural Networks, 18,* 781-789.

Teegavarapu, R. S. V., & Chandramouli, V. (2005). Improved weighting methods, deterministic and stochastic data-driven models for estimation of missing precipitation records. *Journal of Hydrology, 312,* 191-206.

Tim, T. N., & Marwala, T. (2006). Computational intelligence methods for risk assessment of HIV. *Proceedings of the IFMBE, 14,* 3581-3585.

Vivarelli, F., & Williams, C. K. I. (2001). Comparing Bayesian neural network algorithms for classifying segmented outdoor images. *Neural Networks, 14,* 427-437.

Wan, H. B., & Song, Y. H. (1998). Hybrid supervised and unsupervised neural network approach to voltage stability analysis. *Electric Power Systems Research, 47,* 115-122.

Wayman, J. C. (2003). *Multiple imputation for missing data: What is it and how can I use it?* Paper presented at the 2003 Annual Meeting of the American Educational Research Association, Chicago, IL.

Yim, J., & Mitchell, H. (2005). Comparison of country risk models: hybrid neural networks, logit models, discriminant analysis and cluster techniques. *Expert Systems with Applications, 28*, 137-148.

Yuan, K. H., & Bentler, P. M. (2000). Three likelihood-based methods for mean and covariance structure analysis with non-normal missing data. *Sociological Methodology, 30*, 165-200.

Zhang, H., & Ishikawa, M. (2004). A solution to combinatorial optimization with time-varying parameters by a hybrid genetic algorithm. *International Congress Series, 1269*, 149-152.

Zhong, M., Lingras, P., & Sharma, S. (2004). Estimation of missing traffic counts using factor, genetic, neural, and regression techniques. *Transportation Research Part C: Emerging Technologies, 12*, 139-166.

Chapter IV
Maximum Expectation Algorithms for Missing Data Estimation

ABSTRACT

Two sets of hybrid techniques have recently emerged for the imputation of missing data. These are, first, the combination of the Gaussian Mixtures Model and the Expectation Maximization algorithms (the GMM-EM) and second, the combination of Auto-Associative Neural Networks with Evolutionary Optimization (the AANN-EO). In this chapter, the evolutionary optimization method implemented is the particle swarm optimization method (the AANN-PSO). Both the GMM-EM and AANN-EO techniques have been discussed individually and their merits discussed at length in the available literature. This chapter provides a comparison between these techniques, using datasets from an industrial power plant, an industrial winding process and an HIV sero-prevalence survey. The results show that GMM-EM method is suitable and performs better in cases where there is little or no interdependency between the input variables, whereas the AANN-PSO combination is suitable when there are inherent nonlinear relationships between some of the given variables.

INTRODUCTION

Databases, such as those that store measurement or medical data may become subject to missing values in either the data acquisition or data-storage process. Problems in a sensor, a break in the data transmission line or non-response to questions posed in a questionnaire are prime examples of how data can go missing. The problem of missing data creates a difficulty in the analysis and decision-making processes that depend on the data to be in a complete form and, thereby, they require methods of estimation that are accurate and efficient. Various techniques exist as a solution to this problem, ranging from data deletion to methods employing statistical and artificial intelligence techniques for the imputation of missing variables. However, some statistical methods, like mean substitution have a high likelihood of producing biased estimates (Tresp, Neuneier, & Ahmad, 1995) or make assumptions about the data that may not be true, affecting the quality of decisions based on these data.

The estimation of missing data in real-time applications requires a system that possesses knowledge of characteristics such as correlations between variables, which are inherent in the input space. Computational intelligence techniques and maximum likelihood techniques do possess such characteristics and as a result are important for the imputation of missing data.

This chapter now compares the above two approaches to the problem of missing data estimation. The first technique is based on the combined use of Gaussian Mixture Models with Expectation Maximization, the GMM-EM (Schafer, 1997; Schafer & Olsen, 1998; Schafer & Graham, 2002). The second approach is the use of a system based on the missing data estimation error equation made out of an Auto-Associative Neural Network (Adbella & Marwala, 2005) and solved using Particle Swarm Optimization, the AANN-PSO. A genetic algorithm was used in Chapter II to solve this equation instead of particle swarm optimization. The estimation abilities of both of these techniques will be compared, based on three datasets and conclusions are then drawn.

Stoica, Xu, and Li (2005) observed that the EM algorithm can be rather slow to converge in some problems. Consequently, they introduced a new algorithm called equalization-maximization algorithm for estimating parameters with missing data. They derived an equalization-maximization algorithm in a generalized fashion and implemented this in the case of a Gaussian auto-regressive time series with a varying number of missing observations. They observed that equalization-maximization did out-perform the EM algorithm in terms of computational speed, but did not necessarily do the same in terms of estimating missing data.

Park, Qian, and Jun (2007) proposed an algorithm for estimating parameters that is based on the likelihood function which accounts for the missing information. They assumed a binomial response and normal exploratory model for the missing data and fitted the model by using the Monte Carlo EM algorithm. They derived the Expectation step using the Metropolis-Hastings algorithm to generate a sample for missing data, and for the Maximization step, they maximized the likelihood function using the Newton-Raphson method. They also derived the asymptotic variances and the standard errors of the maximum likelihood estimates by using the observed Fisher information. Furthermore, M'hiri, Cammoun, and Ghorbel (2007) used the EM algorithm in logistic linear models for non-ignorable missing data estimation.

Zhong, Lingras, and Sharma (2004) developed and used genetically designed neural network and regression models and factor models for missing data estimation. They applied these techniques to traffic counts and found that genetically designed regression models give the most accurate results. The average errors for refined models that they obtained were lower than 1% and the 95th percentile errors were below 2% for counts with stable patterns. Furthermore, they found that even for counts with unstable patterns, the average errors were still lower than 3% in the cases considered.

Teegavarapu and Chandramouli (2005) observed that distance-weighted and data-driven methods had been extensively used for estimating missing rainfall data. Furthermore, they observed that the inverse distance weighting method was one of the most applied methods for estimating the missing rainfall data using data that were recorded in other available recording gages. These researchers realized that this method suffered from major conceptual limitations and so they proposed a data-driven model that uses an artificial neural network and a stochastic interpolation technique. They tested these methods by estimating missing precipitation data from 20 rain-gauging stations. The results they obtained showed that the conceptual revisions improved the estimation of missing precipitation records.

Muteki, MacGregor, and Ueda, (2005) proposed to use auxiliary data related to missing data to estimate missing data. They proposed latent variable approaches that used both auxiliary data and measured data for missing data estimation and applied this to the material properties of rubbers used in industrial polymer blends. They observed that using auxiliary information was advantageous when the percentage of missing data was high and for the case when certain combinations of missing data exhibited little correlation with observed data. They proposed a multi-block approach and a novel two-stage projection approach, both of which both incorporated auxiliary data. The novel two-stage projection approach was found to offer slightly better estimates in two industrial polymer-blending problems and, therefore, auxiliary information improved the estimates.

BACKGROUND

Missing Data

Real time processing applications that are highly dependent on the data often suffer from the problem of missing input variables. Various heuristics for missing data imputation such as mean substitution and hot deck imputation also depend on the knowledge of how the data points become missing. There are several reasons why the data may be missing. As a result, missing data may follow an observable pattern. Exploring the pattern is important and may lead to the possibility of being able to identify cases and variables that affect the missing data (Schafer, 1997). Having identified the variables that predict the pattern, a proper estimation method can be selected. In this chapter, the performances of a combination of neural networks and particle swarm optimization as well as a combination of Gaussian mixture models and EM algorithm are both used for missing data estimation and then their performances are explained in terms of the nature of the data.

Auto-Associative Neural Networks

As explained in earlier chapters, auto-associative neural networks, also known as autoencoders, are neural networks trained to recall the input space. Thompson, Marks, and Choi (2002) distinguished two primary features of auto-associative networks, namely the auto-associative nature of the network and the presence of a bottleneck that occurs in the hidden layers of the network, resulting into a butterfly-like structure. In cases where it is necessary to recall the input, auto-associative networks are preferred due to their remarkable ability to learn certain linear and nonlinear inter-relationships such as correlation and covariance inherent in the input space. Auto-associative networks project the input onto some smaller set by intensively squashing it into smaller details. Though dependent on the type of application, the optimal number of hidden nodes of auto-associative neural networks must be smaller than that of the input layer (Frolov *et al.*, 1995; Thompson, Marks, & Choi, 2002).

Auto-associative networks have been used in various applications including the treatment of missing data problem by a number of researchers (Frovolov et al., 1995; Dhlamini, Nelwamondo & Marwala, 2006; Mohamed & Marwala, 2005). The research by Leke, Marwala, and Tettey (2006) successfully used an auto-associative network to model the risk of HIV. Daszykowski, Walczak, and Massart (2003) used an auto-associative neural network for data compression and visualization. They found that auto-associative networks could deal with linear and nonlinear correlation among variables and, thereby, were very useful in data analysis. They showed that auto-associative networks were able to nonlinearly compress and visualize chemical datasets. Shimizu et al. (1997) applied autoassociative networks to fault diagnosis in the optimal production of yeast with a controllable temperature. They eliminated high frequency noise in the data by using the wavelet transform. They observed that the diagnosis system accurately detected faults that could not be detected by a linear principal component analysis.

In this chapter, autoassociative networks are constructed using Multi-Layer Perceptrons (MLP) networks and trained using back-propagation (Bishop, 1995). As explained in earlier chapters, the MLPs are feed-forward neural networks with an architecture made up of an input layer, a hidden layer and an output layer. Each layer is formed from smaller units known as *neurons*. Neurons in the input layer receive the input vector $\{x\}$ and distribute it forward to the network. In the next layers, each neuron receives a signal, which is a weighted sum of the outputs of the nodes in the previous layer. Inside each neuron, an activation function is used to control the input. Such a network determines a nonlinear mapping from an input vector to the output vector, parameterized by a set of network weights, which is referred to as a *vector of weights* $\{w\}$. The MLP is used here because it has been found it can model very complex data (Marwala, 2001), and in Chapter II was found to perform better than the radial basis function.

The first step to approximate the weight parameters of the model is to find the approximate architecture of the MLP, where the architecture is characterized by the number of hidden units, the type of activation function, as well as the number of input and output variables. The second step is to estimate the weight parameters using the training set (Japkowicz, 2002). The training estimates the weight vector $\{w\}$ to ensure that the output is as close to the target vector as possible. As explained in the Chapter III, the problem of identifying the weights in the hidden layers is solved by maximizing the probability of the weight parameter, given the data, P($\{w\}$|[D]), and using Bayes' rule as follows (Marwala, 2001):

$$P(\{w\}\,|\,[D]) = \frac{P([D]\,|\,\{w\})P(\{w\})}{P([D])}$$

(4.1)

In equation 4.1, [D] is the training data, P([D]|$\{w\}$) is called the *evidence term* that balances between fitting the data well and helps to avoid overly complex models, and $P(\{w\})$ is the prior probability of $\{w\}$. In this chapter, the autoassociative network is trained using the hybrid Monte Carlo method as was previously done in Chapter III.

Lampinen and Vehtari (2001) gave a short review on the Bayesian approach to neural networks and demonstrated the advantages of this method on three applications. They explained the fundamentals of a Bayesian framework with emphasis on the significance of prior knowledge in Bayesian models and compared it to the traditional error minimization approaches as implemented in Chapter II. They concluded that the generalization capability of these models depends on prior assumptions. They found that the Bayesian method ensures propagation of errors in quantities that are unknown to other assumptions in the model. The applications they considered included a regression, a classification and an inverse problem. For regression problems, models with less restrictive priors were found to be the best. They found that the Bayesian approach prevented guessing attributes that were unknown, e.g., the degrees of freedom in the model. For these reasons, Bayesian training of MLP autoassociative networks is pursued in this chapter.

Particle Swarm Optimization (PSO)

In Chapters II and III, genetic algorithms were used to optimize the missing data-estimation error equation. This chapter uses the PSO to optimize the missing data-estimation error equation. The PSO is a stochastic, population-based evolutionary algorithm that is widely used for optimization. It is based on socio-psychological principles that are inspired by swarm intelligence. It offers us an understanding of social behavior, and it has contributed to engineering applications. Society enables an individual to maintain cognitive robustness through influence and learning. We know that individuals learn to tackle problems by communicating and interacting with other individuals and, thereby, develop generally similar ways of tackling problems (Engelbrecht, 2005). Likewise, swarm intelligence is driven by two factors. These are group knowledge and individual knowledge. Each member of a swarm always acts by balancing between individual knowledge and group knowledge.

To solve optimization problems using the PSO, a fitness function is constructed to describe a measure of the desired outcome. In order to reach an optimum solution, a social network representing a population of possible solutions is defined and randomly generated. The individuals within this social network are assigned neighbors with whom to interact. These individuals are called *particles*, and thus the name Particle Swarm Optimization. Thereafter, a process to update these particles is initiated. This is conducted by evaluating a *fitness* measure of each particle. Each particle can remember the location where it had its greatest success as measured by the fitness function. The best solution for the particle is named the *local best* and each particle makes this information on the local best accessible to their neighbors who in turn also observe their neighbors' success. The process of moving in the search space is guided by these successes and the population ultimately converges by the end of the simulation to an optimum solution.

The PSO was developed by Kennedy and Eberhart (1995) and this procedure was inspired by algorithms that model the "flocking behavior" seen in birds. Researchers in artificial life (Reynolds, 1987; Heppner & Grenander, 1990) developed simulations of bird flocking. In the context of optimization, the concept of birds finding a roosting place is analogous to the process of finding an optimal solution. PSO has been very successful in optimizing many complex problems. Marwala (2005) used PSO to improve finite element models to reflect the measured data better. He compared this method to a finite element model updating approach, which used simulated annealing and genetic algorithm. The proposed methods were tested on a simple beam and an unsymmetrical H-shaped structure. On average, it was observed that the PSO method gave the most accurate results followed by simulated annealing and then the genetic algorithm. Dindar and Marwala (2004) successfully used PSO to optimize the structure of the committee of neural networks. Ransome et al. (2005) as well as Ransome (2006) successfully used PSO to optimize the position of a patient during radiation therapy.

Arumugam and Rao (2008) successfully applied a multiple-objective, particle swarm optimization method for molecular docking. The multi-objective, particle swarm optimization method is a technique that is aimed at solving more than one objective function. For example, in this chapter as in Chapter III, the missing data estimation process is aimed at solving a single objective function to minimize the missing data-estimation error equation, which was described in earlier chapters.

Arya et al. (2007) used PSO and the singular value decomposition method to design neuro-fuzzy networks. This process was aimed at identifying the optimal neuro-fuzzy networks whose objective is to model the data accurately. Berlinet and Roland (2008) introduced a particle swarm optimization algorithm for scheduling problems. Scheduling problems have design variables that are integers. If there are many of these variables, the combinatorial nature of these problems makes the process of finding solutions to these problems extremely difficult. Jarboui et al. (2008) introduced and applied the particle swarm optimization technique, where a distribution vector is used in the update of the velocities of particles. Other successful applications of PSO include Jiang et al. (2007), who successfully applied particle swarm optimization to conceptual design, Kathiravan and Ganguli (2007) to power systems, Lian, Gu, and Jiao (2008) to training a radial basis function to identify chaotic systems, Lin, Chang, and Hsieh (2008) in inverse radiation problems, Qi et al. (2008) in scheduling problems and Guerra and dos S. Coelho (2008) in modeling nonlinear filters.

The PSO is implemented by finding a balance between searching for a good solution and exploiting other particles' successes. If the search for a solution is too limited, the simulation will converge to the first solution encountered, which may be a local optimum. If the successes of others are not exploited, then the simulation will never converge. The PSO approach has the advantages in that it is computationally efficient, simple to implement, has few adjustable parameters when compared to other competing evolutionary programming methods such as genetic algorithms and can adapt to the local and global exploratory ability.

When implementing the PSO, the simulation is initialized with a population of random candidates, conceptualized as particles. Each particle is assigned a random velocity and is iteratively moved through the particle space. At each step, the particle is attracted towards a region of the best fitness function by the location of the best fitness achieved so far in the population.

On implementing the standard PSO, each particle is represented by two vectors: $p_i(k)$ the *position* and $v_i(k)$ the *velocity* at step k. The initial positions and velocities of particles are randomly generated. The subsequent positions and velocities are calculated using the position of the best solution that a particular particle has encountered during the simulation called $pbest_i(k)$ and the best particle in the swarm, which is called $gbest(k)$. The subsequent velocity of a particle i can be calculated using the following equation:

$$v_i(k+1) = \gamma v_i(k) + c_1 r_1 (pbest_i(k) - p_i(k)) + c_2 r_2 (gbest(k) - p_i(k)) \qquad (4.2)$$

The subsequent position of a particle i can be calculated using the equation:

$$p_i(k+1) = p_i(k) + v_i(k+1) \qquad (4.3)$$

Here γ is the inertia of the particle, and it controls the impact of the previous velocity of the particle on the current velocity. These parameters control the exploratory properties of the simulation. A high value of the inertia encourages global exploration, while a low value of the inertia encourages local exploration. The parameters c_1 and c_2 are called *'trust' parameters* and they determine the relative weight of individual and group experiences. The trust parameter c_1 indicates how much confidence the current particle has in itself, while the trust parameter c_2 indicates how much confidence the current particle has on the population. The parameters r_1 and r_2 are random numbers between 0 and 1, which determine the degree to which the simulation should explore the space. In equation 4.3, it can be seen that the PSO makes use of the velocity to update the position of the swarm.

The position of the particle is updated, based on the social behavior of population of particles and it adapts to the environment by continually coming back to the most promising region identified. This process is stochastic and can thus be summarized as follows:

1. Initialize a population of particles' positions and velocities. The positions of the particles must be randomly distributed in the updating parameter space.
2. Calculate the velocity for each particle in the swarm using equation 4.2.
3. Update the position of each particle using equation 4.3.
4. Repeat steps 2 and 3 until convergence results.

To improve the performance of the particle swarm optimization method presented above, several additions and modifications of the PSO algorithm have been proposed and implemented. Liu, Liu, and Duan (2007) proposed a combination of the PSO algorithm and the evolutionary algorithm and applied this method to train recurrent neural networks for time-series prediction. Janson, Merkle, and Middendorf (2007) combined PSO with a gradient-based optimization method and used this to design a composite beam which has the highest possible strength, while Yisu et al. (2008) improved the PSO by introducing cross-over functionality from the genetic algorithm and then applying this to the control of hybrid systems.

Brits, Engelbrecht, and van den Bergh (2007) presented a new particle swarm optimization technique aimed at locating and refining multiple solutions to problems with multi-modal characteristics. Their method extended the uni-modal nature of the standard PSO approach by using many swarms from the initial population. A different solution was represented by a sub-swarm and was individually optimized. Thereby each set of particles in the swarm represented a possible solution. When implemented experimentally it was found to locate all optima successfully.

Sha and Hsu (2006) proposed a hybrid particle swarm optimization for the job shop problem, which operated in a discrete space rather than in a continuous space. Due to the nature of the discrete space, the PSO algorithm was improved through the particle position representation, particle movement and particle velocity. The particle position was then represented, based on a preference list, particle movement on the swap operator, and particle velocity on the taboo list. The heuristic of Giffler and Thompson (1960) was implemented to decode a particle position into a schedule while the taboo search improved the solution's quality. It was observed that the modified PSO was better than the original while the hybrid PSO was observed to be better than the conventional meta-heuristic method. Based on these successes the PSO is used in the estimation of missing data in this chapter.

Neural Networks and PSO for Missing Data Imputation

The method used in this chapter combines the use of auto-associative neural networks with particle swarm optimization to approximate missing data as was previously done in Chapter II. This method has been used to approximate missing data in a database by Abdella and Marwala (2005). The PSO is used now to estimate the missing values by optimizing an objective function known as the *missing-data estimation error equation* and this is presented briefly in this section. The complete vector combining the estimated and the observed values is fed into the auto-associative network the input as shown in Figure 4.1.

Symbols X_k and X_u represent the known variables and the unknown or missing variable, respectively. The combination of X_k and X_u represent the full input space. The missing data estimation error equation can be written as is done in Chapter II, in terms of the weight vectors, observed vector $\{X_k\}$, unknown vector $\{X_u\}$ and the auto-associative neural network function f as follows:

Figure 4.1. An auto-associative network and particle swarm optimization missing data estimator structure

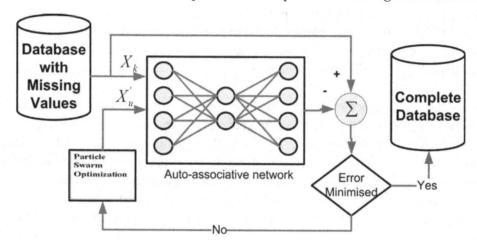

$$\varepsilon = \left\| \left(\left\{ \begin{matrix} \{X_k\} \\ \{X_u\} \end{matrix} \right\} - f\left(\left\{ \begin{matrix} \{X_k\} \\ \{X_u\} \end{matrix} \right\}, \{W\} \right) \right) \right\|$$

(4.4)

Here $\| \ \|$ is the Euclidean norm. It should be noted here that the auto-associative function implemented in this chapter is the MLP. The missing data estimation Equation 4.4 is used as the objective function that is minimized using the PSO method that was discussed in the previous section. The combination of X_k and X_u represent the full input space.

MAXIMUM LIKELIHOOD APPROACH

The maximum likelihood approach to approximating missing data is a very popular technique and is based on a precise statistical model of the data (Hartley, 1958; Dempster, Laird, & Rubin, 1977; Allison, 2002; Tanner, 1996). The model most commonly used is the multivariate Gaussian mixture model while the maximum likelihood method is applied for the task of imputing the missing values. Likelihood methods may be categorized into 'Single Imputations' and 'Multiple Imputations (Little & Rubin, 1987; Schafer & Olsen, 1998). This chapter only considers Single Imputations where the EM algorithm is used and this is discussed below.

Expectation Maximization for Missing Data

The expectation-maximization algorithm was originally introduced by Dempster, Laird, and Rubin (1977). It was aimed at overcoming problems associated with the Maximum Likelihood methods. The Expectation Maximization combines statistical methodology with algorithmic implementation and has gained much attention recently for various missing data problems. Expectation Maximization has also been proven to work better than methods such as list-wise, pair-wise data deletion, and mean substitution because it assumes that incomplete cases have data missing at random rather than missing completely at random (Allison, 2002; Rubin, 1978).

Fraley (1999) computed missing information using the EM algorithm and demonstrated that improved estimates are important for accurate inference. The power method for eigen-computation was used to attain better estimates efficiently because it computes only the largest eigenvalue and eigenvector of a matrix. The obtained results showed that this approach becomes more efficient as the dimensions of the data increase.

Atkinson and Cheng (2000) used robust regression to extend the missing data techniques to data with several outliers. The EM algorithm is a general technique for fitting models to incomplete data. The Expectation Maximization capitalizes on the relationship between missing data and the unknown parameters of a data model. If the missing values were known, then estimating the model parameters would be straightforward. Similarly, if parameters of the data model were known, then it would be possible to obtain unbiased predictions for the missing values. This inter-dependence between model parameters and missing values suggests an iterative method where the missing values are first predicted, based on assumed values for the parameters, then these predictions are used to update the parameter estimates, and then the process is repeated.

The sequence of parameters converges to maximum-likelihood estimates that implicitly average over the distribution of the missing values. It should be noted here that the missing data method that uses a PSO and a neural network first identifies the "correct" model through neural network training and then uses the PSO to identify the missing values. This differs from the EM algorithm in the sense that both missing values and the "correct" model are identified simultaneously.

EM Algorithm

Bordes, Chauveau, and Vandekerkhove (2007) applied the EM algorithm to image reconstruction. They found that the results were within 10% of the experimental data. Cai et al. (1997) successfully applied the EM algorithm to missing data, while Han, Ko, and Seo (2007) combined the EM algorithm with a Laplace method and successfully applied this to modeling of economic data. Ingrassia and Rocci (2007) generalized the EM algorithm to semi-parametric mixture models, which, when tested on real data showed that the proposed method was easy to implement and computationally efficient. Kauermann, Xu, and Vaida (2007) used the EM algorithm to recognize polymorphism in pharmaco-kinetic (pharmaco dynamic) (PK (PD)) phenotypes, while Wang and Hu (2007) improved the EM algorithm's computational load and successfully applied this to brain tissue segmentation. Another successful implementation of the EM algorithm includes binary text classification (Park, Qian, & Jun, 2007). Other improvements to the EM algorithm include accelerating the computational speed by Patel, Patwardhan, and Thorat (2007). Further information on the implementation of EM algorithm can be found in Wang, Schumitzky, and D'Argenio (2007), McLachlan and Krishnan (1997) as well as Neal and Hinton (1998).

In the EM algorithm, the maximum likelihood estimation of the parameter vector is iteratively identified by repeating the following steps (Little & Rubin, 1987):

- *The Expectation E-step:* In the presence of a set of parameter estimates, for example, a mean vector and covariance matrix for a multivariate normal distribution, the E-step estimates the conditional expectation of the complete-data log-likelihood given the observed data and the parameter estimates.

- *The Maximization M-step:* The M-step identifies the parameter estimates that maximize the complete-data log-likelihood from the E-step given a complete-data log-likelihood.
- These steps are iterated until convergence.

In the EM algorithm, $\{y\}$ indicates incomplete data, which is composed of values of observable variables, and $\{z\}$ indicates the missing data. Then $\{y\}$ and $\{z\}$ can be combined to form a complete dataset. If it is assumed that p is the joint probability density function of the complete data with parameters given by the vector $\{\theta\}$: $p(\{y\},\{z\}|\{\phi\})$, then this function can also be viewed as the complete data likelihood, which is a function of $\{\phi\}$. In addition, given the observed data, the conditional distribution of the missing data can be expressed as (Little & Rubin, 1987):

$$p\left(\{z\}|\{y\},\{\phi\}\right)=\frac{p(\{y\},\{z\}|\{\phi\})}{p(\{y\}|\{\phi\})}$$

$$=\frac{p(\{y\}|\{z\},\{\phi\})p(\{z\}|\{\phi\})}{\int p(\{y\}|\{z\},\{\phi\})p(\{z\},\{\phi\})d\{z\}} \tag{4.5}$$

Equation 4.5 is obtained by using Bayes' rule and the law of total probability. This process needs knowledge of the observation likelihood, given the unobservable data $p(\{y\}|\{z\},\{\phi\}$ and the probability of the unobservable data $p(\{z\}|\{\phi\})$. Therefore, the EM algorithm iteratively re-estimates the missing values and thus improves the initial estimate of the missing value by constructing new estimates for $\{\phi_n\}$, $\{\phi_{n+1}\}$ etc. This process can be written as follows (Nelwamondo, 2008):

$$\{\phi\}_{n+1} = \arg\max_{\phi} G(\{\phi\}) \tag{4.6}$$

In equation 4.6, $G(\{\phi\})$ is the expected value of the log-likelihood and is given by:

$$G(\{\phi\}) = \sum_z p(\{z\}|\{y\},\{\phi\}_n) \log p(\{y\},\{z\}|\{\phi\}) \tag{4.7}$$

Simply put, $\{\phi\}_{n+1}$ is the value that maximizes the conditional expectation of the complete data log-likelihood given the observed variables.

To use the EM algorithm to solve for the missing data, Gaussian Mixture Models are used in this chapter. Marwala, Mahola, and Chakraverty, (2007) as well as Nelwamondo, Marwala, and Mahola (2006) extracted features that were used to classify faults in bearings using Gaussian Mixture Models (GMM). The results obtained showed that GMM performs well and is computationally efficient. Chen, Chen, and Hou, (2004) successfully used Gaussian mixture models with the Karhunen–Loeve transform for the identification of speakers. Choi, Park, and Lee (2004) successfully used a Gaussian mixture model through a principal component analysis and a discriminant analysis for process monitoring. Because of these successes, Gaussian Mixture Models are used in this chapter together with expectation maximization to estimate missing data.

Given a sample of m data points $\{y\}_1,...,\{y\}_m$ that is drawn from one of the n Gaussian distributions, and z_i which denotes which Gaussian y_i is from, the probability of $\{y\}$ can be written as follows (Little & Rubin, 1987):

$$P(\{y\}|z=i,\{\phi\})=(2\pi)^{-D/2}|\sigma_i|\exp\{-1/2(\{y\}-\{\mu\}_i)^T\sigma_i^{-1}(\{y\}-\{\mu\}_i)\} \qquad (4.8)$$

Here D is the dimension of the Gaussian distribution while $\{\mu\}_i$ and $\{\sigma\}_i$ are the mean and the variance of the ith Gaussian distribution vector. The aim of the EM algorithm is to estimate the unknown parameters $\{\phi\}=[\{\mu\}_1,...,\{\mu\}_n,\{\sigma\}_1...\{\sigma\}_n,P(\{z\}_{i=1}),...P(\{z\}_{i=n})\}$, representing the expected value and variance of each Gaussian as well as the probability of drawing any Gaussian at any given time. Here the parameter n is the number of Gaussian distribution functions under consideration.

In the E-step, therefore, the unobserved z's may be estimated, conditional on the observed values, using the last values as per Little & Rubin (1987):

$$P\left(z_j=i|\{y\}_j,\{\phi\}_t\right)=\frac{P\left(z_j=i,\{y\}_j|\{\phi\}_t\right)}{P\left(\{y\}_j|\{\phi\}_t\right)}$$

$$=\frac{P\left(\{y\}_j|z_j=i,\{\phi\}_t\right)P\left(z_j=i|\{\phi\}_t\right)}{\sum\limits_{k=1}^{n}P\left(\{y\}_j|z_j=k,\{\phi\}_t\right)P\left(z_j=k|\{\phi\}_t\right)} \qquad (4.9)$$

In the M-step the expected value of the log-likelihood of the joint event needs to be maximized. This expected value of the log-likelihood to be maximized may be written as follows (Dempster, Laird, & Rubin 1977):

$$G(\{\phi\})=E_z\left[\ln\prod\limits_{i=1}^{m}P(\{y\}_j,\{z\}|\{\phi\})|\{y\}_j\right]$$

$$=\sum\limits_{j=1}^{m}\sum\limits_{i=1}^{n}P(z_j=i|\{y\}_j,\{\phi\}_t)\ln P(z_j=i,\{y\}_j|\{\phi\}) \qquad (4.10)$$

The maximization of equation 4.10 must be attained in the presence of the following constraint:

$$\sum\limits_{i=1}^{n}P(z_j=i|\{\phi\})=1 \qquad (4.11)$$

As it is customary, the objective function in equation 4.10 can be combined with the constraint in equation 4.11 by calculating the Lagrangian, L, as indicated below (Snyman, 2005):

$$L(\{\phi\})$$
$$=\left(\sum\limits_{j=1}^{m}\sum\limits_{i=1}^{n}P(z_j=i\begin{vmatrix}\{y\}_j,\{\phi\}_t\end{vmatrix}\left(-\frac{D}{2}\ln(2\pi)-\frac{1}{2}\ln|\sigma_i|-\frac{1}{2}(\{y\}_j-\{\mu\}_i)^T\sigma_i^T(\{y\}_j-\{\mu\}_i)+\ln(P(z_j=1|\{\phi\}))\right)\\-\lambda\left(\sum\limits_{i=1}^{n}P(z_j=i|\{\phi\})-1\right)\right)$$

$$(4.12)$$

Here λ is a parameter known as the *Lagrangian multiplier*. To solve for the new estimate $\{\phi\}_{n+1}$ equation 4.12 can be maximized by finding the gradient of the Lagrangian with respect to parameter $\{\phi\}$ and then equating this to zero as per Snyman (2005):

$$\frac{\partial L(\{\phi\})}{\partial\{\phi\}} = 0$$

$$(4.13)$$

By calculating equation 4.13, the following expression is obtained (Dempster, Laird, & Rubin, 1977):

$$
\begin{aligned}
P\big(\{z\}_j \big|\{\phi\}\big) &= \frac{1}{\lambda}\sum_{j=1}^{m} P\big(z_j = i\big|\{y\}_j, \{\phi\}_t\big) \\
&= \frac{\sum_{j=1}^{m} P\big(z_j = i\big|\{y\}_j, \{\phi\}_t\big)}{\sum_{k=1}^{n}\sum_{j=1}^{m} P\big(z_j = i\big|\{y\}_j, \{\phi\}_t\big)} \\
&= \frac{1}{m}\sum_{j=1}^{m} P\big(z_j = i\big|\{y\}_j, \{\phi\}_t\big)
\end{aligned}
$$

$$(4.14)$$

From this expression ϕ_{n+1} can thus be estimated. This process is repeated until convergence is reached.

EXPERIMENTAL EVALUATION

This section reports on experimental validation of the above method on various sets of data.

Data Analysis

The Gaussian mixture model trained using the EM algorithms and the auto-associative neural network coupled with particle swarm optimization approaches are used for approximating missing data and are then compared using three different datasets. The datasets used are briefly described below.

Power Plant Data

The first dataset used is data from a 120 MW power plant in France (Nelwamondo, 2008), under normal operating conditions. This dataset contains five inputs, namely: *gas flow, turbine valves opening, super heater spray flow, gas dampers* and *airflow*. Sampling of the data was every 1228.8 seconds and 200 instances were recorded. An extract of the data without any missing values is shown in Table 4.1.

The dataset was split into a training dataset and a testing dataset. Due to the limited data available, one seventh of the dataset was kept as the test set, with the remaining data being used for training. For easy comparison with the combined Neural Network and PSO approach (NN-PSO), the training and testing data for the EM algorithm were combined into a single file, with the testing data appearing at the end of the file. This separation ensured that for both the EM and the NN-PSO approaches, testing results are compared using the same amount of testing data and that their respective models are built from the same amount of "training" data. The dataset was transformed using a min-max normalization to [0, 1] before use, to ensure that the dataset was within the active range of the activation function of the neural network.

HIV Database

The data used in this test were obtained from the South African antenatal sero-prevalence survey of 2001. The data for this survey was obtained from questionnaires answered by pregnant women visiting selected public clinics in South Africa. Only women participating for the first time in the survey were eligible to answer the questionnaire.

Data attributes used in this chapter were the *HIV status*, *Education level*, *Gravidity*, *Parity*, *Age Group* and *Age Gap*. The *HIV status* was represented in a binary form, where 0 and 1 represent negative and positive, respectively. The *Education Level* was measured using integers representing the highest grade successfully completed, with 13 representing tertiary education. *Gravidity* was the number of pregnancies, complete or incomplete, experienced by the female, and this variable was represented by an integer between 0 and 11. *Parity* was the number of times the individual had given birth with multiple births being counted as one birth event. Both *parity* and *gravidity* are important, as they show reproductive activity as well as the reproductive health state of the woman. *Age gap* is a measure of the age difference between the pregnant woman and the prospective father of the child. A sample of this dataset is shown in Table 4.2. The data used in this chapter consisted of 5776 instances. The data was divided into two subsets, namely, the training and testing datasets. Testing was conducted with 776 instances.

Data from an Industrial Winding Process

The third dataset used here represents a test setup of an industrial winding process. The data can be found in De Moor (2006). The main part of the plant was composed of a plastic web that unwound from the first reel (the unwinding reel), went over the traction reel and finally rewound on the rewinding reel, as shown in Figure 4.2 (Bastogne et al., 2002).

Table 4.1. Set of power plant measurements under normal operating conditions

Gas Flow	Turbine	Heater	Gas Dampers	Airflow
0.1184	0.0894	0.11387	0.62615	0.07699
0.1086	0.0825	0.11284	0.62617	0.01502
0.0997	0.1992	0.14079	0.62233	0.06197
0.0928	0.1916	0.12733	0.62617	0.05913
0.0889	0.3002	0.13768	0.62611	0.02817
0.0876	0.6318	0.07483	0.63050	0.07981

Table 4.2. Extract of the HIV database used without missing values

HIV	Education	Gravidity	Parity	Age	Age Gap
0	7	10	9	35	5
1	10	2	1	20	2
1	10	6	5	40	6
0	5	4	3	25	3

Figure 4.2. Graphic representation of the winding plot system

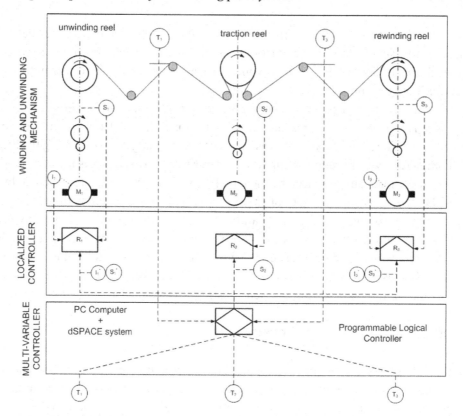

As is shown in Figure 4.2, reels 1 and 3 were coupled with a DC-motor that was controlled with input set-point currents I_1 and I_3. The angular speed of each reel (S_1, S_2 and S_3) and the tensions in the web between reel 1 and 2 (T_1) and between reel 2 and 3 (T_3) were measured using dynamo tachometers and tension meters.

The full dataset has 2500 instances, sampled every 0.1 seconds. In this chapter, testing was done with 500 instances, while the training set and the validation set for the neural network consisted of 1500 and 500 instances. The inputs to the winding system were the angular speeds of reel 1 (S_1), reel 2 (S_2), reel 3 (S_3), the set point current at motor 1 (I_1) and at motor 2 (I_3) as shown in Figure 4.2. A more detailed description of the data can be found in Bastogne et al. (2002).

Performance Analysis

The effectiveness of the missing data system was evaluated using the correlation coefficient and the relative prediction accuracy. The correlation coefficient was used as a measure of similarity between the prediction and the actual data. The correlation coefficient, r was computed as:

$$r = \frac{\sum_{i=1}^{n}(x_i - \overline{x}_i)(\hat{x}_i - \overline{\hat{x}}_i)}{[\sum_{i=1}^{n}(x_i - \overline{x}_i)^2 \sum_{i=1}^{n}(x_i - \overline{x}_i)(\hat{x}_i - \overline{\hat{x}}_i)^2]^{1/2}}$$

(4.15)

In equation 4.15, $\hat{x}$ represents the approximated data, x is the actual data, while $\bar{x}$ represents the mean of the data. The relative prediction accuracy was defined as:

$$Error = \frac{n_\tau}{N} \times 100\%$$

(4.16)

In equation 4.16, n_τ is the number of predictions within a certain tolerance percentage of the missing value. In this chapter, a tolerance of 10% was used. The 10% was chosen arbitrarily, with an assumption that it is the maximum acceptable margin for error in the applications considered. This error analysis can be interpreted as a measure of how many of the missing values are predicted within the tolerance but the tolerance can be made to be any value, depending on the sensitivity of the application.

EXPERIMENTAL RESULTS AND DISCUSSION

This section presents the experimental results obtained by using both of the approaches described above. The predictability was evaluated within 10% of the target value. The evaluation was computed by determining how much of the test sample was estimated within the given tolerance. The results of the test are first presented for the power plant dataset.

For the experiment with the power plant data, the NN-PSO system was implemented using an auto-associative multi-layer perceptron network trained with 4 hidden nodes and for the HIV data, 6 hidden nodes. In both cases, the hyperbolic tangent function was used in the hidden layer and a linear activation function in the outer layer. The GMM architecture used a diagonal covariance matrix with 3 centers for the power plant data and 5 centers for the HIV data. The correlation coefficient and the accuracy within 10% of the actual values are given in Table 4.3.

In this paragraph, the variables in this example were analyzed concerning the missing data mechanism. As discussed in earlier chapters, and as explained in the literature (Little & Rubin, 2000; Burk, 1960) there are four types of missing data mechanisms. These are MCAR, MAR, MNAR and MBND. In the examples in this chapter, the variables are missing through the first three mechanisms. MBND is a missing data mechanism where the values are missing because they cannot be measured physically (or naturally).

It can be seen from the results that the GMM-EM algorithm fails to make a prediction for column 1 in this dataset. The reason is that for the GMM-EM algorithm to make a prediction, the prediction matrix needs to be a positive definite (Dempster, Laird, & Rubin, 1977). The major cause of this happens when one variable is linearly dependent on another variable. This linear dependency may sometimes exist not between the variables themselves, but between elements of moments such as the mean, variances, co-variances and correlations.

Other reasons for this failure include errors on reading the data, in initial values and many more. This problem can be solved by deleting variables that are linearly dependent on each other or by using Principal Components to replace a set of collinear variables with orthogonal components. Seminal work on dealing with "not positive definite matrices" was undertaken by Wothke (1993).

The results show that the NN-PSO method can impute missing values with a higher accuracy of prediction for most cases as shown in Figure 4.3. However, the lack of high accuracy predictions for

86 *Marwala*

Table 4.3. Results of comparative testing using power plant data

	Correlation		10%	
Variable	GMM-EM	NN-PSO	GMM-EM	NN-PSO
Gas Flow	-	0.9790	-	21.43
Turbine	0.7116	0.8061	14.29	14.29
Heater	0.7218	0.6920	7.14	28.57
Gas dumper	-0.4861	0.5093	3.57	10.71
Air Flow	0.6384	0.8776	10.71	7.14

Figure 4.3. Graphical comparison of estimation accuracy using 10% tolerance (power plant data)

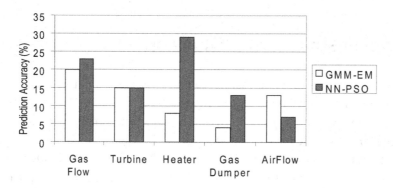

both estimation techniques suggests some degree of difficulty in estimating the missing variables, based on the given set of data.

For neural networks, it is observable that the quality of estimation in each input variable depends on the existence of some form of correlation between variables in the input space such that this linear or nonlinear relationship can be discovered by the neural networks and used to give higher accuracy imputations. The GMM-EM algorithm also requires that the data must not be linearly dependent on variables within the input space, as demonstrated by the need for positive definite matrices. Before commencing with the experiment, data are tested for correlation. This testing process involves finding out if any variable in the data is somehow strongly related to any other variable in the data. Figure 4.4 shows the results obtained when predicting missing variables on the HIV dataset within the 10% tolerance.

The results obtained using the HIV database are presented in Table 4.4 and Figure 4.4. Table 4.4 presents the correlation coefficients between the actual and the predicted data. Results here clearly show that the GMM-EM performs better than the NN-PSO method for the prediction of variables such as *Education*, *Parity*, *Age* and *Age Gap*. Unlike the power plant database, the results here show that the GMM-EM algorithm is better than the NN-PSO for the prediction of variables in the HIV dataset of this study. Since this is a social science database, the reason for the poor performance of the NN-PSO can be either that the variables are not sufficiently representative of the problem space to produce an accurate imputation model, or that people are not very honest in answering the questions in the questionnaire, leading to less dependability of variables on each other. The results obtained from the industrial winding

Table 4.4. Correlation coefficients between actual and predicted data for the HIV database

	Education	Gravidity	Parity	Age	Age Gap
GMM-EM	0.12	0.21	0.91	0.99	1
NN-PSO	0.10	0.71	0.67	0.99	0.99

Figure 4.4. Prediction within 10 % tolerance of missing variable in the HIV data

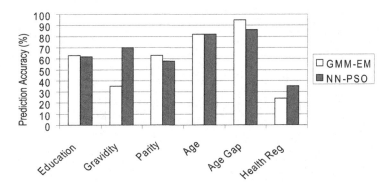

process are shown in Figure 4.5 and the GMM-EM and NN-PSO approaches are compared. Again, the results obtained in this section show that for some variables the GMM-EM algorithm produces better-imputed results, while in others the NN-PSO system can produce a better imputation result. From the observed data, the predicted values are not well correlated with the actual missing variables. A possible explanation is that the missing value is not interdependent with itself, but to other variables in the data. Table 4.5 shows the correlation coefficients. As for the other datasets, the problem of the non-positive definite matrix when imputing values for column 1 prevents the GMM-EM algorithm from being used to estimate the missing data.

CONCLUSION

This chapter investigated and compared the GMM-EM algorithm and the NN-PSO approach for missing data approximation. In one approach, an auto-associative neural network was trained to predict its own input space. Particle swarm optimization and neural networks were used to approximate the missing data. On the other hand, the expectation maximization was implemented with Gaussian mixture models for the same problem. The results show that for some variables the GMM-EM algorithm can produce better imputation accuracy, while for the other variables the NN-PSO gives better results. Thus, the imputation ability of one method over another seems to be highly problem dependant. Findings also show that the GMM-EM algorithm seems to perform better in cases where there is very little dependency among the variables, which is contrary to the performance of the NN-PSO approach.

Table 4.5. Correlation coefficients between actual and predicted data for the winding process

	S1	S2	S3	I1	I2
GMM-EM	-	-0.003	0.009	-0.05	-0.001
NN-PSO	0.203	0.229	0.159	0.038	0.117

Figure 4.5. Prediction within the 10% tolerance of missing variables in the industrial winding process

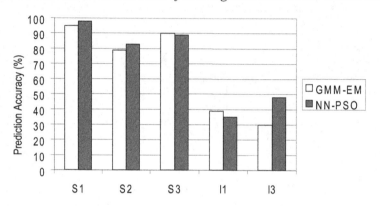

FUTURE RESEARCH

This chapter compared the Gaussian Mixture model trained using the EM algorithm to the hybrid multi-layer perceptron auto-associative neural network and particle swarm optimization for missing data estimation. For future work, the neural network approach should be directly trained using the EM algorithm. Furthermore, other neural network architectures and evolutionary optimization methods should be considered to tackle the missing data problem.

REFERENCES

Abdella, M., & Marwala, T. (2005). The use of genetic algorithms and neural networks to approximate missing data in database. *Computing and Informatics*, *24*, 577-589.

Allison, P. D. (2002). *Missing data: Quantitative applications in the social sciences*. London: Sage Publications.

Arumugam, M. S., & Rao, M. V. C. (2008). Molecular docking with multi-objective particle swarm optimization. *Applied Soft Computing*, *8*(1), 666-675

Arya, L. D., Choube, S. C. Shrivastava, M., & Kothari, D. P. (2007). The design of neuro-fuzzy networks using particle swarm optimization and recursive singular value decomposition. *Neurocomputing*, *71*(1-3), 297-310.

Atkinson, A. C., & Cheng, T. (2000). On robust linear regression with incomplete data. *Computational Statistics & Data Analysis*, *33*(4), 361-380.

Bastogne, T. , Noura, H, Richard A., & Hittinger J. M. (2002). Application of subspace methods to the identification of a winding process. In *Proceedings of the 4th European Control Conference.*

Berlinet, A., & Roland, C. (2008). A novel particle swarm optimization algorithm for permutation flow-shop scheduling to minimize makespan. *Chaos, Solitons & Fractals, 35*(5), 851-861.

Bishop, C. M. (1995). *Neural networks for pattern recognition.* Oxford, UK: Oxford University Press.

Bordes, L., Chauveau, D., & Vandekerkhove, P. (2007). Comparison of ML–EM algorithm and ART for reconstruction of gas hold-up profile in a bubble column. *Chemical Engineering Journal, 130*(2-3), 135-145.

Brits, R., Engelbrecht, A. P., & van den Bergh, F. (2007). Locating multiple optima using particle swarm optimization. *Applied Mathematics and Computation, 189*(2), 1859-1883

Burk, S. F. (1960). A method of estimation of missing values in multivariable data suitable for use with an electronic computer. *Journal of the Royal Statistical Society, B22,* 302-306.

Cai, X., Zhang, N. Venayagamoorthy, G. K., & Wunsch II, D. C. Dempster, A. Laird, N., & Rubin, D. (1977). Maximum likelihood from incomplete data via the EM algorithm. *Journal of the Royal Statistical Society, Series B, 39*(1), 1–38.

Chen, C. T., Chen, C., & Hou, C. (2004). Speaker identification using hybrid Karhunen–Loeve transform and Gaussian mixture model approach. *Pattern Recognition, 37*(5), 1073-1075.

Choi, S. W., Park, J. H., & Lee, I. (2004). Process monitoring using a Gaussian mixture model via principal component analysis and discriminant analysis. *Computers & Chemical Engineering, 28*(8), 1377-1387.

Daszykowski, M., Walczak, B., & Massart, D. L. (2003). A journey into low-dimensional spaces with autoassociative neural networks. *Talanta, 59*(6), 1095-1105.

De Moor, B. L. R. (2006). DaISy: Database for the Identification of Systems. Department of Electrical Engineering, ESAT/SISTA, K.U. Leuven, Belgium, Retrieved August 15, 2007, from http://www.esat.kuleuven.ac.be/sista/daisy

Dempster, A. P, Laird, N. M., & Rubin, D. B. (1977). Maximum likelihood for incomplete data via the EM algorithm. *Journal of the Royal Statistical Society, B39,* 1-38.

Dhlamini, S. M., Nelwamondo, F. V., & Marwala, T. (2006). Condition monitoring of hv bushings in the presence of missing data using evolutionary computing. *WSEAS Transactions on Power Systems, 1,* 296-302.

Dindar, Z. A., & Marwala, T. (2004). Option pricing using a committee of neural networks and optimized networks. In *IEEE 2004 International Conference on Systems, Man and Cybernetics* (pp. 434–438).

Engelbrecht, A. P. (2005). *Fundamentals of computational swarm intelligence.* New York: John Wiley & Sons.

Fraley, C. (1999). On computing the largest fraction of missing information for the EM algorithm and the worst linear function for data augmentation. *Computational Statistics & Data Analysis, 31*(1), 13-26.

Frolov, A., Kartashov, A. Goltsev, A., & Folk, R. (1995). Quality and efficiency of retrieval for Willshaw-like autoassociative networks. *Computation in Neural Systems, 6*, 535-549.

Giffler, B., & Thompson, G. L. (1960). Algorithms for solving production scheduling problems. *Operations Research, 8*, 487-503.

Guerra, F. A., & dos S. Coelho, L. (2008). A particle swarm optimization approach to nonlinear rational filter modeling. *Expert Systems with Applications, 34*(2), 1194-1199.

Han, H., Ko, Y., & Seo, J. (2007). Stacked Laplace-EM algorithm for duration models with time-varying and random effects. *Computational Statistics & Data Analysis, (in press)*

Hartley, H. (1958). Maximum likelihood estimation from incomplete data. *Biometrics, 14*, 174–194.

Heppner, F., & Grenander, U. (1990). A stochastic nonlinear model for coordinated bird flocks. 1990. In S. Krasner (Ed.) *The ubiquity of chaos* (pp. 233-238). Washington DC: AAAS Publications.

Ingrassia, S., & Rocci, R. (2007). A stochastic EM algorithm for a semiparametric mixture model. *Computational Statistics & Data Analysis, 51*(11), 5429-5443.

Janson, S., Merkle, D., & Middendorf, M. (2007). Strength design of composite beam using gradient and particle swarm optimization. *Composite Structures, 81*(4), 471-479.

Japkowicz, N. (2002). Supervised learning with unsupervised output separation. *In: International Conference on Artificial Intelligence and Soft Computing*, 321-325.

Jarboui, B., Damak, N, Siarry, P and Rebai, A. (2008). The landscape adaptive particle swarm optimizer. *Applied Soft Computing, 8*(1), 295-304.

Jiang, Y., Hu, T. Huang, C., & Wu, X. (2007). Particle swarm optimization based on dynamic niche technology with applications to conceptual design. *Advances in Engineering Software, 38*(10), 668-676.

Kathiravan, R., & Ganguli, R. (2007). Particle swarm optimization for determining shortest distance to voltage collapse. *International Journal of Electrical Power & Energy Systems, 29*(10), 796-802.

Kauermann, G., Xu, R., & Vaida, F. (2007). Nonlinear random effects mixture models: Maximum likelihood estimation via the EM algorithm. *Computational Statistics & Data Analysis, 51*(12), 6614-6623.

Kennedy, J. E, & Eberhart, R. C. (1995). Particle swarm optimization. In *Proceedings of the IEEE International Conference on Neural Networks* (pp. 942-1948).

Lampinen, J., & Vehtari, A. (2001). Bayesian approach for neural networks—Review and case studies. *Neural Networks, 14 (3)*, 257-274

Leke, B. B., Marwala, T., & Tettey, T. (2006). Autoencoder networks for HIV classification. *Current Science. 91*(11), 1467-1473.

Lian, Z., Gu, X., & Jiao, B. (2008). Multi-step ahead nonlinear identification of Lorenz's chaotic system using radial basis neural network with learning by clustering and particle swarm optimization. *Chaos, Solitons & Fractals, 35*(5), 967-979.

Lin, Y., Chang, W., & Hsieh, J. (2008). Application of multi-phase particle swarm optimization technique to inverse radiation problem. *Journal of Quantitative Spectroscopy and Radiative Transfer, 109*(3), 476-493.

Little, R. J. A., & Rubin, D. B. (1987). *Statistical analysis with missing data.* New York: John Wiley & Sons.

Little, R. J. A., & Rubin, D. B. (2000). *Statistical analysis with missing data.* 2nd edition, New York: John Wiley & Sons.

Liu, X., Liu, H., & Duan, H. (2007). Time series prediction with recurrent neural networks trained by a hybrid PSO–EA algorithm. *Neurocomputing, 70*(13-15), 2342-2353.

Marwala, T. (2001). Probabilistic fault identification using vibration data and neural networks. *Mechanical Systems and Signal Processing, 15,* 1109-1128.

Marwala, T. (2005). Finite element model updating using particle swarm optimization. *International Journal of Engineering Simulation, 6*(2), 25-30.

Marwala, T., Mahola, U., & Chakraverty, S. (2007). Fault classification in cylinders using multi-layer perceptrons, support vector machines and Gaussian mixture models. *Computer Assisted Mechanics and Engineering Sciences, 14*(2), 307-316.

McLachlan, G., & Krishnan, T. (1997). *The EM algorithm and extensions.* New York: John Wiley & Sons.

M'hiri, S., Cammoun, K., & Ghorbel, F. (2007). Monte Carlo EM algorithm in logistic linear models involving non-ignorable missing data. *Applied Mathematics and Computation*, (in press).

Mohamed, S., & Marwala, T. (2005). Neural network based techniques for estimating missing data in databases. In *Proceedings of the 16th Annual Symposium of the Patten Recognition Association of South Africa*, (pp. 7-32).

Muteki, K., MacGregor, J. F., & Ueda, T. (2005). Estimation of missing data using latent variable methods with auxiliary information. *Chemometrics and Intelligent Laboratory Systems, 78*(1-2), 41-50.

Neal, R., & Hinton, G. (1998). A view of the EM algorithm that justifies incremental, sparse, and other variants. In M. Jordan (Ed.), *Learning in graphical models* (pp. 355-368). Dordrecht: Kluwer Academic Press.

Nelwamondo, F. V. (2008). *Computational intelligence techniques for missing data imputation.* Unpublished doctoral dissertation, University of the Witwatersrand, Johannesburg.

Nelwamondo, F. V., Marwala, T., & Mahola, U. (2006). Early classifications of bearing faults using hidden Markov models, Gaussian mixture models, mel-frequency cepstral coefficients and fractals. *International Journal of Innovative Computing, Information and Control, 2*(6), 1281-1299.

Park, J., Qian, G. Q., & Jun, Y. (2007). Using the revised EM algorithm to remove noisy data for improving the one-against-the-rest method in binary text classification. *Information Processing & Management, 43*(5), 1281-1293.

Patel, A. K., Patwardhan, A. W., & Thorat, B. N. (2007). Acceleration schemes with application to the EM algorithm. *Computational Statistics & Data Analysis, 51*(8), 3689-3702.

Qi, H., Ruan, L. M. Shi, M. An, W., & Tan, H. P. (2008). A combinatorial particle swarm optimization for solving multi-mode resource-constrained project scheduling problems. *Applied Mathematics and Computation, 195*(1), 299-308.

Ransome, T. (2006). *Automatic minimisation of patient setup errors in proton beam therapy.* Unpublished master thesis, University of the Witwatersrand, Johannesburg.

Ransome, T. M., Rubin, D. M., & Marwala, T., & de Kok, E. A. (2005). Optimising the verification of patient positioning in proton beam therapy. In *Proceedings of the IEEE 3rd International Conference on Computational Cybernetics* (pp. 279-284).

Reynolds, C. W. (1987). Flocks, herds and schools: A distributed behavioral model. *Computer Graphics, 2*, 25-34.

Rubin, D. B. (1978). Multiple imputations in sample surveys–a phenomenological Bayesian approach to nonresponse. In *Proceedings of the Survey Research Methods Section* (pp. 20-34).

Schafer, J. L. (1997). *Analysis of incomplete multivariate data.* New York: Chapman & Hall

Schafer, J. L., & Olsen, M. K. (1998). Multiple imputation for multivariate missing-data problems: A data analyst's perspective. *Multivariate Behavioural Research, 33*, 545-571.

Schafer, J. L., Graham, J. W. (2002). Missing data: Our view of the state of the art. *Psychological Methods, 7*(2), 147-177.

Sha, D. Y., & Hsu, C. (2006). A hybrid particle swarm optimization for job shop scheduling problem. *Computers & Industrial Engineering, 51*(4), 791-808.

Shimizu, H., Yasuoka, K. Uchiyama, K., & Shioya, S. (1997). Online fault diagnosis for optimal rice α-amylase production process of a temperature-sensitive mutant of Saccharomyces cerevisiae by an autoassociative neural network. *Journal of Fermentation and Bioengineering, 83*(5), 435-442.

Snyman, J. A. (2005). *Practical mathematical optimization: An introduction to basic optimization theory and classical and new gradient-based algorithms.* New York: Springer-Verlag.

Stoica, P., Xu, L., & Li, J. (2005). A new type of parameter estimation algorithm for missing data problems. *Statistics & Probability Letters, 75*(3), 219-229.

Tanner, M. (1996). *Tools for statistical inference.* New York: Springer-Verlag.

Teegavarapu , R. S. V., & Chandramouli, V. (2005). Improved weighting methods, deterministic and stochastic data-driven models for estimation of missing precipitation records. *Journal of Hydrology, 312*(1-4), 191-206.

Tresp, V., Neuneier, R. Ahmad, S. (1995). Efficient methods of dealing with missing data in supervised learning. In G. Tesauro, D. Touretzky and T. Leen (Eds.) *Advances in Neural Information Processing Systems* (pp. 689-696). Cambridge MA: MIT Press.

Thompson, B. B., Marks, R. J., & Choi, J. J. (2002). Implicit learning in autoencoder novelty assessment. In *Proceedings of the IEEE International Joint Conference on Neural Networks*, (pp. 2878-2883).

Wang, H., & Hu, Z. (2007). Speeding up HMRF_EM algorithms for fast unsupervised image segmentation by Bootstrap resampling: Application to the brain tissue segmentation. *Signal Processing, 87*(11), 2544-2559.

Wang, X., Schumitzky, A., & D'Argenio, D. Z. (2007). Constrained monotone EM algorithms for finite mixture of multivariate Gaussians. *Computational Statistics & Data Analysis, 51*(11), 5339-5351.

Wothke, W. (1993). Non-positive matrices in structural modelling. In K.A. Bollen and J.S. Long (Eds.) *Testing structural equation models* (pp. 256-293). Newbury Park, CA: Sage Publications.

Yisu, J., Knowles, J. Hongmei, L. Yizeng, L., & Kell, D. B. (2008). On the improved performances of the particle swarm optimization algorithms with adaptive parameters, cross-over operators and root mean square (RMS) variants for computing optimal control of a class of hybrid systems. *Applied Soft Computing, 8*(1), 324-336.

Zhong, M., Lingras, P., & Sharma, S. (2004). Estimation of missing traffic counts using factor, genetic, neural, and regression techniques. *Transportation Research Part C: Emerging Technologies, 12*(2), 139-166.

Chapter V
Missing Data Estimation Using Rough Sets

ABSTRACT

A number of techniques for handling missing data have been presented and implemented. Most of these proposed techniques are unnecessarily complex and, therefore, difficult to use. This chapter investigates a hot-deck data imputation method, based on rough set computations. In this chapter, characteristic relations are introduced that describe incompletely specified decision tables and then these are used for missing data estimation. It has been shown that the basic rough set idea of lower and upper approximations for incompletely specified decision tables may be defined in a variety of different ways. Empirical results obtained using real data are given and they provide a valuable insight into the problem of missing data. Missing data are predicted with an accuracy of up to 99%.

INTRODUCTION

There are a number of general ways that have been used to approach the problem of missing data in databases (Little & Rubin, 1987: Rubin, 1976; Little & Rubin, 1989; Collins, Schafer Kam, 2001; Schafer & Graham, 2002). One of the simplest of these methods is the 'list-wise deletion', which simply deletes instances with missing values (Scheuren, 2005; King et al., 2001; Abdella, 2005). The major disadvantage of this method is the dramatic loss of information in data sets (King et al., 1988). Also, Enders and Peugh (2004) demonstrated that when there is a group of missing data, list-wise deletion gives biased parameters and standard errors. Another approach is 'pair-wise deletion' (Marsh, 1998).

Tsikritis (2005) observed that appropriately dealing with missing data has been underestimated by the operation management authors, unlike in other fields such as marketing, organizational behavior analysis, economics, statistics and psychometrics that have intensely attended to the issue. Tsikritis found from a review of 103 survey articles appearing in the *Journal of Operations Management* between 1993 and 2001 that list-wise deletion was the most widely used technique to deal with missing data. This is in spite of the fact that list-wise deletion is usually the least accurate method for handling missing data. This indicates, therefore, that missing data estimation methods that are simple to understand and implement still need to be developed and then promoted for wider usage.

Kim and Curry (1997) found that when 2% of the features are missing and the complete observation is deleted then up to 18% of the total data may be lost. The second common technique imputes the data by finding estimates of the values and missing entries are then replaced with these estimates. Various estimates have been used and these estimates include zeros, means and other statistical calculations. These estimations are then used as if they were the observed values. Xia et al. (1999) estimated missing data in climatological time and investigated six methods for imputing missing climatological data including *daily maximum temperature, minimum temperature, air temperature, water vapor pressure, wind speed* and *precipitation*. These researchers used the multiple regression analysis with the five closest weather stations and the results obtained from the six methods showed similar estimates for the averaged precipitation amount.

Another common technique assumes some models for the prediction of the missing values and uses the maximum likelihood approach to estimate the missing values (Nelwamondo, Mohamed, & Marwala, 2007; Dempster, Laird, & Rubin., 1977; Abdella & Marwala, 2006). In Chapter IV, the hybrid auto-associative neural networks and simple genetic algorithms are compared to the Gaussian mixture models trained with the expectation maximization (GMM-EM) algorithm and tested using data sets from an industrial power plant, an industrial winding process and HIV sero-prevalence survey data. The results obtained show that both methods perform well. The GMM-EM method was found to perform well in cases where there was little or no inter-dependency between variables, whereas the hybrid auto-associative neural network and genetic algorithm was found to be suited to problems where there were some inherent nonlinear relationships between some of the given variables.

A great deal of research has been conducted to find new ways of approximating missing values. Among others, Abdella and Marwala (2006), Nelwamondo and Marwala (2007a), Ssali and Marwala (2007) as well as Mohamed and Marwala (2005) have used computational intelligence methods to approximate missing data. Qiao, Gao, and Harley (2005) have used neural networks and particle swarm optimization to keep track of the dynamics of the power plant in the presence of missing data. Nauck and Kruse (1999), Gabrys (2002) as well as Nelwamondo and Marwala (2007c) used fuzzy approaches to deal with missing data. A different approach was taken by Wang (2005) who replaced incomplete patterns with fuzzy patterns. The patterns without missing values were, along with fuzzy patterns, used to train the neural network. In this model, the neural network learned to classify without actually predicting the missing data and this approach is adopted in Chapter VIII. Similar work was also conducted by Arbuckle (1996)

Mitra, Pal, and Saddiqi (2003) described the incorporation of a minimal spanning tree, based on graph-theoretic technique, an expectation maximization algorithm as well as rough set initialization for non-convex clustering. The expectation maximization algorithm was found better able to handle uncertainties and the statistical model of the data. The rough set theory was found to assist in the increased speed of convergence and in avoiding local minima problem.

Hong, Tseng, and Wang (2002) solved the problem of effecting a set of certain and possible rules from incomplete data sets using rough set theory. They introduced a learning algorithm, which concurrently derived rules from incomplete data sets and estimated the missing values during the learning process. They first assumed unknown values as any possible values and then gradually refined these unknown values using the incomplete lower and upper approximations that were derived from the training examples. By so doing, the approximations and examples interacted with each other to give accurate rules and, thereby, inferred appropriate unknown values with greater accuracy. The rules that were identified were thus used as knowledge about the incomplete data set.

The use of neural networks comes with a greater computational cost in that the data must be made available before the missing condition occurs. This chapter implements a rough set based approach for estimating missing data. *Rough sets* are closely related to fuzzy logic and are mathematical and computational tools that create an intelligent look-up table that can relate different data points in terms of how similar these points are to what is in the look-up table (Ziarko, 1998; Yao & Yao, 2002). This becomes possible in large data sets where the probability is high that a set of data will be repeated. It is envisaged that in large databases, it is more likely that the missing values might be correlated with some other variables observed elsewhere in the same dataset. Instead of approximating missing data, it might, therefore, be cheaper to spot similarities between the observed data instances and those that contain missing attributes. The key issue is spotting similarity. The rough set is such a tool that assist in intelligently spotting similarities because of its ability to offer algebraic representation of modeling information (Comer, 1991) and its ability to model uncertainty (Beaubouef & Petry, 2003).

APPLICATIONS OF ROUGH SETS

There are many applications of rough sets reported in the literature (Pattaraintakorn & Cercone, 2007; Golan & Ziarko, 1995; Marwala, 2007; Wong, Ziarko, & Ye, 1986; Crossingham, 2007; Crossingham, & Marwala, 2007a). Most of these applications assume that complete data is available (Grzymala-Busse, 2004; Nelwamondo & Marwala, 2008). Tettey, Nelwamondo, and Marwala (2007) as well as Nelwamondo and Marwala (2007b) who proposed an algorithm based on rough set theory for missing data estimation and applied a rough set technique for missing data estimation to a large and real database. Nishino, Nagamachi, and Tanaka (2006) applied a variable precision Bayesian-based rough set model to human evaluation data. In large databases, it is more likely that the missing values may be correlated with other variables observed elsewhere in the same data. Instead of approximating missing data, it was found to be cheaper to identify indiscerniblity relations between the observed data instances and those that contain missing attributes. The results obtained using the HIV database had accuracies ranging from 74.7% to 100%. However, the drawback of this method was that it made no extrapolation or interpolation and as a result, can only be used if the missing case is similar or related to another case with more observations. Pe-a, Ltourneau, and Famili (1999) successfully applied rough sets algorithms to the problem of predicting aircraft component failure, while Ohrn and Rowland (2000) successfully used rough sets as a knowledge discovery technique for multi-factorial medical outcomes. Furthermore, Hassanien (2007) used fuzzy rough sets for the detection of cancer.

Marwala and Crossingham (2007a) proposed an approach to training rough set models using a Bayesian framework trained using the Markov Chain Monte Carlo (MCMC) method. The prior probabilities were constructed from the prior knowledge that good rough set models have fewer rules. Markov Chain Monte Carlo sampling was conducted through sampling in the rough set granule space and Metropolis-Hastings algorithm was used as an acceptance criteria. The proposed method was tested to estimate the risk of HIV given the demographic data. The results obtained showed that the proposed approach could achieve an average accuracy of 58% with the accuracy varying up to 66%. In addition, the Bayesian rough set gave the probabilities of the estimated HIV status as well as the linguistic rules describing how the demographic parameters drive the risk of HIV.

Tettey, Nelwamondo, and Marwala (2007) presented an analysis of HIV data obtained from a survey performed on pregnant women by the Department of Health in South Africa. The HIV data was analyzed by formulating a rough set approximation of the six demographic variables analyzed. Unlike the variables in Chapter IV, this time the variables were *Race, Age of Mother, Education, Gravidity, Parity* and *Age of Father*. It was found that of the 4096 possible subsets in the input space, the data only represented 225 of those cases with 130 cases being discernible and 96 cases indiscernible. The rough sets analysis was suggested as a quick way of analyzing data and extracting rules that define inter-relationships over neuro-fuzzy models when it came to data driven identification. This was because neuro-fuzzy models give rules that are not completely expressed in linguistic fashion and, thereby, are still difficult to interpret. Comparisons of rule extraction using rough sets and using neuro-fuzzy were conducted and the results were found to be in favor of the rough sets.

Marwala and Crossingham (2007b) proposed a neuro-rough model for modeling the risk of HIV from demographic data. The model was also formulated using a Bayesian framework (Pawlak, 2003) and trained using the Markov Chain Monte Carlo method and the Metropolis-Hastings criterion. When the model was tested given the demographic data and asked to estimate the risk of HIV infection, it was found to give an accuracy of 62%. This is to be contrasted with 58% obtained from a Bayesian formulated rough set model which was trained using the Markov Chain Monte Carlo method and 62% obtained from a Bayesian formulated multi-layered perceptron model, trained using hybrid Monte Carlo procedure. The presented model combined the accuracy of the Bayesian multi-layered perceptron model and the transparency of Bayesian rough set model.

There is also a great deal of information about various applications of rough sets to medical data sets. Rough sets have been mostly used in prediction cases and Rowland, Ohno-Machado, and Ohrn (1998) compared neural networks and rough sets for the prediction of *ambulation* following a spinal cord injury. Although rough sets performed somewhat worse than neural networks, they demonstrated that they can be used in prediction problems. They offer linguistic rules that aid medical practitioners to make practical diagnosis.

Rough sets have also been used in learning Malicious Code Detection (Zhang et al., 2006) and in fault diagnosis (Tay & Shen, 2003). Grzymala-Busse and Hu (2001) have presented nine different approaches for estimating missing values in databases. Amongst others, the presented methods included choosing the most common concept and most common attribute, assigning all possible values related to the current concept, deleting cases with missing values, treating missing values as special values and imputing for missing values using other techniques such as neural networks and maximum likelihood approaches. Some of the techniques proposed come with either a heavy computational expense or loss of information. The following section discusses the difference between rough sets and the closely related concept of a fuzzy set.

ROUGH SETS VS. FUZZY SETS

Fuzzy sets are sets that have elements with degrees of membership. In other words, in fuzzy logic an element of a set has a degree of belonging or membership to that particular set. Zadeh (1965) introduced fuzzy sets as an expansion of the classical concept of a set. In *classical set theory*, the membership of elements in a set is evaluated in binary terms in that i.e., either it is a member of that set or it is not a member of that particular set. Fuzzy set theory allows the steady evaluation of the membership of elements in a set and this is illustrated with the help of a *membership function* allowed to fall within the interval [0, 1]. A fuzzy set, is therefore a generalized version of a classical set. Conversely, a classical set is a special case of the membership functions of fuzzy sets which only permit values 0 or 1. Thus far, fuzzy set theory has not generated any results that differ from the results from a probability or classical set theory.

Zhang and Shao (2006) developed a self-learning method to identify a fuzzy model and to extrapolate missing rules. This was done using the modified gradient descent method and confidence measure. The method can simultaneously identify a fuzzy model, revise its parameters and establish optimal output fuzzy sets. When tested on a classical truck control problem, the results obtained show the usefulness and accuracy of the advanced method.

Coulibaly and Evora (2007) investigated the multilayer perceptron network, the time-lagged feed-forward network, the generalized radial basis function network, the recurrent neural network, the time delay recurrent neural network and the counter-propagation fuzzy-neural network along with different optimization methods for estimating daily time series with missing daily total *precipitation* records and extreme *temperatures*. The results obtained revealed that the multi-layer perceptron, the time-lagged feed-forward network and the counter-propagation fuzzy-neural network offered the highest accuracy in estimating missing precipitation values. The multi-layer perceptron was found to be successful at imputing missing daily precipitation values. In addition, the multi-layer perceptron was most suitable for imputing missing daily maximum and minimum temperature values. The counter-propagation fuzzy-neural network was similar to the multi-layer perceptron at imputing missing daily maximum temperatures; sadly, it was less useful for estimating the minimum temperature. The recurrent neural network and time delay recurrent neural network were found to be the least appropriate method for imputing both daily precipitation and the extreme temperature records, while the radial basis function was good at approximating maximum and minimum temperature.

A *rough set* is defined as a proper estimation of a classical (crisp) set through a pair of sets that offer the lower and the upper approximation of the original set. The lower approximations are sets that are similar to sets that have already been observed in the information system while the higher approximation sets are sets that can only be inferred either strongly or weakly from the information system. These lower and upper approximation sets are crisp sets in the classical description of rough set theory (Pawlak, 1991), but in other variants, the approximating sets may also be fuzzy sets.

The distinguishing features between fuzzy sets and rough sets are that while fuzzy sets operate via membership functions, rough sets operate through upper and lower approximation sets (Chanas & Kuchta, 1992). The similarity between the two approaches is that they are both designed to deal with the vagueness and uncertainty of the data. For example, Jensen and Shen (2004) introduced dimensionality reduction that retained semantics using both the rough and fuzzy-rough based approaches. Some researchers, such as Deng et al. (2007), introduced a method that combines both rough and fuzzy sets. The reason for this hybridization process is that even though these techniques are similar, each one offers its own unique advantages. Such hybrid fuzzy rough sets scheme has been applied to model completed problems such as breast cancer detection. The following section describes rough sets, which is a method that will be adopted in this chapter.

ROUGH SET THEORY

The *Rough set theory* was introduced by Pawlak (1991). It is a mathematical tool to deal with vagueness and uncertainty. It is based on set of rules that are expressed in terms of linguistic variables. Rough sets are of fundamental importance to computational intelligence and in cognitive science. They are highly applicable to the tasks of machine learning and decision analysis, especially in the analysis of decisions where there are inconsistencies. Because they are rule-based, rough sets are very transparent. However, they are not as accurate in their predictions, and most certainly are not universal approximators, as other machine learning tools such as neural networks are. It can thus be concluded that in machine learning there is always a trade-off between prediction accuracy and transparency.

Crossingham and Marwala (2007b) presented an approach to optimize rough set partition sizes using various optimization techniques. Three optimization techniques were implemented to perform the granularization process, namely the genetic algorithm, hill climbing and simulated annealing. These optimization methods maximize the classification accuracy of the rough sets. The presented rough set partition method was tested on a set of demographic properties of individuals. The three techniques were compared for computational time, accuracy and number of rules produced when applied to the HIV data set. The optimized methods results were compared with a non-optimized discretization method known as equal-width-bin partitioning. The accuracies achieved after optimizing the partitions using genetic algorithm, hill climbing and simulated annealing were 66.89%, 65.84% and 65.48%, respectively, compared to an accuracy of equal-width-bin partitioning of 59.86%. In addition to rough sets providing the plausability of the estimated HIV status, they also provided the linguistic rules describing how the demographic parameters drive the risk of HIV.

Rough sets theory provides a technique of reasoning from vague and imprecise data (Goh & Law, 2003). The technique is based on the assumption that some information is associated somehow with some information of the universe of the discourse (Komorowski et al., 1999; Yang & John, 2006; Kondo, 2006). This, therefore, implies that if some aspects of the data are missing then they can be estimated from the part of the information in the universe of discourse that is similar to the observed part of that particular data. Objects with the same information are indiscernible in the view of the available information. An elementary set consisting of indiscernible objects forms a basic granule of knowledge. A union of elementary sets is referred to as a *crisp set* otherwise the set is considered a rough set. The next few sub-sections briefly introduce concepts that are common to rough set theory.

Information System

An information system (Λ), is defined as a pair (U, A) where U is a finite set of objects called the universe and A is a non-empty finite set of attributes as shown in equation 5.1 (Yang & John, 2006; Pawlak & Skowron, 2007a).

$$\Lambda = (U,A) \tag{5.1}$$

Every attribute $a \in A$ has a value, which must be a member of a value set V_a of the attribute a (Dubois, 1990):

$$a: U \to V_a \tag{5.2}$$

A rough set is defined with a set of attributes and the indiscernibility relation between them. Indiscernibility is discussed in the next section.

Indiscernibility Relation

Indiscernibility relation is one of the fundamental ideas of rough set theory (Grzymala-Busse & Siddhaye, 2004; Zhao, Yao, & Luo, 2007; Pawlak & Skowron, 2007b). Indiscernibility simply implies similarity (Goh & Law, 2003) and therefore these sets of objects are indistinguishable. Given an information system Λ and subset $B \subseteq A$, B the indiscernibility determines a binary relation $I(B)$ on U such that (Pawlak, Wong, & Ziarko, 1988; Ohrn, 1999; Wu, Mi, & Zhang, 2003; Pawlak & Skowron, 2007c):

$(x,y) \in I(B)$
if and only if
$a(x) = a(y)$ (5.3)

for all $a \in A$, where $a(x)$ denotes the value of attribute a for element x. Equation 5.3 implies that any two elements that belong to $I(B)$ should be identical from the point of view of a. Suppose U has a finite set of N objects $\{x_1, x_2, ..., x_N\}$. Let Q be a finite set of n attributes $\{q_1, q_2, ..., q_n\}$ in the same information system Λ, then according to Pawlak (1991); Inuiguchi & Miyajima (2007):

$$\Lambda = \langle U, Q, V, f \rangle$$ (5.4)

where f is the total decision function called the *information function*. From the definition of the indiscernibility relation given in this section, any two objects have a similarity relation to attribute a if they have the same attribute values everywhere except for the missing values.

Information Table and Data Representation

An *information table* is used in rough sets theory as a way of representing the data. The data in the information table are arranged based on their condition attributes and decision attribute (D). The condition attributes and decision attributes are analogous to the independent variables and dependent variable (Goh & Law, 2003; Questier et al., 2002). These attributes are divided into $C \cup D = Q$ and $C \cup D = 0$.

An information table can be classified into *complete* and *incomplete classes*. All objects in a complete class have known attribute values B. On the other hand, an information table is considered incomplete if at least one attribute variable has a missing value. An example of an incomplete information table is given in Table 5.1. Data is represented in a table where each row represents an *instance*, sometimes referred to as an *object*. Every column represents an *attribute*, which can be a measured variable. This kind of a table is also referred to as *Information System* (Komorowski, et al., 1999; Leung, Wu, & Zhang, 2006).

Decision Rules Induction

Rough sets also involve generating decision rules for a given information table. The rules are normally determined based on condition attributes values (Goh & Law, 2003; Bi, Anderson, & McClean, 2003;

Table 5.1. An example of an Information Table with missing values

	B_1	B_2	B_3	D
1	1	1	0.2	P
2	1	2	0.3	A
3	0	1	0.3	P ·
4	?	?	0.3	A
5	0	3	0.4	A
6	0	2	0.2	P
7	1	4	?	A

Slezak & Ziarko, 2005). The rules are presented in an *if CONDITION(S)-then DECISION* format. Stefanowski (1998) successfully used rough set based approaches for inference of decision rules. Wang et al. (2006) used rough set theory to deal with vagueness and uncertainty and thereby reduce the redundancy in assessing the degree of malignancy in brain glioma, based on magnetic resonance imaging (MRI) findings as well as the clinical data before the operation. The data included inappropriate features at the same time as uncertainties and missing values. The rough set rules that were extracted from this dataset were used to forecast the degree of malignancy. Rough set based feature selection algorithms were used to choose features so that the classification accuracy based on decision rules could be improved. These chosen feature subsets were used to produce decision rules for the classification task. A method based on particle swarm optimization reduced the attributes of the rough set and it was compared with other conventional rough set reduction schemes. The results obtained demonstrated that the method identifies *reducts* (see below) that generate decision rules with higher classification rates than conventional approaches.

Lower and Upper Approximation of Sets

Lower and upper approximations are defined based on an indiscernibility relation as it was discussed above. The lower approximation is defined as the collection of cases whose equivalent classes are contained in the cases that need to be approximated, whereas the upper approximation is defined as the collection of classes that are partially contained in the set that needs to be approximated (Rowland, Ohno-Machado, & Ohrn, 1998; Degang, et al., 2006; Witlox & Tindemans, 2004). Let a concept X be defined as a set of all cases defined by a specific value of the decision. Any finite union of an elementary set associated with B is called a *B - definable set* (Grzymala-Busse & Siddhaye, 2004). The set X is approximated by two *B - definable sets*, referred to as the *B-lower approximation,* denoted as $\underline{B}X$ and the *B-upper approximation* denoted as $\overline{B}X$. The B-lower approximation is defined as (Grzymala-Busse & Siddhaye, 2004; Bazan, Nguyen, & Szczuka, 2004):

$$\underline{B}X = \left\{ x \in U \middle| [x]_B \subseteq X \right\} \qquad (5.5)$$

and the B-upper approximation is defined as:

$$\overline{B}X = \left\{ x \in U \middle| [x]_B \cap X \neq 0 \right\} \tag{5.6}$$

Other methods have been reported in the literature for defining the lower and upper approximations for completely specified decision tables. Some of the common ones include approximating the lower and upper approximation of X using Equations 5.7 and 5.8, respectively as follows (Grzymala-Busse, 2004):

$$\cup \left\{ [x]_B \middle| x \in U, [x]_B \subseteq X \right\} \tag{5.7}$$

$$\cup \left\{ [x]_B \middle| x \in U, [x]_B \cap X \neq 0 \right\} \tag{5.8}$$

The definition of definability is modified in cases of incompletely specified tables. In this case, any finite union of characteristic sets of B is called a B - *definable set*. Three different definitions of approximations have been discussed by Grzymala-Busse and Siddhaye (2004). Again letting B be a subset of A of all attributes and $R(B)$ be the characteristic relation of the incomplete decision table with characteristic sets $K(x)$, where $x \in U$, the following are defined:

$$\underline{B}X = \{ x \in U \middle| K_B(x) \subseteq X \} \tag{5.9}$$

and

$$\overline{B}X = \left\{ x \in U \middle| K_B(x) \cap X \neq 0 \right\} \tag{5.10}$$

Equations 5.9 and 5.10 are referred to as *singletons*. The subset lower and upper approximations of incompletely specified data sets are then defined as:

$$\cup \left\{ K_B(x) \middle| x \in U, K_B(x) \subseteq X \right\} \tag{5.11}$$

and

$$\cup \left\{ K_B(x) \middle| x \in U, K_B(x) \cap X = 0 \right\} \tag{5.12}$$

More information on these methods can be found in (Grzymala-Busse, 2004; Grzymala-Busse & Hu, 2001; Grzymala-Busse, 1992; Grzymala-Busse & Siddhaye, 2004). It follows from the properties that a crisp set is only defined if $\underline{B}(X) = \overline{B}(X)$. Roughness, therefore, is defined as the difference between the upper and the lower approximation.

Set Approximation

There are various properties of rough sets that have been presented in (Pawlak, 1991) and (Pawlak, 2002). One property of a rough set is the *definability* of a rough set (Quafafou, 2000), which is discussed briefly above and is when the lower and upper approximations are equal. Otherwise, if this is not the case, then the target set is indefinable. Some of the special cases of definability are (Pawlak, Wong, & Ziarko, 1988):

- *Internally definable* set: Here $\underline{B}X \neq 0$ and $\overline{B}X = U$. Here the attribute set B has objects that are certainly elements of the target set X, even though there are no objects that can be definitively excluded from set X.
- *Externally definable* set: Here $\underline{B}X = 0$ and $\overline{B}X \neq U$. Here the attribute set B has no objects that are certainly elements of the target set X, even though there are objects that can be definitively excluded from set X.
- *Totally un-definable* set: Here $\underline{B}X = 0$ and $\overline{B}X = U$. Here the attribute set B has no objects that are certainly elements of the target set X, though there are no objects that can definitively be excluded from set X.

Other properties of rough sets are the *reduct* and *core*. A fascinating inquiry is whether there are attributes B in the information system that are more essential to the knowledge represented in the equivalence class structure than other attributes. It will be interesting to find out if there is a subset of attributes, which by itself can completely describe the knowledge in the database, and this attribute set is known as a *reduct*.

Beynon (2001) observed that the elementary feature of the variable precision rough sets model entailed an exploration for subsets of condition attributes that give identical information for classification functions as the complete set of given attributes. Beynon labeled these subsets *approximate reducts* and described these for an identified classification error represented by β and then identified specific anomalies and fascinating implications for identifying β-reducts which guaranteed a general knowledge similar to that obtained from the full set of attributes. Terlecki and Walczak (2007) described the relations between rough set reducts and emerging patterns. From this study, practical usage of these observations to the minimal reduct problem was established. Also, using these relations to test the differentiating factor of an attribute set was established.

Ziarko and Shan (1995) defined a *reduct* formally as a subset of attributes $RED \subseteq B$ such that:

- $[x]_{RED} = [x]_B$ i.e., the equivalence classes that were induced by reducing the attribute set RED is identical to the similar class structure that was induced by full attribute set B.
- The attribute set RED is minimal because of the fact that $[x]_{(RED-A)} \neq [x]_B$ for any attribute $A \in RED$. Simply, there is no attribute that can be taken away from the set RED without altering the equivalence classes $[x]_B$.

Therefore a reduct can be visualized as an adequate set of features that can sufficiently express the category structure. One characteristic of a reduct of an information system is that it is not unique because there may be other subsets of attributes that may still preserve the equivalence class structure conveyed in the information system. The set of characteristics that are common in all reducts is called a *core*.

Boundary Region

The boundary region, which can be written as the difference $\overline{B}X - \underline{B}X$, is a region that is composed of objects that cannot be included or excluded as members of the target set X. Simply, the lower approximation of a target set is an approximation that consists of only those objects which can be positively identified as members of the set. The upper approximation is a loose approximation and includes objects that may be members of the target set. The *boundary region* is the region in between the upper approximation and the lower approximation.

Rough Membership Functions

A *rough membership function* is a function $\mu_A^x : U \rightarrow [0,1]$ that when applied to object x, quantifies the degree of overlap between set X and the indiscernibility set to which x belongs. The rough membership function is used to calculate the *plausibility* and can be defined per (Pawlak, 1991):

$$\mu_A^x(X) = \frac{\left| [x]_B \cap X \right|}{\left| [x]_B \right|}$$

(5.13)

Now that rough sets have been discussed, the next step is to discuss how rough sets can be used for missing data estimation, which is the subject of the remainder of this chapter.

MISSING DATA IMPUTATION BASED ON ROUGH SETS

For any missing data process, it is important that there be adequate information in the universe of discourse, in advance, that can accurately characterize the dynamics of the system that the particular data set represents. For example, Gold and Bentler (2000) characterized this universe of discourse using regression models trained by the Monte Carlo method. The algorithm implemented in this chapter (shown in Figure 5.1), belongs to the hot-deck approach to missing data, and estimates the missing values by presenting a list of all possible values based on the observed data.

As mentioned earlier, the hypothesis here is that in some finite database, a case similar to the missing data case could have been observed before. Therefore it should be computationally cheaper to use such values instead of computing missing values with complex methods such as neural networks or other multivariate analysis (Little, Schnabel, & Baumert, 2000; Graham and Hofer, 2006). The algorithm implemented in this chapter is shown in Algorithm 1, followed by a worked out example demonstrating how the missing values are imputed. In our example, the degree of *belongingness* $\kappa \left(o[x_{14}] \right) = o \ [x_{14}] = 1 / \left| dom(x_{14}) \right|$ where $o \neq o'$ and $dom(x_{14})$ denotes the domain of the attribute x_{14} which is the fourth instance of x_1 and $\left| dom(x_{14}) \right|$ which is the cardinality of $dom(x_{14})$. If the missing values are to be interpreted possibilistically, all attributes have the same possibilistic degree of being the actual one.

The algorithm in this study is fully dependent on the available data and makes no additional assumptions about the data or the distribution thereof. As presented in the algorithm, a list of possible values is given in a case where a crisp set could not be found. It is from this list that possible values may be heuristically chosen. A justification to this is sometimes the exact value needs to be known. As a result, it may be cheaper to have a *rough* value. The possible imputable values are obtained by collecting all the entries that lead to a particular decision D. The algorithm used in this application is a simplified version of the algorithm of Hong, Tseng and Wang (2002).

There are two approaches to reconstructing the missing values. The missing values can be *probabilistically* interpreted or be *possibilistically* interpreted (Nakata & Sakai, 2006). The algorithm will now be illustrated using an example. Missing values will be denoted by a question mark (?) symbol. Attribute values of the attribute a are denoted as V_a. Using the notation defined in Gediga and Duntsch

Figure 5.1. Algorithm of using rough sets to impute missing data

Algorithm 1: Rough sets based missing data imputation algorithm

input:	Incomplete data set Λ with a attributes and i instances. All these instances should belong to a decision D

output: A vector containing possible missing values

assumption: D and some attributes will always be known

for all *i* **do**

→Partition the input space according to D
→Arrange all attributes according to order of availability, with D being first.
end

for each *attribute* **do**

→Without directly extracting the rules, use the available information to extract relationships to other instances i in the $\wedge$.
→ Compute the family of equivalent classes $e(a)$ containing each object o_i for all input attributes.
→ The degree of *belongingness* $k(o[A]I/|dom(a_{i_{missing}})|$ *where* $o \neq o$ and $dom(x_{1_4})$ denotes the domain of attribute x_{1_4}, which is the fourth instance of x_1, and $|dom(x_{1_4})|$ is the cardinality of $dom(x_{1_4})$ while extracting relationships do

If i has the same attribute values with a_j everywhere except for the missing value, replace the missing value, a, with the value v_j, from a_j, where j is an index to another instance. Otherwise proceed to the next step

end
→ Complete the lower approximation of each attribute, given the available data of the same instance with the missing value.
while doing this **do**

If more than one v_j value is suitable for the estimation, postpone the replacement for later when it will be clear which value is appropriate.

end
→ Compute the incomplete upper approximations of each subset partition.
→ Do the computation and imputation of missing data as was done with the lower approximation.
→ Either crisp sets will be found, otherwise, rough sets can be used and missing data can be heuristically selected from the obtained rough set.
end

(2003) we let $rel_Q(x)$ represent a set of all *Q-relevant* attributes of x. It is assumed that an information table such as that presented in Table 5.1, where x_1 is in binary form, $x_2 \in [1:5]$ being integers and x_3 either 0.2, 0.3 or 0.4.

The algorithm firstly seeks relationships between variables. Since this is a small database, it is assumed that the only variable that will always be known is the decision, D. The first step is to partition the data according to this decision, as follows (Nelwamondo, 2008):

$$\varepsilon(D) = \{o_1, o_3, o_6\}, \{o_2, o_4, o_5, o_7\} \tag{5.14}$$

Two partitions are obtained due to the binary nature of the decision in the given example. The next step is to extract indiscernible relationships within each attribute. For x_1, the following is obtained:

$$IND(x_1) = \left\{ \begin{array}{l} (o_1,o_1),(o_1,o_2),(o_1,o_4),(o_1,o_7),(o_2,o_2),(o_2,o_4), \\ (o_2,o_7),(o_3,o_3),(o_3,o_4),(o_3,o_5),(o_3,o_6),(o_4,o_4), \\ (o_4,o_5),(o_4,o_6),(o_4,o_7),(o_5,o_5),(o_5,o_6),(o_6,o_6),(o_7,o_7) \end{array} \right\} \tag{5.15}$$

The family of equivalent classes $\varepsilon(x_1)$ containing each object o_i for all input variables is computed as follows:

$$\varepsilon(x_1) = \{o_1, o_2, o_4, o_7,\}, \{o_3, o_4, o_5, o_6,\} \tag{5.16}$$

Similarly,

$$\varepsilon(x_2) = \{o_1, o_3, o_4\}, \{o_2, o_4, o_6,\}, \{o_4, o_5\}\{o_4, o_7,\}\{o_4\}\{o_7\} \tag{5.17}$$

and

$$\varepsilon(x_3) = \{o_1, o_6, o_7\}, \{o_2, o_3, o_4, o_7\}, \{o_5, o_7\} \tag{5.18}$$

The lower approximation is defined as:

$$\underline{A}(X_{miss}, \{X_{avail}, D\}) A(X_{miss}, \{X_{avail}, D\}) = \{E(X_{miss}) | \exists(X_{avail}, D), E(X) \subseteq (X_{avail}, D)\} \tag{5.19}$$

whereas the upper approximation is defined as:

$$\overline{A}(X_{miss}, \{X_{avail}, D\}) A(X_{miss}, \{X_{avail}, D\}) = \{E(X_{miss}) | \exists(X_{avail}, D), E(X) \cap X_{avail} \cap D\} \tag{5.20}$$

Using $IND(x_1)$, the families of all possible classes containing o_4 which is given by:

$$Poss\varepsilon(x_1)o_i = \{o_1, o_2, o_7\}, \{o_1, o_2, o_4, o_7\}, i = 1, 2, 7$$
$$Poss\varepsilon(x_1)o_i = \{o_3, o_5, o_6\}, \{o_3, o_4, o_5, o_6\}, i = 3, 5, 6$$
$$Poss\varepsilon(x_1)o_4 = \{o_4, o_1, o_2, o_7\}, \{o_3, o_4, o_5, o_6\} \tag{5.21}$$

The probabilistic degree to which the chosen value is the right one is given by (Nakata & Sakai, 2006):

$$k((\{o_i\}) \in \varepsilon(x_1)) = \frac{1}{2}, i = 1, 2, 7$$

$$k((\{o_i\}) \in \varepsilon(x_1)) = \frac{1}{2}, i = 3, 5, 6$$

$$k((\{o_i\}) \in \varepsilon(x_1)) = \frac{1}{2}, i = 4 \tag{5.22}$$

else:

$$k((\{o_i\}) \in \varepsilon(x_1)) = 0 \tag{5.23}$$

The else part applies to all other conditions, such as $k(\{o_1, o_2, o_3\}) \in \varepsilon(x_1) = 0$. A family of weighted equivalent classes can now be computed as follows:

$$\varepsilon(x_1) = \left\{ \left\{ o_1, o_2, o_4, o_7 \right\}, \left\{ \frac{1}{2} \right\} \right\}, \left\{ \left\{ o_3, o_4, o_5, o_6 \right\} \left\{ \frac{1}{2} \right\} \right\} \tag{5.24}$$

The values $\varepsilon(x_2)$ and $\varepsilon(x_3)$ are computed in a similar way. Then the families of weighted equivalent classes can be used to obtain the lower and upper approximations as presented above. The degree to which the object o has the same value as object o' of the attributes is referred to as the *degree of belongingness* and is defined in terms of the binary relation for indiscernibility as (Nakata & Sakai, 2006):

$$IND(X) = \left\{ \left((o,o'), \kappa \left(o[X] = o'[X] \right) \right) \mid \left(\kappa \left(o[X] = o'[X] \right) \neq 0 \right) \wedge (o \neq o') \right\} \cup \left\{ ((o,o'),1) \right\} \tag{5.25}$$

where $\kappa \left(o[X] = o'[X] \right)$ is the indiscernibility degree of the objects o and o' and this is equal to the degree of belongingness:

$$\kappa \left(o[X] = o'[X] \right) = A_i \overset{\otimes}{\in} X \, \kappa \left(o[A_i] = o'[A_i] \right) \tag{5.26}$$

where the operator $\otimes$ depends on whether the missing values are possibilistically or probabilistically interpreted. For a probabilistic interpretation, the parameter is a product denoted by $\times$, otherwise the operator *minimize* is used.

EXPERIMENTAL EVALUATION

The data used for testing was obtained from the South African antenatal sero-prevalence survey of 2001 and was explained in Chapter IV. The data for this survey were obtained from questionnaires answered by pregnant women visiting selected public clinics in South Africa. Only women participating for the first time in the survey were eligible to answer the questionnaire. Data attributes used in this chapter are the *HIV status, education level, gravidity, parity, age, age of the father, race* and *region*. The difference between the data used in this chapter and those used in Chapter IV is that in this chapter the additional variables *region* and *race* are used. The HIV status is the decision and is represented in a binary form, where 0 and 1 represent negative and positive, respectively. Race is measured on the scale 1 to 4 where 1, 2, 3, and 4 represent African, Mixed, European and Asian, respectively. The data used was obtained in three regions, referred to as region A, B and C in this chapter. Using these values for region and race rather than binary numbers is not problematic because the nature of the rough set is discontinuous.

The *education level* was measured using integers representing the highest grade successfully completed, with 13 representing tertiary education. *Gravidity* was the number of pregnancies, complete or incomplete, experienced by a female, with this variable being an integer between 0 and 11. *Parity* was the number of times the individual had given birth, with multiple births being counted as one. Both *parity* and *gravidity* are important, as they show the degree of reproductive activity and the reproductive health state of the women. *Age gap* measured the age difference between the pregnant woman and the prospective father of the child. A sample of this data set is shown in Table 5.2.

Data Preprocessing

As mentioned in a previous section, the HIV/AIDS data used in this work was obtained from a survey performed on pregnant women. Like all data in raw form, several steps need to be taken to ensure that the data is in a usable form. Several types of outliers were identified in the data. Firstly, some of the data records were not complete. This is probably because the people being surveyed omitted certain information and errors were made by the person who manually recorded the surveys onto a spreadsheet. For instance, *Gravidity* is defined as the number of times a woman has been pregnant and *parity* is defined as the number of times a woman has given birth. Any instance where the value of parity is greater than that of gravidity is wrong. The justification for this is that is it not possible for a woman to give birth more than she has been pregnant if multiple births are in a pregnancy are counted as 1. The whole observation was considered an outlier and was removed. It should be emphasized that rough sets are not very good at handling outliers as compared to neural networks. Therefore, in this chapter it was vital that these outliers were removed from the dataset.

Variable Discretization

The degree of *discretization* defines the granularity with which the universe of discourse can be analyzed. If one chooses to discretize the variables into a large number of categories, then the rules extracted are more complex to analyze. Therefore, if one would like to use the rough sets for rule analysis and interpretation rather than for classification it is advisable that the number of categories be as small as possible. For the purposes of this work, the input variables were discretized into four categories. A description of the categories and their definitions is shown in Table 5.3. Table 5.4 shows the simplified version of the information system shown in Table 5.2.

Table 5.2. Extract of the HIV database used, with missing values. Key: Reg=Region, Educ=Education, Gravid=Gravidity, Fat=Father's age

Race	Reg	Educ	Gravid	Parity	Age	Fat	HIV
1	C	?	1	2	35	41	0
2	B	13	1	0	20	22	0
3	?	10	2	0	?	27	1
2	C	12	1	?	20	33	1
3	B	9	?	2	25	28	0
?	C	9	2	1	26	27	0
2	A	7	1	0	15	?	0
1	C	?	4	?	25	28	0
4	A	7	1	0	15	29	1

Table 5.3. The discretized variables

Race	Age	Education	Gravidity	Parity	Father's Age	HIV
1	≤ 19	Zero (0)	Low (≤ 3)	Low (≤ 3)	≤ 19	0
2	[20 – 29])	P (1 – 7)	High (> 3)	High (> 3)	([20 - 29])	1
3	[30 – 39])	S (8 - 12)	-	-	([30 - 39])	-
4	≥ 40	T (13)		-	≥ 40	-

Table 5.4. Extract of the HIV database used with missing values after discretization. Key: Educ.=Education, Par.=Parity

Race	Region	Educ	Gra.	Par.	Age	Father's Age	HIV
1	C	?	≤ 3	≤ 3	[31:40]	[41:50]	0
2	B	T	≤ 3	≤ 3	≤ 20	[21:30]	0
3	?	S	≤ 3	≤ 3	?	[21:30]	1
2	C	S	≤ 3	?	≤ 20	[21:30]	1
3	B	S	?	≤ 3	[21:30]	[21:30]	0
?	C	S	≤ 3	≤ 3	[21:30]	[21:30]	0
2	A	P	≤ 3	≤ 3	≤ 20	?	0
1	C	?	>3	>	[21:30]	[21:30]	0
4	A	P	≤ 3	≤ 3	≤ 20	[21:30]	1
1	B	S	≤ 3	≤ 3	≤ 20	[21:30]	1

RESULTS AND DISCUSSION

The experimentation was performed using both the original and the simplified data sets. The results obtained in both cases are summarized in Table 5.5. It can be seen that the prediction accuracy is much higher for the generalized data set. Furthermore, instead of being exact, the likelihood of being correct is even higher if one has to give a rough estimate. For instance, instead of saying that someone has a level of education of 10, it is much safer to specify that the person has a secondary education. Although this approach leaves out detail, it is often the case that the left-out details are not required. In a decision system such as the one considered in this chapter, knowing that the prospective father is 19 years old may carry the same significance as saying that the father is a teenager.

CONCLUSION

Rough sets have been used for missing data imputation and characteristic relations were introduced to describe incompletely specified decision tables. It was shown that the basic rough set idea of lower and upper approximations for incompletely specified decision tables could be defined in a variety of different ways. The technique was tested with a real database and the results with the HIV database were acceptable with accuracies ranging from 74.7% to 100%. One drawback of this method is that it

Table 5.5. Missing data estimation results for both the original data and the generalized data

	Education	Gravidity	Parity	Father's age
Original	83.1	86.5	87.8	74.7
Generalized	99.3	99.2	99	98.5

makes no extrapolation or interpolation. As a result, it can only be used if the missing case is similar or related to another case associated with more observations or full observation.

FUTURE WORK

This chapter introduced rough sets for modeling missing data problem. Rough sets work by partitioning variables into bins. As a future study, a process that optimizes the *sizes* of these bins should be implemented. Another issue in rough sets that ought to be addressed is the issue of reducts, particularly with respect to how reducts can be used to improve the missing data estimation problem. In a future study, the impact of reducts on missing data estimation must be taken into account.

REFERENCES

Abdella, M. (2005). *The use of genetic algorithms and neural networks to approximate missing data in database.* Unpublished master's thesis, University of the Witwatersrand, Johannesburg.

Abdella, M., & Marwala, T. (2006). The use of genetic algorithms and neural networks to approximate missing data in database. *Computing and Informatics, 24,* 1001–1013.

Arbuckle, J. L. (1996). Full information estimation in the presence of incomplete data. In G.A. Marcoulides and R.E. Schumacker (Eds.) *Advanced structural equation modeling: Issues and techniques.* Mahwah, NJ: Lawrence Erlbaum Associates.

Bazan, J., Nguyen, H. S., & Szczuka, M. (2004). A view on rough set concept approximations. *Fundamenta Informaticae, 59,* 107–118.

Beaubouef, T., & Petry, F. E. (2003). Rough set uncertainty in an object oriented data model. *Intelligent Systems for Information Processing, 9,* 37-46.

Beynon, M. (2001). Reducts within the variable precision rough sets model: A further investigation. *European Journal of Operational Research, 134*(3), 592-605.

Bi, Y., Anderson, T., & McClean, S. (2003). A rough set model with ontologies for discovering maximal association rules in document collections. *Knowledge-Based Systems, 16*(5-6), 243-251.

Chanas, S., & Kuchta, D. (1992). Further remarks on the relation between rough and fuzzy sets. *Fuzzy Sets and Systems, 47*(3), 391–394.

Comer, S. D. (1991). An algebraic approach to the approximation of information. *Fundamenta Informaticae, 14*(4), 495–502.

Crossingham, B. (2007). *Rough set partitioning using computational intelligence approach.* Unpublished master's thesis, University of the Witwatersrand, Johannesburg.

Crossingham, B., & Marwala, T. (2007a). Using genetic algorithms to optimise rough set partition sizes for HIV data analysis. *Studies in Computational Intelligence, 78,* 245-250.

Crossingham, B., & Marwala, T. (2007b). Using optimisation techniques to granulise rough set partitions. *Computational Models for Life Sciences, 952,* 248-257.

Collins, L. M, Schafer, J. L., & Kam, C. M. (2001). A comparison of inclusive and restrictive strategies in modern missing data procedures. *Structural Equation Modelling, 6,* 330-351.

Coulibaly, P., & Evora, N. D. (2007). Comparison of neural network methods for infilling missing daily weather records. *Journal of Hydrology, 341*(1-2), 27-41.

Degang, C., Wenxiu, Z., Yeung, D., & Tsang, E. C. C. (2006). Rough approximations on a complete completely distributive lattice with applications to generalized rough sets. *Information Sciences, 176*(13), 1829-1848.

Dempster, A. P, Laird, N. M., & Rubin, D. B. (1977). Maximum Likelihood for Incomplete Data via the EM Algorithm. *Journal of the Royal Statistical Society, B39,*1-38.

Deng, T., Chen, Y., Xu, W., & Dai, Q. (2007). A novel approach to fuzzy rough sets based on a fuzzy covering. *Information Sciences, 177*(11), 2308-2326.

Dubois, D. 1990, Rough fuzzy sets and fuzzy rough sets. *International Journal of General Systems, 17,* 191–209.

Enders, C. K., & Peugh, J. L. (2004). Using an EM covariance matrix to estimate structural equation models with missing data: Choosing an adjusted sample size to improve the accuracy of inferences. *Structural Equation Modeling, 11,* 1-19.

Gabrys, B. (2002). Neuro-fuzzy approach to processing inputs with missing values in pattern recognition problems. *International Journal of Approximate Reasoning, 30,* 149–179.

Gediga, G., & Duntsch, I. (2003). Maximum consistency of incomplete data via non-invasive imputation. *Artificial Intelligence Review, 19,* 93–107.

Goh, C., & Law, R. (2003). Incorporating the rough sets theory into travel demand analysis. *Tourism Management, 24,* 511–517.

Golan, R. H., & Ziarko, W. (1995). A methodology for stock market analysis utilizing rough set theory. In *Proceedings of Computational Intelligence for Financial Engineering* (pp. 32–40).

Gold, M. S., & Bentler, P. M. (2000). Treatments of missing data: A Monte Carlo comparison of RB-HDI, iterative stochastic regression imputation, and expectation-maximization. *Structural Equation Modeling, 7,* 319-355.

Graham, J. W., & Hofer, S. M, (2006). Multiple imputation in multivariate research. In T.D. Greco, B. Matarazzo and R. Slowinski (Eds.), *Rough membership and Bayesian confirmation measures for parameterized rough sets* (Vol. 6104, pp. 314-324). Proceedings of SPIE: The International Society for Optical Engineering.

Grzymala-Busse, J. W. (1992). LERSA system for learning from examples based on rough sets. In R. Slowinski, *Handbook of applications and advances of the rough sets theory* (pp. 3-13). Dordrecht: Kluwer Academic Publishers.

Grzymala-Busse, J. W. (2004). Three approaches to missing attribute values - A rough set perspective. In *IEEE 4th International Conference on Data Mining* (pp. 57–64).

Grzymala-Busse, J. W., & Hu, M. (2001). A comparison of several approaches to missing attribute values in data mining. *Lecture Notes in Artificial Intelligence, 205,* 378-385.

Grzymala-Busse, J. W., & Siddhaye, S. (2004). Rough set approaches to rule induction from incomplete data. In *Proceedings the 10th International Conference on Information Processing and Management of Uncertainty in Knowledge-Based Systems* (pp. 923–930).

Hassanien, A. (2007). Fuzzy rough sets hybrid scheme for breast cancer detection. *Image and Vision Computing, 25*(2), 172-183.

Hong, T., Tseng, L., & Wang, S. (2002). Learning rules from incomplete training examples. *Expert Systems with Application, 22,* 285–293.

Inuiguchi, M., & Miyajima, T. (2007). Rough set based rule induction from two decision tables. *European Journal of Operational Research, 181*(3), 1540-1553.

Jensen, R., & Shen, Q. (2004). Semantics-preserving dimensionality reduction: Rough and fuzzy-rough based approaches. *IEEE Transactions on Knowledge and Data Engineering, 16*(12), 1457–1471.

Kim, J., & Curry, J. (1997). The treatment of missing data in multivariate analysis. *Sociological Methods and Research, 6,* 215–241.

King, G., Honaker, J., Joseph, A., & Scheve, K. (2001). Analyzing incomplete political science data: An alternative algorithm for multiple imputation. *95, American Political Science Review* 49-69.

King, G., Honaker, J., Joseph, A., & Sheve, K. (1998). *Listwise deletion is evil: What to do about missing data in political science?* Paper presented at the annual meeting of the American Political Science Association, Boston, MA.

Komorowski, J., Pawlak, Z., Polkowski, L., & Skowron, A. (1999). *A rough set perspective on data and knowledge: The handbook of data mining and knowledge discovery.* Oxford: Oxford University Press.

Kondo, M. (2006). On the structure of generalized rough sets. *Information Sciences, 176*(5), 589-600.

Little, R. J. A., & Rubin, D. B. (1987). *Statistical analysis with missing data.* New York: John Wiley & Sons.

Little, R. J. A., & Rubin, D. B. (1989). The analysis of social science data with missing values. *Sociological Methods and Research, 18*, 292-326.

Little, T. D., Schnabel, K.U., & Baumert, J. (2000). *Modeling longitudinal and multilevel data: Practical issues, applied approaches, and specific examples.* Erlbaum, New Jersey: Psychology Press.

Leung, Y., Wu, W., & Zhang, W. (2006). Knowledge acquisition in incomplete information systems: A rough set approach. *European Journal of Operational Research, 168*(1), 164-180.

Marsh, H. W. (1998). Pairwise deletion for missing data in structural equation models with missing data: Nonpositive definite matrices, parameter estimates, goodness of fit, and adjusted sample sizes. *Structural Equation Modeling, 5*, 22-36.

Marwala, T. (2007). *Computational intelligence for modelling complex systems.* New Delhi, India: Research India Publications.

Marwala, T., & Crossingham, B. (2007a). Bayesian approach to rough set. *ArXiv: 0704.3433.*

Marwala, T., & Crossingham, B. (2007b). Bayesian approach to neuro-rough models for modelling HIV. *ArXiv: 0705.0761.*

Mitra, P., Pal, S. K., & Saddiqi, M. A. (2003). Non-convex clustering using expectation maximization algorithm with rough set initialization. *Pattern Recognition Letters, 24*(6), 863-873.

Mohamed, S., & Marwala, T. (2005). Neural network based techniques for estimating missing data in databases. In *Proceedings of the 16th Annual Symposium of the Pattern Recognition Association of South Africa,* Langebaan, South Africa (pp. 27–32).

Nakata, M., & Sakai, H. (2006). Rough sets approximations to possibilistic information. In *Proceedings of the IEEE International Conference on Fuzzy Systems,* Vancouver, Canada (pp. 10804–10811).

Nauck, D., & Kruse, R. (1999). Learning in neuro-fuzzy systems with symbolic attributes and missing values. In *Proceedings of the IEEE International Conference on Neural Information Processing,* Perth (pp. 142–147).

Nelwamondo, F.V. (2008). *Computational intelligence techniques for missing data imputation.* Unpublished doctoral dissertation, University of the Witwatersrand, Johannesburg.

Nelwamondo, F.V., & Marwala. (2008). Techniques for handling missing data: applications to online condition monitoring. *International Journal of Innovative Computing, Information and Control 4*(6), 2008, 1507-1526

Nelwamondo, F.V., & Marwala, T. (2007a). Rough sets computations to impute missing data. *ArXiv: 0704.3635.*

Nelwamondo, F.V., & Marwala, T. (2007b) Rough set theory for the treatment of incomplete data. In *Proceedings of the IEEE Conference on Fuzzy Systems* (pp. 338-343).

Nelwamondo, F. V., & Marwala, T. (2007c). Fuzzy ARTMAP and neural network approach to online processing of inputs with missing values. *SAIEE Africa Research Journal, 98*(2), 45-51.

Nelwamondo, F.V., Mohamed, S., & Marwala, T. (2007). Missing data: A comparison of neural network and expectation maximization techniques. *Current Science, 93*(11), 1514-1521.

Nishino, T., Nagamachi, M., & Tanaka, H. (2006). Variable precision Bayesian rough set model and its application to human evaluation data. In *Proceedings of SPIE - The International Society for Optical Engineering* (pp. 294-303).

Ohrn, A. (1999). *Discernibility and rough sets in medicine: Tools and applications.* Unpublished doctoral dissertation, Norwegian University of Science and Technology, Norway.

Ohrn, A., & Rowland, T. (2000). Rough sets: A knowledge discovery technique for multifactorial medical outcomes. *American Journal of Physical Medicine and Rehabilitation, 79*, 100-108.

Pattaraintakorn, P., & Cercone, N. (2007). Integrating rough set theory and medical applications. *Applied Mathematics Letters*, (in press).

Pawlak, Z. (1991). *Rough sets: Theoretical aspects of reasoning about data.* Dordrecht: Kluwer Academic Publishers.

Pawlak, Z. (2002). Rough sets and intelligent data analysis. *Information Science, 147*, 1–12.

Pawlak, Z. (2003). A rough set view on Bayes' theorem. *International Journal of Intelligent Systems, 18*(5), 487-498.

Pawlak, Z., & Skowron, A. (2007a). Rough sets: Some extensions. *Information Sciences, 177*(1), 28-40.

Pawlak, Z., & Skowron, A. (2007b). Rough sets and Boolean reasoning. *Information Sciences, 177*(1), 41-73

Pawlak, Z., & Skowron, A. (2007c). Rudiments of rough sets information sciences. *Information Science, 177*(1), 3-27.

Pawlak, Z., Wong, S. K. M., & Ziarko, W. (1988). Rough sets: Probabilistic versus deterministic approach. *International Journal of Man-Machine Studies, 29*, 81–95.

Pe-a, J., Ltourneau, S., & Famili, A. (1999). Application of rough sets algorithms to prediction of aircraft component failure. *Lecture Notes in Computer Science 1642/1999*, 473-484.

Questier, F. Arnaut-Rollier, I. Walczak, B., & Massart, D. L. (2002). Application of rough set theory to feature selection for unsupervised clustering. *Chemometrics and Intelligent Laboratory Systems, 63*(2), 155-167.

Qiao, W., Gao, Z., & Harley, R. G. (2005). Continuous online identification of nonlinear plants in power systems with missing sensor measurements. In *Proceedings of the 2005 IEEE International Joint Conference on Neural Networks*, Montreal, Canada (pp. 1729-1734).

Quafafou, M. (2000). α-RST: A generalization of rough set theory, *Information Sciences, 124*(1-4), 301-316.

Rowland, T., Ohno-Machado, L., & Ohrn, A. (1998). Comparison of multiple prediction models for ambulation following spinal chord injury. *In Chute, 31*, 528–532.

Rubin, D.B. (1976). Inference with missing data. *Biometrika, 63*, 581-592.

Schafer, J.L., & Graham, J.W. 2002. Missing data: Our view of the state of the art. *Psychological Methods, 7*, 147-177.

Scheuren, F. (2005). Multiple imputation: How it began and continues. *The American Statistician, 59*, 315-319.

Slezak, D., & Ziarko, W. (2005). The investigation of the Bayesian rough set model. *International Journal of Approximate Reasoning, 40*(1-2), 81-91.

Ssali, G., & Marwala, T. (2007). Estimation of missing data using computational intelligence and decision trees. *arXiv:0709.1640.*

Stefanowski, J. (1998). On rough set based approaches to induction of decision rules. In L. Polkowski and A. Skowron (Eds.), *Rough sets in knowledge discovery 1: Methodology and applications* (pp. 500–529), Heidelberg: Physica-Verlag.

Tay, F. E. H., & Shen, L. (2003). Fault diagnosis based on rough set theory. *Engineering Applications of Artificial Intelligence, 16*, 39–43.

Terlecki, P., & Walczak, K. (2007). On the relation between rough set reducts and jumping emerging patterns. *Information Sciences, 177*(1), 74-83.

Tettey, T., Nelwamondo, F.V., & Marwala, T. (2007). HIV Data analysis via rule extraction using rough sets. In *Proceedings of the 11th IEEE International Conference on Intelligent Engineering Systems,* Budapest, Hungary (pp. 105-110).

Tsikritis, N. (2005). A review of techniques for treating missing data in OM survey research. *Journal of Operations Management, 24*(1), 53-62.

Wang, S. (2005). Classification with incomplete survey data: a Hopfield neural network approach. *Computers and Operations Research, 24*, 53–62.

Wang, W., Yang, J., Jensen, R., & Liu, X. (2006). Rough set feature selection and rule induction for prediction of malignancy degree in brain glioma. *Computer Methods and Programs in Biomedicine, 83*(2), 147-156.

Witlox, F., & Tindemans, H. (2004). The application of rough sets analysis in activity based modelling: Opportunities and constraints. *Expert Systems with Applications, 27*, 585-592.

Wong, S. K. M., Ziarko, W., & Ye, R. L. (1986). Comparison of rough-set and statistical methods in inductive learning. *International Journal of Man-Machine Studies, 24*, 53–72.

Wu, W., Mi, J., & Zhang, W. (2003). Generalized fuzzy rough sets. *Information Sciences, 151*, 263-282.

Xia, Y., Fabian, P., Stohl, A., & Winterhalter, M. (1999). Forest climatology: Estimation of missing values for Bavaria, Germany. *Agricultural and Forest Meteorology, 96*(1-3), 131-144.

Yang, Y., & John, R. (2006). Roughness bound in set-oriented rough set operations. In *Proceedings of the 2006 IEEE International Conference on Fuzzy Systems,* Vancouver, Canada (pp. 1461–1468).

Yao, J. T., & Yao, Y. Y. (2002). Induction of classification rules by granular computing. In *Proceedings of the Third International Conference on Rough Sets and Current Trends in Computing (TSCTC'02),* London, UK (pp. 331–338).

Zadeh, L. A. (1965). Fuzzy sets. *Information and Control, 8,* 338–353.

Zhang, B., Yin, J., Tang, W., Hao, J., & Zhang, D. (2006). Unknown malicious codes detection based on rough set theory and support vector machine. *Proceedings of the 2006 IEEE International Joint Conference on Neural Networks* (pp. 4890–4894). Vancouver, Canada.

Zhang, L., & Shao, C. (2006). Designing fuzzy inference system based on improved gradient descent method. *Journal of Systems Engineering and Electronics, 17*(4), 853-857.

Zhao, Y., Yao, Y., & Luo, F. (2007). Data analysis based on discernibility and indiscernibility. *Information Sciences, 177*(22), 4959-4976.

Ziarko, W. (1998). Rough sets as a methodology for data mining. In L. Polkowski and A. Skowron (Eds.), *Rough sets in knowledge discovery 1: Methodology and applications* (pp. 554–576). Heidelberg: Physica-Verlag.

Ziarko, W, & Shan, N. (1995). Discovering attribute relationships, dependencies and rules by using rough sets. In *Proceedings of the 28th Annual Hawaii International Conference on System Sciences (HICSS'95)* (pp. 293–299).

Chapter VI
Support Vector Regression for Missing Data Estimation

ABSTRACT

This chapter develops and compares the merits of three different data imputation models by using accuracy measures. The three methods are auto-associative neural networks, a principal component analysis and support vector regression all combined with cultural genetic algorithms to impute missing variables. The use of a principal component analysis improves the overall performance of the auto-associative network while the use of support vector regression shows promising potential for future investigation. Imputation accuracies up to 97.4% for some of the variables are achieved.

INTRODUCTION

The problem with data collection in surveys is that the data invariably suffers from some loss of information. This may for example be a consequence of problems such as incorrect data entry, or unfilled fields in surveys. This chapter explores three different methods for data imputation. These are the combination of cultural genetic algorithms with three learning methods i.e., neural networks, a principal component analysis and support vector regression.

The general approach pursued in this chapter is to use regression models to model the inter-relationships between data variables using neural networks (Chang and Tsai, 2008), a principal component analysis (Adams et al., 2002) and a support vector regression (Cheng, Yu, & Yang, 2007). Thereafter, a controlled and planned approximation of missing data is conducted using an optimization method, in this chapter a cultural genetic algorithm (Yuan and Yuan, 2006) is selected.

Data imputation using auto-associative neural networks as a regression model has been conducted as explained in earlier chapters by Abdella and Marwala (2006); Abdella (2005); Leke, Marwala, and Tettey (2006); Nelwamondo, Mohamed, and Marwala (2007), while other variations include expectation maximization (Nelwamondo, 2008); rough sets as is described in Chapter V; decision trees (Barcena and Tussel, 2002). The use of auto-associative networks comes with a trade-of between computational complexity and time. However, the advantage of using auto-associative networks is that it does give good results as observed in Chapters III and IV.

Auto-associative networks are used in this chapter because they have been found to be successful when they were applied to many problems including fault diagnosis in the optimal production of yeast with a controllable temperature by Shimizu et al. (1997). The auto-associative network based system was able to accurately detect faults in real-time. When the same problem was solved using linear principal component analysis, it could not detect these faults.

Shen, Fu, and Lu (2005) presented a support vector regression based color image watermarking scheme that operates by using the information supplied by the reference positions and the watermark which was adaptively embedded into the blue channel of the host image (taking into account the human visual system). Other successful implementations of support vector machine include Marwala, Chakraverty, and Mahola (2006) who successfully applied it to fault classification in mechanical systems and Msiza, Nelwamondo, and Marwala (2007) who used support vector machines for water demand time-series forecasting.

Pan, Flynn, and Cregan (2007) successfully applied principal component analysis and sub-space principal component analysis for monitoring of a combined cycle gas turbine while Marwala and Hunt (2001) successfully applied principal component analysis and neural networks for damage detection in a population of cylindrical shells. Other successful applications of principal component analysis include the classification of pasteurized milk (Horimoto & Nakai, 1998) as well as Brito et al. (2006) who used principal component analysis for classifying heat-treated liver pastes according to container type, using heavy metal content and manufacturer's data.

This chapter investigates the use of different regression methods that offer a solution to the data imputation problem. The chapter first gives a short background on missing data, neural networks and the other regression methods used in this chapter. The dataset that was used to evaluate these three methods is then introduced and explained. Thereafter, the methods are presented and then implemented. The results obtained in this chapter are given and discussed. Thereafter, conclusions are drawn and possibilities for further work are identified. The methods investigated in this chapter are applied to an HIV database (Department of Health - South Africa, 2000). As indicated before, Acquired Immunodeficiency Syndrome (AIDS) is a collection of symptoms and infections resulting from the specific damage to the immune system caused by the Human Immunodeficiency Virus (HIV) in humans (Marx, 1982). The world has seen an increase in HIV infection rates in recent years. Research into this subject is ongoing, particularly in trying to identify ways of dealing with the virus. Thus demographic data are used often to classify people living with HIV and how they are affected. Proper data collection needs to be done to understand where and how the virus is spreading. By identifying factors that are viewed as making certain people or population groups a higher risk group, governments can then deploy strategies and plans within those groups to help the people. This database is often full of missing values and this chapter proposes three approaches to deal with the missing data.

BACKGROUND

Data collection forms the backbone of most research projects and applications. To accurately use the data for decision making, all information is normally required to be available.

Missing Data

In practice, most data collection exercises suffer from missing values or even data variables. For example, the source of this 'missingness' can be from unfilled fields in a survey or from data entry mistakes. Simply removing all entries with missing values, as explained in earlier chapters, is not always the best solution, particularly if this would result in a substantial loss of information, leading to incomplete or erroneous conclusions.

Methods are needed to impute the missing data. Numerous ways have been used to do this. The approach taken in this chapter is to use regression methods to find the inter-relationships between the data and then to verify the approximations that were made. The next sub-sections discuss the different regression methods used in this chapter.

Neural Networks

As described before, neural networks are computational models that have the ability to learn and model complicated systems (Bishop, 1995; Fu and Yan, 1995; Daszykowski, Walczak, & Massart, 2003). The neural network architecture used in this chapter is the multi-layer perceptron network (Bishop, 1995) and was described in detail in Chapter II. Multi-layer perceptrons (MLP) have been successfully used in modelling engineering systems. Marwala and Hunt (2000) successfully used probabilistic multi-layer perceptrons for fault identification using vibration data in cylindrical shells while Marwala and Lagazio (2004) used the multi-layered perceptron for modelling and controlling inter-state conflict. Furthermore, Leke and Marwala (2005) used multi-layer neural networks for stock market prediction while Patel and Marwala (2006) used multi-layer perceptron neural networks to identify a viable strategy to predict stock prices. From the observations above, it is evident that the multi-layer perceptron can model complex systems ranging from the financial markets to mechanical engineering. In addition, Chapter II demonstrated that the multi-layer perceptron performs better than radial basis function. Consequently, MLP is used to construct an autoassociative network to be used to estimate missing data in this chapter.

The multi-layer perceptron has two layers of weights which connect the input layer to the output layer. The middle of the network is made up of a hidden layer. This layer can further be made up of a number of hidden nodes. This number has to be optimized so that the network can model systems better (Bishop, 1995; Krose and van der Smagt, 1996). An increase in hidden nodes translates into an increase in the complexity of the input-output mapping function. In this chapter, the outer layer activation function of the MLP is a linear function while the inner layer activation function is the hyperbolic tangent function. The MLP is used to construct an auto-associative neural network, which is the subject of the next section.

Auto-Associative Networks

Auto-associative neural networks are neural networks that are trained to recall their inputs. Thus the number of inputs is equal to the number of outputs. Auto-associative neural networks have a bottleneck that results from the structure of the hidden nodes (Sherrington and Wong, 1989; Rios and Kabuka, 1995; Thompson, Marks, & Choi, 2002). There are usually fewer hidden nodes than input nodes, thereby resulting in a butterfly-like structure. An auto-associative network is preferred in recall applications as it can map linear and nonlinear relationships between all of the inputs. The auto-associative structure results in the compression of data into smaller dimensions and then decompresses them into the output (Abdella and Marwala, 2005; Nelwamondo and Marwala, 2007a&b).

Marais and Marwala (2007) used multi-layer auto-associative networks to determine the presence of a worm, based on routing information available from Internet routers. The auto-associative network was used to detect anomalies in normal routing behavior, was trained using information from a single router, and could detect both global instability caused by worms and localized routing instability.

Aldrich (1998) successfully used auto-associative neural networks for the visualization of transformed multivariate datasets, while Huang, Shimizu, and Shioya (2002) used an auto-associative neural network model for data pre-processing and output evaluation for online fault detection in the production of virginiamycin.

Other successful implementations of auto-associative networks include in understanding memory retrieval and spontaneous activity bumps in small-world networks by Anishchenko and Treves (2006) as well as Zhao and Xu (2005) who used this for multivariate statistical process monitoring. Further information on auto-associative neural networks can be found in Guan, Lam, and Chen (2000), Kropas-Hughes et al. (2000), Kramer (1992) as well as Marseguerra and Zoia (2005a&b, 2006)

In this chapter an auto-associative network was constructed using a multi-layer perceptron neural network that was trained using scaled conjugate gradient method (Bishop, 1995). The HIV data was fed into the network and the networks were trained to recall their own inputs. The structure of the auto-associative network is shown in Figure 6.1. The next section describes the support vector regression that was also used to construct an auto-associative network.

Support Vector Regression

Support vector machines constitute a supervised learning method used mainly for classification, are derived from statistical learning theory and were first introduced by Vapnik (1998). They have also been extended to regression, thus resulting in the term Support Vector Regression (SVR) (Gunn, 1998; Chang, Tsai and Young, 2007; Chuang, 2008).

Figure 6.1. Auto-associative neural network

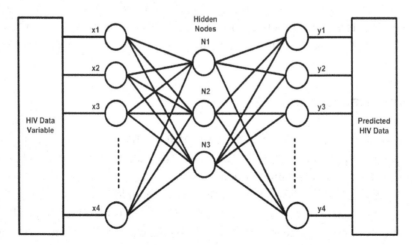

Pires and Marwala (2004; 2005) successfully used support vector machines for option pricing and further extended these to a Bayesian framework while Gidudu, Hulley, and Marwala (2007) successfully used support vector machines for image classification. Jayadeva and Chandra (2007) successfully used the regularized least squares fuzzy support vector regression for financial time-series forecasting while Zhang, Sato, and Iai (2006) successfully used support vector regression for online health monitoring of large-scale structures. Thissen et al. (2004) successfully used support vector machines for spectral regression applications. Xi, Poo, and Chou (2007) used support vector regression model successfully for predictive control of an air conditioned plant.

One of the problems with support vector regression is the computational load needed to train them. Researchers such as Guo and Zhang (2007) have developed methods for accelerating support vector regression while Üstün, Melssen, and Buydens (2007) visualized and interpreted support vector regression models. Other applications of support vector regression include nonlinear time series prediction (Lau and Wu, 2007), the identification of people (Palanivel and Yegnanarayana, 2008), in analyzing chemical compounds (Zhou et al., 2006), response modeling (Kim, Lee, & Cho, 2008), the estimation of software project effort (Oliveira, 2006) and the real-time prediction of order flow times (Alenezi, Moses, and Trafalis, 2008). In a drive to improve the performance of support vector regression, some innovative approaches were introduced including those by Üstün, Melssen, and Buydens (2006) who included a Pearson VII function based kernel and Üstün et al. (2005) who used a genetic algorithm and a simplex method.

The basic idea behind support vector regression is to map the input space into an output space. Suppose there is the training dataset with one input and one output being considered: $\{(x_1,y_1),...,(x_i,y_i)\} \subset \chi \times \Re$, where χ is the space of the input parameters and $\Re$ denotes the real number set. It is desired to find a function $f(x)$ that will map the training inputs to the training outputs. In support vector regression it is intended to find this function that has at most ε deviation from the actual training targets y_i. Several kinds of functions $f(x)$ can be fitted to map training inputs into training outputs. These functions are known as *kernel functions* but these cannot just be any functions because kernel functions have to adhere to some criteria (Joachims, 1999). For the purposes of explanation a linear kernel function is considered.

$$f(x) = \langle w,x \rangle + b \quad with \quad w \in \chi \quad , \quad b \in \Re \tag{6.1}$$

where $\langle .,. \rangle$ denotes the dot product.

It is desired to find small values for w. One way to do this is to minimize the Euclidean norm $\|w\|^2$ (Drezet and Harrison, 2001). The slack variables ξ_i, ξ_i^* are then included so that certain infeasible constraints in the minimization of the Euclidean norm can be used and the minimization problem then becomes (Xie, Liu, & Tang, 2007):

$$\min \quad \frac{1}{2}\|w\|^2 + C\sum_{i=1}^{l}(\xi_i + \xi_i^*) \tag{6.2}$$

$$subject \quad to \quad \begin{cases} y_i - \langle w,x_i \rangle - b \leq \varepsilon + \xi_i \\ \langle w,x_i \rangle + b - y_i \leq \varepsilon + \xi_i^* \\ \xi_i, \xi_i^* \geq 0 \end{cases} \tag{6.3}$$

where l is the number of training points used. The constraints above deal with the ε-insensitive loss function used to penalize certain training points that are outside of the bound given by ε, which is a value chosen by the user. There are various other loss functions such as the Huber loss function which can also be used, but the most common one is the ε-insensitive loss function (Gunn, 1998). This loss function is given by:

$$|\xi|_\varepsilon = \begin{cases} 0 & if \quad |\xi| \leq \varepsilon \\ |\xi| - \varepsilon & otherwise \end{cases} \tag{6.4}$$

The value for C in equation 6.2 is understood as being the degree to which deviations from ε are tolerated (Trafalis and Ince, 2000). It can be seen as measuring the over-fitting a function too well to its training points. If the value of C is set too high then the function found ($f(x)$) will be too well fitted to the training data and will not predict data well that has not been seen in the training of the function. It means that points lying outside of the bounds given by ε are not penalized enough. This results in the function being too well fitted to the training data (Trafalis and Ince, 2000). A sketch of a linear function being fitted to training data can be seen in Figure 6.3 with the bounds being shown (Gunn, 1998).

The function on the right of Figure 6.2 is used to penalize those points that lie outside of the bounds shown on the left (Gunn, 1998). The more a point lies outside of one of the bounds (either below or above), the more the point is penalized and thus plays a smaller role in determining the function. Those points that fall within the bounds of the function are not penalized at all and their corresponding slack variable values (ξ_i, ξ_i^*) are given a zero and thus these points will play a major contribution in the determination of the function $f(x)$.

The optimization problem of equation 6.4 is then set up as a quadratic programming problem by first finding the Lagrangian multiplier and applying the Karush-Kuhn Tucker (KKT) conditions (Joachims, 1999). Then the values for w and b can be found so that the linear function that fits the training data of equation 6.1 can be found explicitly. Note that this example using the constrained optimization prob-

Figure 6.2. Linear support vector regression for a set of data (left) and the ε-insensitive loss function (right)

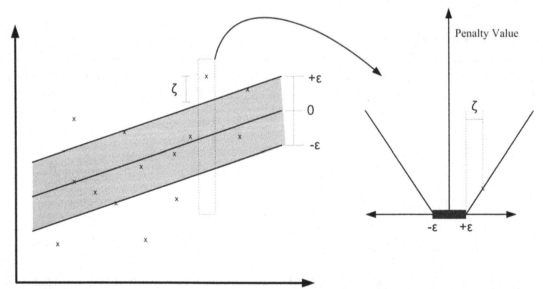

lem is only for a linear kernel function and the constrained optimization problem is different for other kernel functions.

In the same manner, a nonlinear model can be applied to model the data sufficiently well. This can be done by using a nonlinear mapping function to map the data into a high dimensional feature space where linear regression is performed. Once again, the kernel approach was implemented to deal with the issue of the curse of dimensionality.

For nonlinear problems the ε-insensitive loss function can be used to give (Gunn, 1998):

$$\max_{\alpha,\alpha^*} W(\alpha, \alpha^*) = \max_{\alpha,\alpha^*} \sum_{i=1}^{l} \alpha_i^*(y_i - \varepsilon) - \alpha_i(y_i + \varepsilon) - \frac{1}{2}\sum_{i=1}^{l}\sum_{j=1}^{l}(\alpha_i^* - \alpha_i)(\alpha_j^* - \alpha_j)K(x_i, x_j) \tag{6.5}$$

subject to:

$$0 \le \alpha_i, \alpha_i^* \le C, \ i = 1,\dots,l$$

$$\sum_{i=1}^{l}(\alpha_i - \alpha_i^*) = 0 \tag{6.6}$$

Here K is the kernel function and α and α^* are Lagrangian multipliers. Solving equations 6.5 and 6.6 gives the Lagrangian multipliers and the resulting regression equation can be written as follows (Gunn, 1998):

$$f(x) = \sum_{SVs}(\bar{\alpha}_i - \bar{\alpha}_i^*)K(x_i, x) + \bar{b} \tag{6.7}$$

$$b = -\frac{1}{2}\sum_{i=1}^{l}(\alpha_i - \alpha_i^*)(K(x_i, x_r) + K(x_i, x_s)) \tag{6.8}$$

The least squares support vector toolbox was used for the investigation (Suykens et al., 2002).

Principal Component Analysis

A Principal Component Analysis (PCA) (Shlens, 2005; Xie, Liu, & Tang, 2007) is a statistical technique that is commonly used to find patterns in data with high dimensions (Smith, 2002). The data can then be expressed in a way that highlights its similarities and differences. Another property is that, after finding patterns in the data, the data can be compressed without much data loss. This is advantageous for artificial neural networks applications because it results in a reduction of the number of nodes needed, thus increasing computational speeds.

Linkens and Vefghi (1997) successfully applied a principal component analysis for the recognition of patient anaesthetic levels while Yap et al. (1996) applied a principal component analysis to sinogram recovery. Ko, Zhou, and Ni (2002) used a principal component analysis and the frequency response functions for modeling seismic damage to a 38-storey building while Mirme, Minkkinen, and Ruuskanen (1996) used a principal component analysis for analyzing the behavior of aerosol, black carbon and gaseous pollutants in urban air. Further successes of the principal component analysis include the fields of water quality (Sârbu and Pop, 2005) and environmental science (Johnson et al., 2007).

As explained in Chapter IV, a principal component analysis is implemented in the following manner. First, data is gathered and the mean of each dimension is subtracted from the data. Second, the covariance matrix of the data is calculated. Third, the eigenvalues and eigenvectors of the covariance matrix are calculated. The highest eigenvalue corresponds to the eigenvector that is the principal component. This then is where the notion of data compression comes in. Using the chosen eigenvectors, the dimension of the data can be reduced whilst still retaining a large amount of information. Through using only the largest eigenvalues and their corresponding eigenvectors, compression can be attained as well as a simple transformation used. The data compression or transformation is:

$$[P] = [D] \times [PC] \tag{6.9}$$

where $[D]$ is the original dataset, $[PC]$ is the principal component matrix and $[P]$ is the transformed data. The principal component analysis multiplication results in a dataset that emphasizes the relationships between the data whether of a smaller or the same dimension. To return to the original data, the following equation is used:

$$[D'] = [P] \times [PC]^{-1} \tag{6.10}$$

Here $[D']$ is the re-transformed data. We find that $[D] \approx [D']$ if all of the principal components data are used from the covariance matrix. The transformed data ($[D]$) can be used in conjunction with an Artificial Neural Network (ANN) to increase the efficiency of the ANN by reducing its complexity. This results from the property of the PCA to extract linear relationships between the data variables; thus the ANN only needs to extract the nonlinear relationships. Overall, this results in less training cycles that are needed. Thus ANNs can be built more efficiently. Figure 6.3 illustrates this concept.

Cultural Algorithms

Cultural algorithms are a type of evolutionary computation with a knowledge module known as the *belief space* inserted on top of the population module. Therefore, cultural algorithms are modification

Figure 6.3. PCA Auto-associative neural network

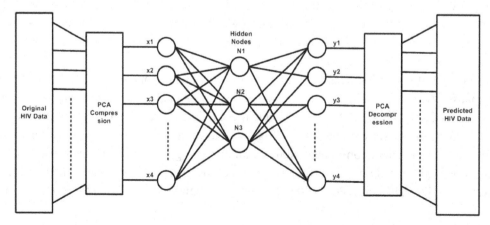

of the simple genetic algorithm. Cultural algorithms were proposed by Kobti, Reynolds, and Kohler (2003) and Alami, Imrani, and Bouroumi (2007) who developed a multi-population cultural algorithm using fuzzy clustering.

Yuan and Yuan (2006) applied a cultural algorithm for the scheduling of hydrothermal systems, while Digalakis and Margaritis (2002) applied multi-population cultural algorithm for the scheduling of electricity generation. In this chapter, a cultural genetic algorithm was applied. Therefore, in addition to the standard genetic algorithm, the belief space of a cultural algorithm was implemented. This belief space was divided into different categories that represent the different domains of knowledge that the population has of the search space. After each iteration step the belief space was revised using the best individuals of the population. There are many belief categories and these include (Kobti, Reynolds, & Kohler, 2003):

- *Normative knowledge,* which is a set of desired value ranges for the individuals in the population.
- *Domain specific knowledge,* which is some knowledge of the problem at hand, called a *prior* in Bayesian statistics.
- *Situational knowledge,* which is knowledge of vital incidents in the search space.
- *Spatial knowledge,* which is information on the landscape of the search space.

In this chapter, normative knowledge is applied. These contain bounds defined by the standard deviations of the observed data and spatial knowledge of the estimated gradient of the missing data estimation equation with respect to the missing variable. The Genetic algorithm (GA) is defined as a population based model that uses selection and recombination operators to generate new sample points in search space (Whitley, 1994). Details of genetic algorithms were explained in the earlier chapters.

In this chapter, the cultural genetic algorithm was used to find the input into regression model that will result in the most accurate missing data value. The genetic algorithm is generally good when used for approximating nonlinear functions. A genetic algorithm starts with the creation of a random population of *"chromosomes"*. These chromosomes are normally in binary format. From this random population, an *evaluation function* is used to find which of the chromosomes is the *fittest*. Those which are deemed to be fit are then used for the *selection* stage. A recombination of the chromosomes is performed by taking the fittest chromosomes and choosing bits from each that will be swapped. This process is called *crossover*. These crossovers then result in a new population of chromosomes. The final stage is *mutation* where bits are randomly changed, but within the chromosomes. From this new population, the *fitness operation* begins again until a preset number of iterations are reached. The genetic algorithm toolbox that was used in this chapter is by Houck, Joines, and Kay (1995).

The cultural algorithms were implemented in this chapter as follows:

- Select initial population
- Create a belief space (in this chapter, the bounds and the estimated gradient of the fitness function form the belief space)
- Reiterate until the termination state is achieved:
  - o Execute the cross-over and mutation of the individuals in population space
  - o Assess each individual by using the missing data estimation equation (fitness)
  - o Choose the parents to reproduce a fresh generation of offspring

    o   Allow the belief space i.e., bounds and gradient, alter the genome of the offspring

    o   Allow the best individuals to change the belief space

DATA COLLECTION AND PRE-PROCESSING

The dataset that was used for this investigation was the HIV data from antenatal clinics from South Africa, described in Chapters IV and V. The difference between the data used in this chapter and the previous chapter is that in this chapter three more variables are used that were not used in the previous chapter. These are *region* which has information as to whether the individual in question is from a rural or urban area, *rapid plasma regain* (RPR) a blood test for syphilis that detects an antibody found in the bloodstream when a patient has syphilis and *income to expenditure ratio* (WTREV).

The dataset was collected by the Department of Health of South Africa. The dataset contains multiple input fields that resulted from the survey. The information is in a number of different formats. For example, *provinces of origin*, *region* and *race* are strings while *age*, *gravidity* and *parity* are integers. Thus conversions were needed. The strings were converted to integers by using a lookup table e.g., there are only 9 provinces so 1 was substituted as the code for Gauteng Province.

Data collected from surveys and other data collection methods normally contain outliers. These are normally removed from the dataset. In this investigation, however, a dataset that had outliers had the outlier removed and the dataset was then classified as incomplete. This then means that the data can still be used in the final survey results if the missing values are imputed. The data with missing values was not used for the training of the computational methods. The data variables and their ranges are shown in Table 6.1.

As explained before, there are three different types of missing data mechanisms (Little and Rubin, 2000; Abdella and Marwala, 2005; Leke and Marwala, 2006; Nelwamondo and Marwala, 2008): Missing Completely at Random (MCAR) – This is when the probability of the missing value of a variable x is unrelated to itself or any other variables in the dataset; Missing at Random (MAR) – This implies

Table 6.1. HIV data variables

Variable	Type	Range
HIV Status	Binary	[0, 1]
Education	Integer	0 – 13
Age Group	Integer	14 – 60
Age Gap	Integer	1 – 7
Gravidity	Integer	0 – 11
Parity	Integer	0 – 40
Race	Integer	1 – 5
Province	Integer	1 – 9
Region	Integer	1 -36
RPR	Integer	0 – 2
WTREV	Continuous	0.64 - 1.28

that the probability of missing data of a particular variable x depends on other variables but not itself; and Non-ignorable – This is when the missing value of variable x depends on itself even though other variables are known.

Table 6.1 shows the different variables. Each variable is made missing following a specific mechanism. *Age group* and *Education* were made missing following the MNAR mechanism, while *Parity* was made missing following the MAR mechanism and the rest of the variables were made to follow the MCAR mechanism. Despite the mechanisms by which the data variables were made missing, the inter-relationships that exist amongst the variables do not change and, therefore, the methods presented in this chapter are valid.

The pre-processed data resulted in a reduction of training data from the original 16500 to 12750 records. To use the data for training must be normalized. This ensures that all data variables can be used in the training process. If the dataset is not normalized, some of the data variables with larger variances will influence the result more than others e.g., if only *WTREV* and *Age Group* data are used, the age data will be influential as it has large values. Thus all of the data were normalized to be between 0 and 1. The training data was then split into 3 partitions. 60% was used for training, 15% for validation and the last 25% was used for the testing stages.

METHODOLOGY

The approach taken in this chapter was to use a combined regression and optimization techniques. As indicated before, the optimization technique chosen in this chapter was a cultural genetic algorithm. Figure 6.4 illustrates the way in which the regression methods and the optimization technique are used to impute data.

First the regression methods have to be trained before being used for data imputation. The following subsections discuss the training procedures used for the regression methods.

Figure 6.4. Data imputation configuration with a cultural genetic algorithm

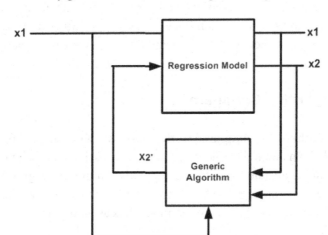

Artificial Neural Network (ANN) Training and Validation

To train the ANN, the optimum number of hidden nodes was needed. To find it, a simulation was constructed to calculate the average error using different numbers of hidden nodes. The number of hidden nodes was optimized, and was found to be 10. This was using a scaled conjugate gradient method (Møller, 1993) which was discussed in Chapter II, a linear outer activation function and a hyperbolic tangent function as the inner activation function.

Then the optimal number of training cycles had to be found. This was found by analyzing the validation error as the training cycles increased. This analysis was used both to avoid the possibility of over-training the ANN and to use the fastest way to train the ANN without compromising on accuracy. Validation was done with a dataset that was not used for training. This then resulted in an unbiased error check that indicated if the network was well trained or not.

PCA ANN Training and Validation

Training data was first used to extract the principal components. After the extraction process, the training dataset was multiplied by the principal components and the resulting dataset was used to train a new ANN. This was then labelled a PCA-ANN. Two PCA-ANNs were trained. One PCA-ANN has no compression and was just a transformation; the other PCA-ANN compressed the data from 11 dimensions to 10. The number of hidden nodes and training cycles were optimized as in the previous sub-section. The number of hidden nodes for the PCA-ANN-11 was 10 and was 9 for the PCA-ANN-10. The inner and outer activation functions were in the ANN and were explained above. Validation was also carried out with an unseen dataset. This also ensured that the ANN was trained well but not over-trained.

Support Vector Regression (SVR) Training and Validation

To train the support vector regression model less training dataset is needed. Only 3000 data records were used in this case. This was due to time constraints, as the training takes a considerable amount of time. Even though a smaller training set was used, the validation error was small. A radial basis function kernel function was used (Vapnik, 1995). The bias point and the regularization have to be optimized. To optimize the two, a genetic algorithm was used, a technique used by Chen and Wang (2007) with good results. The cultural GA used a validation set to find the parameters that result in the minimum error in a support vector regression validation dataset. Validation was carried out after training with an unseen set and the SVR performed well. This meant that the SVR can be used with the cultural GA to impute missing data.

Cultural Genetic Algorithm Configuration

The cultural genetic algorithm was configured in the model shown in Figure 6.5. The inputs $x1$ to $x5$ were known, $x6$ was unknown and was found by using the regression method and the cultural GA. The cultural genetic algorithm put a value from its initial population into the regression model. The model recalls the value and it becomes an output. The cultural GA tries to minimize the error between its approximated value and the value that the regression model has as an output.

Figure 6.5. Cultural genetic algorithm configuration

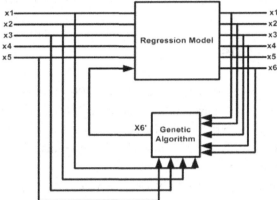

This minimization was conducted using the fitness function. As discussed in Chapters II, III and IV, the fitness evaluation function for a missing data estimation method can be written as follows and is normally a reduction of missing data estimation error function:

$$e = \left\| \left(\begin{Bmatrix} \{x_u\} \\ \{x_k\} \end{Bmatrix} - f \begin{pmatrix} \{x_u\} \\ \{x_k\} \end{pmatrix} \right) \right\| \tag{6.11}$$

Here $\| \ \|$ is Euclidean norm, subscript u indicates unknown while k indicates known. It is intended that the cultural GA will locate a global minimum value. In this chapter the cultural genetic algorithm uses a normalized geometric selection, along with a simple crossover for recombination and non-uniform mutation (Goldberg, 1989). The cultural GA was used to approximate the values of the missing data and the auto-associative network or the SVR mechanism then used the evaluation function in equation 6.11 to calculate the fitness.

When using the ANN and SVR, inputs are recalled, and $\{x_u\}$ which is the unknown parameter is approximated by the cultural GA, given $\{x_k\}$, the known data variables. The function f is the regression model and was changed for each of the models previously discussed. Figure 6.5 shows the configuration of the cultural GA with the regression methods. The cultural GA tries to reduce the error between the regression method and the data inputs, resulting in a data variable that is likely to be the missing value. However, for completeness, all of the outputs were used to reduce the error of the approximated value.

The regression methods discussed in the preceding section were combined with the cultural genetic algorithm, as shown in Figure 6.5. This results in multiple data imputation mechanisms. These are the:

- ANNGA, which is a combination of the ANN and cultural GA.
- PCANNGA, which is a combination of PCA, ANN and cultural GA.
- SVRGA, which is a combination of SVR and cultural GA.

The cultural genetic algorithm was set up with 50 initial population and 50 generation cycles. As mentioned earlier, the cultural GA used a simple crossover, geometric selection, non-uniform mutation (Goldberg, 1989), missing data bounds defined as within two standard deviations of the previously observed values as well as numerically approximated gradient for the belief space.

TESTING OF THE METHODS

The testing set for the data imputation methods contained 1000 sets. These were complete datasets that had some of their data removed to ascertain the accuracy of the imputation methods. The testing set was made up of data that the imputation methods had not yet seen (i.e., data that were not part of the training or validation set). This dataset was also chosen randomly from the initial dataset that was outlined in the previous sections.

The variables to be imputed were chosen to be the *HIV status, Age, Age Group, Parity* and *Gravidity*. These were taken as being the most important data variables that need to be imputed. The testing sets were composed of 3 different datasets made up of a 1000 random records each. This offered an unbiased result as testing with only 1 test can yield results which are the best but may be biased due to the data used.

Different measures of accuracy were used for the evaluating the effectiveness of the imputation methods. This was intended to offer a better understanding of the results as well as to measure the effectiveness of the presented method. The accuracy measures are discussed below.

Mean Square Error

The mean square error was used for the regression and classification data. It was used to measure the error between the imputed data and the real data value. The mean square error was calculated after the imputation by the cultural GA. This was before de-normalization and rounding and as a result did not carry over any rounding errors.

Classification Accuracy

For the classification value of the HIV data, the only accuracy used was the number of correct hits. This means the number of times the method estimated the correct status. This was done after de-normalization and rounding.

Prediction within Years / Unit Accuracy

Prediction within year was used as a useful and easy-to-understand measure of accuracy. This for example would be expressed as 80% accuracy within 1 year for *age* data. This means that for *age* data, the values that are found are 80% accurate within a tolerance of 1 year.

RESULTS

All of the results shown in the tables are in percentages of accuracy. The variables *HIV, Gravidity, Parity* and *Age Gap* gave positive match accuracy. *Education Level* and *Age* were all accurate with a tolerance of 1 year.

The ANNGA

The ANNGA was tested with its complete optimized variables and trained network. The results of the ANNGA data imputation are tabulated in Table 6.2. These results indicate that the ANNGA architecture performed well on all variables except the *education level*. The high estimation accuracies are on par with previous research.

The PCANNGA

The PCANNGA architecture was run with two configurations. The first configuration had no compression, thus it was named PCANNGA11, indicating the transformation from 11 inputs to 11 outputs. The second configuration had a compression of 1 value thus was named PCANNGA-10, indicating that the compression and transformation was from 11 inputs to 10 inputs. The results of the test are shown in Table 6.3.

The results for PCANNGA-11 indicate a good estimation for all the variables except *education level*. PCANNGA-10 performs poorly on *Age* and *Age Gap* while giving good results for the other variables.

Table 6.2. ANNGA results

ANNGA(%)	Run 1	Run 2	Run 3	Average
HIV Classification	68.9	68.6	68.0	68.5
Education Level	25.1	25.1	27.2	25.8
Gravidity	82.7	82.0	84.0	82.9
Parity	81.3	81.1	82.1	81.5
Age	86.9	86.4	85.5	82.3
Age Gap	96.6	96.0	95.4	96

Table 6.3. PCANNGA results

PCANNGA−11 (%)	Run 1	Run 2	Run 3	Average
HIV Classification	65.0	61.6	62.8	63.1
Education Level	27.8	27.3	28.2	27.8
Gravidity	87.6	86.5	87.1	87.1
Parity	87.5	86.3	87.7	87.2
Age	94.9	94.8	93.5	95.7
Age Gap	98.1	98.3	96.9	97.4
PCANNGA −10 (%)	Run 1	Run 2	Run 3	Average
HIV Classification	64.2	60.9	67.2	64.1
Education Level	27.0	31.3	30.2	29.5
Gravidity	86.4	86.3	88.2	61.0
Parity	86.2	86.2	87.6	86.7
Age	8.0	8.2	12.1	9.4
Age Gap	23.9	20.0	24.1	22.7

This resulted from the loss of information during the compression stage. This in turn impacted on the regression ability of the network, resulting in poor imputation accuracy for some of the variables.

The SVRGA

The SVRGA imputation model took a long time to run. The results from the SVRGA are presented in Table 6.4. The SVRGA performed badly in the HIV classification. It performed averagely in with *Education level*, *Parity* and *Gravidity*. With *Age* and *Age gap* it performed well.

Comparison of Results

To compare results, the previous accuracies as well as the mean square error of each method were analyzed. This gave an indication of how the errors in the imputation process affect the accuracy and which model produces the best results. The average mean square errors of the imputation methods are shown in Table 6.5.

For mean square error, a smaller value is desirable. It can be seen from Table 6.5 that for *HIV classification*, the SVRGA performed the worst as it had the highest error, but for the *education level* it performed the best as it has the lowest error. Figure 6.6 contains a graph of the average mean square error of the imputation models without HIV classification.

From Figure 6.6 it can be seen that the SVRGA has the smallest average mean square error (if HIV classification is not included) compared to the rest of the methods. This indicates that the SVRGA functioned well on regression parameters and poorly on classification of HIV. The graph in Figure 6.7 makes this clear.

Table 6.4. SVRGA results

SVRGA (%)	Run 1	Run 2	Run 3	Average
HIV Classification	22.5	22.1	21.4	22.0
Education Level	65.4	40.3	45.6	50.4
Gravidity	80.9	63.2	67.4	70.5
Parity	81.4	63.3	66.9	70.5
Age	96.1	89.2	83.5	89.6
Age Gap	92.6	92.7	94.3	93.2

Table 6.5. Average mean square errors

Results	NN	PCANN11	PCANN10	SVRGA
HIV	0.2691	0.3037	0.3016	0.7644
Education	0.1666	0.1322	0.1235	0.0421
Gravidity	0.0019	0.0014	0.0015	0.0031
Parity	0.0026	0.0023	0.0024	0.0044
Age	0.0010	0.0003	0.1574	0.0032
Age Gap	0.0013	0.0005	0.0979	0.0021

The ANNGA performed the best with an average accuracy of 68.5% while the rest of the models fell behind with the SVRGA having the lowest average accuracy of 22 %. For *Education level* accuracy, the SVRGA performed the best. It had an overall accuracy of 50%. This was measured within a tolerance of 1 year. The accuracies of the models are shown in Figure 6.8.

The SVR predicted the education level better than the rest and thus was performing better when combined with a cultural genetic algorithm to impute the missing variables. The last comparison is of the *age* accuracy. The average accuracies with 1 year tolerance are shown in Figure 6.9.

From Figure 6.9, it can be seen that the PCANN10 performed poorly overall. As explained earlier this resulted through the data loss from the compression of the data. The SVRGA performed better than the ANNGA, but the PCANN11 performed better than all. In almost all of the accuracy tests, the PCANN11 performed better than the ANNGA, thus proving that the combination of the PCA and ANN can result in a better method for imputation. The PCANN11 even had a lower average mean square error than the ANNGA. The PCA without compression improved the performance of the ANNGA. From the comparison of all of the imputation models, it can be seen that the PCANN11 performed better, even though it had a worse *HIV classification* rate. The SVRGA only performed well on the *education level* and thus cannot be considered superior to the PCANN11.

Figure 6.6. Comparison of average mean square error without HIV classification

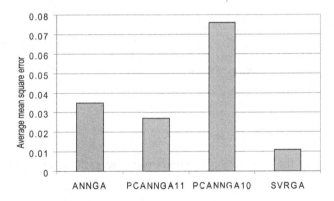

Figure 6.7. Comparison of Average HIV classification accuracy

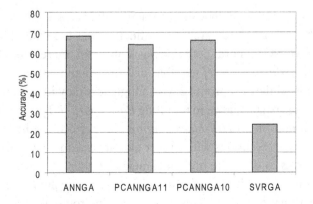

Figure 6.8. Comparison of Average Education Accuracy

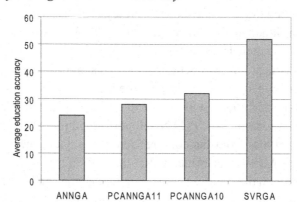

Figure 6.9. Comparison of age accuracy within 1 year

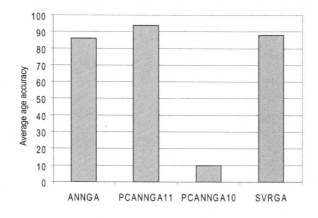

DISCUSSION

The general performance of the imputation methods must be satisfactory to researchers and highly accurate.

General Performance

The high accuracy of the imputation methods in estimating the variables makes them viable solutions for HIV / AIDS research. This gives researchers confidence that the data collected does not have to have a lot discarded. The ANNGA neural network was stable and the results were good. The SVRGA performed the best with the *education level* variable, and this should be further investigated. On average, the PCANNGA11 showed the best promise with a high accuracy in missing data imputation. This resulted from its good average performance in imputing *Parity*, *Gravidity*, *Age* and *Age Gap*, while only lagging behind by a small margin on *HIV classification* and performing better than the ANNGA on predicting the *Education Level*. Solutions with higher tolerances can be attained but the low tolerance used in this investigation was to illustrate high accuracies. Higher tolerances can be used selectively and instead of using years in a variable like *education levels,* it can be put into 3 categories like primary school, high school and tertiary education. This was done using rough set theory in Chapter V.

Further Regression vs. Classification

An investigation into the data required only for classification purposes such as *HIV classification* can yield better results. This came at the price of loss of generalization. Leke and Marwala (2005) investigated a classification based problem of HIV classification only. This cannot be directly used with data imputation without them resulting in highly complex hybrid networks.

CONCLUSION

This chapter investigated and compared the use of three regression methods with a cultural genetic algorithm combination for missing data approximation. An auto-associative neural network was trained to predict its input space, and was reconfigured with a principal component analysis to form a principal component analysis, auto-associative neural network that predicts the principal component transformed input space. Support vector regression was also used in the same manner as in the auto-associative network. The regression methods were combined with cultural genetic algorithms to approximate missing data from an HIV survey dataset. The combination of principal component, auto-associative neural network and cultural genetic algorithm model performed the best overall with the dataset considered, giving accuracies of up to 97.4%, this was followed by the combined auto-associative neural network and cultural genetic algorithm model. The hybrid support vector regression and cultural genetic algorithm model performed well on approximating a missing variable where the rest of the models performed poorly. This suggests future investigations into hybrid systems with combinations of the regression models to get better results and better methods for data imputation in the future.

FURTHER RESEARCH

Further research need to be done on the support vector regression (SVR) by optimizing choices that are made in constructing the SVR. In this chapter, there were cases where the SVR outperformed normal neural networks. A hybrid approach of using the ANNGA and SVRGA or PCANNGA11 and SVRGA together is also a viable investigation area. It is expected that this would increase the performance of the neural network based methods in imputing the *education level* while assisting the SVRGA in imputing the *HIV classification*.

REFERENCES

Abdella, M. (2005). *The use of genetic algorithms and neural networks to approximate missing data in database.* Unpublished master's thesis, University of the Witwatersrand, Johannesburg.

Abdella, M., & Marwala, T. (2005). Treatment of missing data using neural networks. In *Proceedings of the IEEE International Joint Conference on Neural Networks,* Montreal, Canada (pp. 598-603).

Abdella, M., & Marwala, T. (2006). The use of genetic algorithms and neural networks to approximate missing data in database. *Computing and Informatics, 24,* 1001–1013.

Adams, E., Walczak, B., Vervaet, C., Risha, P. G., & Massart, D. L. (2002). Principal component analysis of dissolution data with missing elements. *International Journal of Pharmaceutics*, *234*(1-2), 169-178.

Alami, J., El Imrani, A., & Bouroumi, A. (2007). A multipopulation cultural algorithm using fuzzy clustering. *Applied Soft Computing*, *7*(2), *506-519*.

Aldrich, C. (1998). Visualization of transformed multivariate datasets with autoassociative neural networks. *Pattern Recognition Letters*, *19*(8), 749-764.

Alenezi, A., Moses, S. A., & Trafalis, T. B. (2007) (in press). Real-time prediction of order flowtimes using support vector regression. *Computers and Operations Research*.

Anishchenko, A., & Treves, A. (2006). Autoassociative memory retrieval and spontaneous activity bumps in small-world networks of integrate-and-fire neurons. *Journal of Physiology-Paris*, *100*(4), 225-236.

Barcena, M. J., & Tussel, F. (2002). Multivariate data imputation using trees, *Universidad del País Vasco-Departamento de Economía Aplicada III*. Series BILTOKI, Retrieved February19, 2008, from http://www.et.bs.ehu.es/biltoki/EPS/dt200205.pdf

Bishop, C. M. (1995). *Neural networks for pattern recognition*. Oxford, UK: Oxford University Press.

Brito, G., Andrade, J. M., Havel, J., Díaz, C., García, F. J., & Peña-Méndez, E. M. (2006). Classification of some heat-treated liver pastes according to container type, using heavy metals content and manufacturer's data, by principal components analysis and potential curves. *Meat Science*, *74*(2), 296-302.

Chang, B. R., & Tsai, H. F. (2008). Forecast approach using neural network adaptation to support vector regression grey model and generalized auto-regressive conditional heteroskedasticity. *Expert Systems with Applications*, *34*(2), 925-934.

Chang, B. R. Tsai, H. F., & Young, C-P. (2007) (in press). Diversity of quantum optimizations for training adaptive support vector regression and its prediction applications. *Expert Systems with Applications*.

Chen, K., & Wang, C. (2007). Support vector regression with genetic algorithms forecasting tourism demand. *Tourism Management, 28*, 215–226.

Cheng, J., Yu, D., & Yang, Y. (2007). Application of support vector regression machines to the processing of end effects of Hilbert–Huang transform. *Mechanical Systems and Signal Processing*, *21*(3), 1197-1211.

Chuang, C-C. (2008). Extended support vector interval regression networks for interval input–output data. *Information Sciences*, *178*(3), 871-891.

Daszykowski, M., Walczak, B., & Massart, D. L. (2003). A journey into low-dimensional spaces with autoassociative neural networks. *Talanta*, *59*(6), 1095-1105

Department of Health, South Africa. (2000). HIV/AIDS/STD strategic plan for South Africa.

Digalakis, J. G., & Margaritis, K. G. (2002). A multipopulation cultural algorithm for the electrical generator scheduling problem. *Mathematics and Computers in Simulation*, *60*(3-5), 293-301.

Drezet, P. M. L., & Harrison, R. F. (2001). A new method for sparsity control in support vector classification and regression. *Pattern Recognition, 34*(1), 111-125.

Fu, A. M. N., & Yan, H. (1995). Distributive properties of main overlap and noise terms in autoassociative memory networks. *Neural Networks, 8*(3), 405-410

Guo, G., & Zhang, J. S. (2007). Reducing examples to accelerate support vector regression. *Pattern Recognition Letters, 28*(16), 2173-2183.

Gidudu, A., Hulley, G., & Marwala, T. (2007). Image classification using SVMs: One-against-one vs one-against-all. In *Proceeding of the 28th Asian Conference on Remote Sensing*, Malaysia, CD-Rom.

Goldberg, D. E. (1989). *Genetic algorithms in search, optimization and machine learning.* Boston, MA: Kluwer Academic Publishers.

Guan, Z-H, Lam, J., & Chen, G. (2000). On impulsive autoassociative neural networks. *Neural Networks, 13*(1), 63-69.

Gunn, S. R. (1998). *Support vector machines for classification and regression* (Tech. Rep. 1998) Southampton, England: Image Speech and Intelligent Systems Research Group, University of Southampton.

Horimoto, Y., & Nakai, S. (1998). Classification of pasteurized milk using principal component similarity analysis of off-flavours. *Food Research International, 31*(4), 279-287.

Houck, C., Joines, J., & Kay, M. (1995). *A genetic algorithm for function optimization: A Matlab implementation* (Tech. Rep. No. 95-09). Chapel Hill, NC: North Carolina State University Information Engineering.

Huang, J., Shimizu, H., & Shioya, S. (2002). Data preprocessing and output evaluation of an autoassociative neural network model for online fault detection in virginiamycin production. *Journal of Bioscience and Bioengineering, 94*(1), 70-77.

Jayadeva R. K., & Chandra, S. (2007) (in press). Regularized least squares fuzzy support vector regression for financial time series forecasting. *Expert Systems with Applications.*

Joachims, J. (1999). Making large-scale SVM learning practical. In B. Scholkopf, C. J. C. Burges and A. J. Smola, editors, *Advances in kernel methods-Support vector learning* (pp. 169-184). Cambridge, MA: MIT Press.

Johnson, G. W., Ehrlich, R., Full, W., & Ramos, S. (2007). Principal components analysis and receptor models in environmental forensics. *Introduction to Environmental Forensics (Second Edition)* (pp. 207-272).

Kim, D., Lee, H., & Cho, S. (2008). Response modeling with support vector regression. *Expert Systems with Applications, 34*(2), 1102-1108.

Ko, J. M., Zhou, X. T., & Ni, Y. Q. (2002). Seismic damage evaluation of a 38-storey building model using measured FrF data reduced via principal component analysis. *Advances in Building Technology* (pp. 953-960).

Kobti, Z., Reynolds, R., & Kohler T. (2003). *A multi-agent simulation using cultural algorithms: The effect of culture on the resilience of social systems.* Paper presented at the IEEE Congress on Evolutionary Computation, Canberra, Australia.

Kramer, M. A. (1992). Autoassociative neural networks. *Computers and Chemical Engineering, 16*(4), 313-328.

Kropas-Hughes, C. V., Oxley, M. E., Rogers, S. K., & Kabrisky, M. (2000). Autoassociative–heteroassociative neural networks. *Engineering Applications of Artificial Intelligence, 13*(5), 603-609.

Krose, B., & van der Smagt, P. (1996). *An Introduction to Neural Networks (Book Style).* University of Amsterdam.

Lau, K. W., & Wu, Q. H. (2007) (in press). Local prediction of nonlinear time series using support vector regression. *Pattern Recognition.*

Leke, B. B., & Marwala, T. (2005). Optimization of the stock market input time-window using Bayesian neural networks. In *Proceedings of the IEEE International Conference on Service Operations, Logistics and Informatics,* Beijing, China (pp. 883-894).

Leke, B. B., & Marwala, T. (2006). Ant colony optimization for missing data estimation. In *Proceedings of the Pattern Recognition of South Africa* (pp. 183-188).

Leke, B. B., Marwala, T., & Tettey, T. (2006). Autoencoder networks for HIV classification. *Current Science, 91*(11), 1467-1473.

Linkens, D. A and Vefghi, L. (1997). Recognition of patient anaesthetic levels: neural network systems, principal components analysis, and canonical discriminant variates. *Artificial Intelligence in Medicine, 11*(2), 155-173.

Little, R. J., & Rubin, D. B. (2000). *Statistical analysis with missing data.* 2nd Edition. New York: John Wiley & Sons.

Marais, E., & Marwala, T. (2007). Predicting the presence of internet worms using novelty detection. *ArXiv: 0705.1288.*

Marseguerra, M., & Zoia, A. (2006). The autoassociative neural network in signal analysis: III. Enhancing the reliability of a NN with application to a BWR. *Annals of Nuclear Energy, 33*(6), 475-489.

Marseguerra, M., & Zoia, A. (2005a). The autoassociative neural network in signal analysis: I. The data dimensionality reduction and its geometric interpretation. *Annals of Nuclear Energy, 32*(11), 1191-1206.

Marseguerra, M., & Zoia, A. (2005b). The Autoassociative neural network in signal analysis: II. Application to online monitoring of a simulated BWR component. *Annals of Nuclear Energy, 32*(11), 1207-1223.

Marwala, T., Chakraverty, S., Mahola, U. (2006). Fault classification using multi-layer perceptrons and support vector machines. *International Journal of Engineering Simulation, 7*(1), 29-35.

Marwala, T., & Hunt, H. E. M. (2000). Probabilistic fault identification using vibration data and neural networks. In *Proceedings of the 18ᵗʰ International Modal Analysis Conference,* San Antonio, Texas (pp. 674-680).

Marwala, T., & Hunt, H. E. M. (2001). Maximum likelihood and Bayesian neural networks for damage identification. In *Proceedings of the 19ᵗʰ International Modal Analysis Conference,* Kissimmee, Florida (pp. 355-361).

Marwala, T., & Lagazio, M. (2004). Modelling and controlling interstate conflict. In *Proceedings of the IEEE International Joint Conference on Neural Networks,* Budapest, Hungary (pp. 1233-1238).

Marx, J. L. (1982). New disease baffles medical community. *Science, 217*(4560), 618-621.

Mirme, A., Minkkinen, P., & Ruuskanen, J. (1996). Behaviour of urban aerosol, black carbon and gaseous pollutants in urban air: Exploratory principal component analysis. *Nucleation and Atmospheric Aerosols* (pp. 423-426).

Møller, A. F. (1993). A scaled conjugate gradient algorithm for fast supervised learning. *Neural Networks, 6,* 525-533.

Msiza, I., Nelwamondo, F. V., & Marwala, T. (2007). Artificial neural networks and support vector machines for water demand time series forecasting. *IEEE International Conference on Systems, Man and Cybernetics* (pp. 638-643), Montreal, Canada.

Nelwamondo, F. V. (2008). *Computational intelligence techniques for missing data imputation.* Unpublished doctoral dissertation, University of the Witwatersrand, Johannesburg.

Nelwamondo, F. V., & Marwala, T. (2007a). Rough sets computations to impute missing data. *ArXiv: 0704.3635.*

Nelwamondo, F. V., & Marwala, T. (2007b). Rough set theory for the treatment of incomplete data. In *Proceedings of the IEEE Conference on Fuzzy Systems,* London, UK (pp. 338-343).

Nelwamondo, F. V., & Marwala, T. (2008). Techniques for handling missing data: applications to online condition monitoring. *International Journal of Innovative Computing, Information and Control, 4*(6), 1507-1526.

Nelwamondo, F. V., Mohamed, S., & Marwala, T. (2007). Missing data: A comparison of neural network and expectation maximisation techniques. *Current Science, 93*(11), 1514-1521.

Oliveira, A. L. I. (2006). Estimation of software project effort with support vector regression, *Neurocomputing. 69*(13-15), 1749-1753.

Palanivel, S., & Yegnanarayana, B. (2008). Multimodal person authentication using speech, face and visual speech. *Computer Vision and Image Understanding, 109*(1), 44-55.

Pan, L., Flynn, D., & Cregan, M. (2007). Sub-space principal component analysis for power plant monitoring power. *Plants and Power Systems Control* (pp. 243-248).

Patel, P. B., & Marwala, T. (2006). Forecasting closing price indices using neural networks. In *Proceedings of the IEEE International Conference on Systems, Man and Cybernetics,* Taiwan (pp. 2351-2356).

Pires, M. M., & Marwala, T. (2004). Option pricing using neural networks and support vector machines. In *Proceedings of the IEEE International Conference on Systems, Man and Cybernetics,* The Hague, Nederland (pp. 1279-1285).

Rios, A., & Kabuka, M. (1995). Image compression with a dynamic autoassociative neural network. *Mathematical and Computer Modelling, 21*(1-2), 159-171.

Sârbu, C., & Pop, H. F. (2005). Principal component analysis versus fuzzy principal component analysis: A case study: the quality of danube water (1985–1996). *Talanta, 65*(5), 1215-1220.

Shen, R., Fu, Y., & Lu, H. (2005). A novel image watermarking scheme based on support vector regression. *Journal of Systems and Software, 78*(1), 1-8.

Sherrington, D., & Wong, K. Y. M. (1989). Random Boolean networks for autoassociative memory. *Physics Reports, 184*(2-4), 293-299.

Shimizu, H., Yasuoka, K., Uchiyama, K., & Shioya, S. (1997). Online fault diagnosis for optimal rice α-amylase production process of a temperature-sensitive mutant of Saccharomyces cerevisiae by an autoassociative neural network. *Journal of Fermentation and Bioengineering, 83*(5), 435-442.

Shlens, J. (2005). *A tutorial on principal component analysis.* Retrieved August 26, 2008, from http://www.snl.salk.edu/~shlens/notes.html

Smith, L. I. (2002). *A tutorial on principal component analysis.* Retrieved August 26, 2008, from http://www.cs.otago.ac.nz/cosc453

Suykens, J. A. K., Van Gestel, T., & De Brabanter, J., De Moor, B., & Vandewalle, J. (2000). *Least Squares Support Vector Machines.* Singapore: World Scientific.

Thissen, U., Pepers, M., Üstün, B., Melssen, W. J., & Buydens, L. M. C. (2004). Comparing support vector machines to PLS for spectral regression applications. *Chemometrics and Intelligent Laboratory Systems, 73*(2), 169-179.

Thompson, B. B., Marks, R. J., & Choi, J. J. (2002). Implicit Learning in Autoencoder Novelty Assessment. In *Proceedings of the IEEE International Joint Conference on Neural Networks* (pp. 2878–2883).

Trafalis, T. B., & Ince, H. (2000). Support vector machine for regression and applications to financial forecasting. In *Proceedings of the IEEE International Joint Conference on Neural Networks* (pp. 348–353).

Üstün, B., Melssen, W. J., & Buydens, L. M. C. (2006). Facilitating the application of support vector regression by using a universal Pearson VII function based kernel. *Chemometrics and Intelligent Laboratory Systems, 81*(1), 29-40.

Üstün, B., Melssen, W J., & Buydens, L. M. C. (2007). Visualisation and interpretation of support vector regression models. *Analytica Chimica Acta, 595*(1-2), 299-309.

Üstün, B., Melssen, W. J., Oudenhuijzen, M., & Buydens, L. M. C. (2005). Determination of optimal support vector regression parameters by genetic algorithms and simplex optimization. *Analytica Chimica Acta, 544*(1-2), 292-305.

Vapnik, V. (1995). *The nature of statistical learning theory.* Heidelberg: Springer-Verlag.

Vapnik, V. (1998). *Statistical learning theory.* New York: John Wiley & Sons.

Whitley, D. (1994). A genetic algorithm tutorial. *Statistics and Computing, 4*(2), 65-85.

Xi, X-C., Poo, A-N and Chou, S-K. (2007). Support vector regression model predictive control on a HVAC plant, *Control Engineering Practice, 15*(8), 897-908.

Xie, X., Liu, W. T., & Tang, B. (2007) (In Press). Sinogram Spacebased estimation of moisture transport in marine atmosphere using support vector regression. *Remote Sensing of Environment.*

Yap, J. T., Kao, C-M., Cooper, M., Chen, C. T., & Wernick, M. (1995). Sinogram recovery of dynamic PET using principal component analysis and projections onto convex sets. In R Myers, V Cunningham, D Bailey and T Jones (Eds.) *Quantification of brain function using PET* (pp. 109-112). San Diego, CA: Academic Press.

Yuan, X., & Yuan, Y. (2006). Application of cultural algorithm to generation scheduling of hydrothermal systems. *Energy Conversion and Management, 47*(15-16), 2192-2201.

Zhang, J., Sato, T., & Iai, S. (2006). Support vector regression for online health monitoring of large-scale structures. *Structural Safety, 28*(4), 392-406.

Zhao, S., & Xu, Y. (2005). Multivariate statistical process monitoring using robust nonlinear principal component analysis. *Tsinghua Science & Technology, 10*(5), 582-586.

Zhou, Y-P., Jiang, J-H., Lin, W-Q, Zou, H-Y., Wu, H-L., Shen, G-L., & Yu, R-Q., (2006). Boosting support vector regression in QSAR studies of bioactivities of chemical compounds. *European Journal of Pharmaceutical Sciences, 28*(4), 344-353.

Chapter VII
Committee of Networks for Estimating Missing Data

ABSTRACT

This chapter introduces a committee of networks for estimating missing data. The first committee of networks consists of multi-layer perceptrons (MLPs), support vector machines (SVMs) and radial basis functions (RBFs). The committee was constructed from a weighted combination of these three networks. The second, third and fourth committees of networks were evolved using a genetic programming approach and used the MLPs, RBFs and SVMs, respectively. The committee of networks was collectively implemented with hybrid particle-swarm optimization and a genetic algorithm for missing data estimation. They were tested on an artificial taster as well as HIV datasets and then compared to the individual multi-layer perceptron, radial basis functions and support vector regression for missing data estimation. It was found that the committee of network approach provided improved results over the three methods acting individually. However, this improvement comes with a higher computational load than does using the individual approaches. Furthermore, it is found that evolving a committee method was a good way of constructing a committee.

INTRODUCTION

Several techniques have been introduced for missing data estimation (Abdella, 2005; Abdella & Marwala, 2005, 2006; Allison, 2000). A number of of these methods make use of machine learning approaches to accomplish this mission. In this chapter, committees of machine learning algorithms are used for missing data estimation. The principal incentive for using the committees of networks technique is from the intuitive logic that many 'heads' are better than one and, therefore, using many networks is thus better than using one.

Du, Zhai, and Wan (2007) used a committee of probabilistic radial basis function neural networks to identify palm prints while Anthony (2007) studied the generalization error of fixed combinations of classifiers. Sheikh-Ahmad et al. (2007) used a committee of neural networks for force prediction models in a milling process; Marwala (2000) used a committee of multi-layered perceptrons for damage identification in structures, whereas Marwala (2001a) implemented a probabilistic fault identification process in structures using a committee of neural networks and vibration data. Furthermore, Marwala et al. (2001) used a committee of agents and genetic programming to evolve a stock market prediction system.

In this chapter, the fact that the committee of networks approach will yield superior performance to the stand-alone networks for regression problems is extended to the missing data estimation problem and then demonstrated mathematically. After that, it is tested on an artificial taster and HIV prediction problems that were described in earlier chapters. Two approaches are pursued in constructing committee methods. These are the traditional approach proposed by Perrone and Cooper (1993), and the second approach is evolving the committee method by using a type of evolutionary programming known as *genetic programming*.

MISSING DATA APPROACH

In this chapter, the missing data estimation approach that was adopted entailed the use of a committee consisting of an auto-associative Multi-Layer Perceptron (MLP), Radial Basis Function (RBF) and a Support Vector Machine (SVM). These auto-associative networks were trained to predict their own input vectors and therefore were called *recall networks*. The results obtained from these networks were averaged in a weighted manner. Thus far, the MLP, RBF and SVM, which are the members of the committee of networks presented in this chapter, have each been individually used for missing data estimation (Nelwamondo, 2008) as was shown in earlier chapters. Pelckmans et al. (2005) proposed a technique for handling missing values in support-vector machine classifiers while Junninen et al. (2004) developed missing values imputation methods for air quality data sets. On the other hand, Chandramouli et al. (2007) used artificial neural networks for estimating missing microbial concentrations in a riverine database whereas Zhong, Lingras, and Sharma (2004) estimated missing traffic counts using factor, genetic, neural, and regression techniques. In this chapter, just as in Zhong, Lingras, and Sharma (2004), regression methods are used for estimating the missing data. The missing data estimation procedure adopted in this chapter is composed of two components i.e., regression and optimization components. The missing-data estimation error equation was written in earlier chapters as follows:

$$e = \left\| \left(\left\{ \begin{matrix} \{X_k\} \\ \{X_u\} \end{matrix} \right\} - f\left(\left\{ \begin{matrix} \{X_k\} \\ \{X_u\} \end{matrix} \right\} \right) \right) \right\| \tag{7.1}$$

In equation 7.1 the observed data of the complete dataset $\{X\}$ is $\{X_k\}$, the missing vector to be estimated is $\{X_u\}$, $\| \ \|$ is the Euclidean norm, f is the auto-associative network, which in this chapter is a multi-layer perceptron, a radial basis function or a support vector machine. The missing data estimation objective is, therefore, defined as the desire to identify the missing vector $\{X_u\}$ that would minimize equation 7.1 given the known parameters $\{X_k\}$ and the auto-associative mathematical model f. In this chapter, to approximate the missing input values, equation 7.1 is minimized using some optimization technique. In this chapter the hybrid genetic algorithm and particle swarm optimization method (HGAPSO) is used for this task. This hybrid approach is chosen over the traditional gradient-based approaches because of its simplicity of implementation and its characteristic of ensuring a higher probability of finding the global optimum solution than the traditional optimization methods (Goldberg, 1989). Furthermore, it is found to be robust. To identify the unknown vector $\{X_u\}$ that would minimize equation 7.1, given the known parameters $\{X_k\}$, and for the mathematical model f to be successful, identifying a global optimum solution as opposed to local optimum point is extremely critical. If this global solution is not realized, a wrong estimation of missing data will be achieved. Essentially, what is being done in this chapter is to perform an operation G that entails obtaining the missing data $\{X_u\}$ from the known data $\{X_k\}$ and a mathematical relationship f through the HGAPSO optimization process. This is shown schematically in Figure 7.1 and represented mathematically as follows:

$$\{X_u\} = G\big(\{x_k\}, f\big) \tag{7.2}$$

In equation 7.2, the relationships function f was obtained using regression models. In this chapter, the committee of multi-layer perceptron, radial basis functions and support vector machines as well as the evolved committee of networks are used to identify this relationship. This is the subject of the next section.

COMMITTEE OF NETWORKS

Abdel-Aal (2005a) introduced a three-member committee of networks for improving electric load forecasts and found that the committee reduced forecasting errors when compared to individual networks. However, Abdel-Aal used just one machine-learning algorithm, which was the multi-layer perceptron neural network. The same author extended the application of the committee of networks to the problem of modeling medical data and found that the committee approach offered a reduction in classification errors of up to 20% when compared to the stand-alone networks. In the committee approach, as it will be demonstrated later, it is known that the presence of diversity within the members of the committee improves the performance of the committee. Abdel-Aal (2005b) introduced diversity by training various members of the committee with different data.

Other successful implementations of the committee approach that showed improvement over individual approaches included an application to human face recognition by Zhao, Huang, and Sun (2004), recognition of swallow acceleration signals by Das, Reddy, and Narayanan (2001), selecting salient features by Bacauskiene and Verikas (2004), speaker verification by Reddy and Buch (2003), automatic fire detection by Fernandes et al. (2004), as well as permeability prediction by Chen and Lin (2006). Anthony (2007) studied the generalization error of a fixed size of a committee of networks.

Figure 7.1. The missing data estimation procedure adopted in this chapter

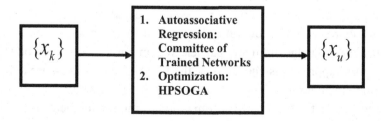

In this section, a committee of networks procedure that is illustrated in Figure 7.2 is introduced. The committee approach, shown in this figure, contains three networks and the output was the weighted average of the outputs of these three networks. The ideas presented in this section are the adaptation and the extension of the work by Perrone and Cooper (1993), who introduced the concept of a committee of networks, which was extended and applied to mechanical systems by Marwala and Hunt (1999). In this section, it is proven that a committee of networks gives results that are more reliable than when using networks separately for missing data estimation.

The mapping of the input vector $\{x_k\}$, which represents the known input to the output vector $\{x_u\}$, may be written as the desired function plus an error as follows:

$$\{x_u\}_1 = \{h_1(\{x_k\})\} + \{e_1(\{x_k\})\} \tag{7.3}$$

$$\{x_u\}_2 = \{h_2(\{x_k\})\} + \{e_2(\{x_k\})\} \tag{7.4}$$

$$\{x_u\}_3 = \{h_3(\{x_k\})\} + \{e_3(\{x_k\})\} \tag{7.5}$$

In equations 7.3 to 7.5, the parameter $\{h_i\}$ is the output vector from the approximated mapping function from network model h_i and $\{e_i\}$ is the mapping error vector for the ith network. The Mean Square Errors (MSE) for these three models in equations 7.3, 7.4 and 7.5 may, therefore, be written as follows:

$$E_1 = \varepsilon\left[\left(\{x_u\}_1 - \{h_1\}\right)^2\right] = \varepsilon\left[e_1^2\right] \tag{7.6}$$

$$E_2 = \varepsilon\left[\left(\{x_u\}_2 - \{h_2\}\right)^2\right] = \varepsilon\left[e_2^2\right] \tag{7.7}$$

$$E_3 = \varepsilon\left[\left(\{x_u\}_3 - \{h_3\}\right)^2\right] = \varepsilon\left[e_3^2\right] \tag{7.8}$$

In equations 7.6 to 7.8, the parameter ε indicates the expected value and corresponds to the integration over the input data, and is defined as follows:

$$\varepsilon[e_1^2] = \int e_1^2 p(\{x_k\})d\{x_k\} \tag{7.9}$$

$$\varepsilon[e_2^2] = \int e_2^2 p(\{x_k\})d\{x_k\} \tag{7.10}$$

Figure 7.2. A three-network committee

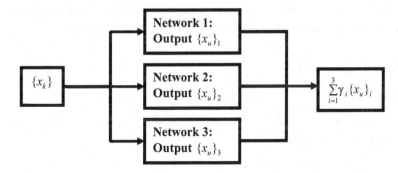

$$\varepsilon[e_3^2] = \int e_3^2 p(\{x_k\})d\{x_k\} \qquad (7.11)$$

In equations 7.9 to 7.11, p is the probability density function and d is the differential operator. The average MSE of the three networks acting individually may thus be written as follows:

$$E_{AV} = \frac{E_1 + E_2 + E_3}{3} = \frac{1}{3}\left(\varepsilon\left(e_1^2\right) + \varepsilon\left(e_2^2\right) + \varepsilon\left(e_3^2\right)\right) \qquad (7.12)$$

Equal Weights

In this section, the concept of a *committee of networks* is explained. The output of the committee is the average of the outputs from the three networks as illustrated in Figure 7.2. By giving equal weighting functions, the committee prediction may thus be written in the following form:

$$\{x_{COM}\} = \frac{1}{3}\left(\{x_k\}_1 + \{x_k\}_2 + \{x_k\}_3\right) \qquad (7.13)$$

The Mean Square Errors (MSE) of the committee of networks can be written as follows:

$$
\begin{aligned}
E_{COM} \\
&= \varepsilon\left[\left(\frac{1}{3}\{\{x_k\}_1 + \{x_k\}_2 + \{x_k\}_3\} - [\{h_1\} + \{h_2\} + \{h_3\}]\right)^2\right] \\
&= \varepsilon\left[\left(\frac{1}{3}\left(\left[(\{x_k\}_1 - \{h_1\})\right] + \left[(\{x_k\}_2 - \{h_2\})\right] + \left[(\{x_k\}_3 - \{h_3\})\right]\right)\right)^2\right] \\
&= \varepsilon\left[\left(\frac{1}{3}(e_1 + e_2 + e_3)\right)^2\right] \\
&= \frac{1}{9}\left(\varepsilon\left[e_1^2\right] + 2\left(\varepsilon[e_1 e_2] + \varepsilon[e_1 e_3] + \varepsilon[e_2 e_3]\right) + \varepsilon\left[e_2^2\right] + \varepsilon\left[e_3^2\right]\right)
\end{aligned}
$$

$$\qquad (7.14)$$

If it is assumed that the errors (e_1, e_2 and e_3) are uncorrelated then:

$$\varepsilon[e_1 e_2] = \varepsilon[e_1 e_3] = \varepsilon[e_2 e_3] = 0 \qquad (7.15)$$

To ensure that the errors are uncorrelated, in this chapter, a *diverse committee* of networks consisting of the multi-layer perceptron, radial basis functions and support vector machines was chosen. By substituting equation 7.15 into equation 7.14 the MSE of the committee becomes:

$$E_{COM} = \frac{1}{9}\left(\varepsilon\left[e_1^2\right] + \varepsilon\left[e_2^2\right] + \varepsilon\left[e_3^2\right]\right) \qquad (7.16)$$

The error of the committee in equation 7.16 can be related to the average error of the networks acting individually (equation 7.12) as follows:

$$E_{COM} = \frac{1}{3} E_{AV}$$ (7.17)

Equation 7.17 shows that the MSE of the committee is one-third of the average MSE of the individual method. From equation 7.17, it can be deduced that the MSE of the committee is always equal to or less than the average MSE of the three methods acting individually.

Variable Weights

The three networks might not necessarily have the same predictive capacity. To accommodate the strength of each member of the committee, the networks should be given appropriate weighting functions. It will be explained later how these weighting functions are evaluated using the prior knowledge of the strength of each approach. The estimated missing data may be defined as the combination of the three independent approaches with approximate weighting functions as [a modification of equation 7.13]:

$$\{x_k\}_{COM} = \gamma_1 \{x_k\}_1 + \gamma_2 \{x_k\}_2 + \gamma_3 \{x_k\}_3$$ (7.18)

In equation 7.18, γ_1, γ_2 and γ_3 are the weighting functions and $\sum_{i=1}^{3} \gamma_i = 1$. The MSE due to the weighted committee can be written as follows:

$$
\begin{aligned}
E_{COM} \\
&= \varepsilon \left[\gamma_1 \{x_u\}_1 + \gamma_2 \{x_u\}_2 + \gamma_3 \{x_u\}_3 - [\gamma_1 h_1 + \gamma_2 h_2 + \gamma_3 h_3] \right] \\
&= \varepsilon \left[\left((\gamma_1 \{x_u\}_1 - h_1) + (\gamma_2 \{x_u\}_2 - h_2) + (\gamma_3 \{x_u\}_3 - h_3) \right)^2 \right] \\
&= \varepsilon \left[(\gamma_1 e_1 + \gamma_2 e_2 + \gamma_3 e_3)^2 \right]
\end{aligned}
$$ (7.19)

Equation 7.19 may be rewritten in Lagrangian form as:

$$E_{COM} = \varepsilon \left[\gamma_1 e_1 + \gamma_2 e_2 + \gamma_3 e_3 \right] + \lambda (1 - \gamma_1 - \gamma_2 - \gamma_3)$$ (7.20)

Here λ is the Lagrangian multiplier. The derivative of error in equation 7.21 with respect to γ_1, γ_2, γ_3 and λ may be calculated and equated to zero as follows:

$$\frac{dE_{COM}}{d\gamma_1} = \varepsilon \left[2(\gamma_1 [e_1] + \gamma_2 [e_2] + \gamma_3 [e_3])[e_1] \right] - \lambda = 0$$ (7.21)

$$\frac{dE_{COM}}{d\gamma_2} = \varepsilon \left[2(\gamma_1 [e_1] + \gamma_2 [e_2] + \gamma_3 [e_3])[e_2] \right] - \lambda = 0$$ (7.22)

$$\frac{dE_{COM}}{d\gamma_2} = \varepsilon \left[2(\gamma_1 [e_1] + \gamma_2 [e_2] + \gamma_3 [e_3])[e_3] \right] - \lambda = 0$$ (7.23)

$$\frac{dE_{COM}}{d\lambda} = (1 - \gamma_1 - \gamma_2 - \gamma_3) = 0$$ (7.24)

Solving equations 7.22 to 7.24, the minimum errors obtained are:

$$\gamma_1 = \frac{1}{1 + \frac{\varepsilon[e_1^2]}{\varepsilon[e_2^2]} + \frac{\varepsilon[e_1^2]}{\varepsilon[e_3^2]}} \tag{7.25}$$

$$\gamma_2 = \frac{1}{1 + \frac{\varepsilon[e_2^2]}{\varepsilon[e_1^2]} + \frac{\varepsilon[e_2^2]}{\varepsilon[e_3^2]}} \tag{7.26}$$

$$\gamma_3 = \frac{1}{1 + \frac{\varepsilon[e_3^2]}{\varepsilon[e_1^2]} + \frac{\varepsilon[e_3^2]}{\varepsilon[e_2^2]}} \tag{7.27}$$

Equations 7.25 to 7.27 may be generalized for a committee with n-trained networks and be expressed for network i as follows:

$$\gamma_i = \frac{1}{\sum_{j=1}^{n} \frac{\varepsilon[e_i^2]}{\varepsilon[e_j^2]}} \tag{7.28}$$

By analyzing equation 7.28, it can be deduced that if the predictive capacity of the three networks are equal, then each method should be assigned equal weights. This conclusion is trivial, but it is deduced in this chapter to confirm the effectiveness of the proposed method.

Since it is not known which network is more accurate at a given instance, the weighting functions are determined from the data that are used for training and validation of the networks process (prior knowledge).

Committee Gives Solution That is More Reliable

Axiom 1: If three independent (uncorrelated) methods are used simultaneously, the reliability of the combination is at least as good as when the methods are used individually. Suppose that the probabilities of success for the approach 1 (m_1), approach 2 (m_2), and approach 3 (m_3) are $P(m_1)$, $P(m_2)$, and $P(m_3)$, respectively. The reliability of the three methods acting in parallel is given by (Marwala, 2001b):

$$P(m_1 \cup m_2 \cup ...m_n) = P(m_1) + P(m_2)... + P(m_n) - [P(m_1 \cap m_2) + P(m_2 \cap m_3) + ... + P(m_{n-1} \cap m_n)]...$$
$$+ P(m_1 \cap m_2...m_n) \tag{7.29}$$

From equation 7.29 it can be deduced that the reliability of the committee is always higher than that of the individual methods.

EVOLVING THE COMMITTEE OF NETWORKS

In the previous section, a method for constructing an optimal committee was developed. This method was mainly based on the expected errors of each member of the committee. Another crucial aspect of this method is that it works if the networks that form a committee are uncorrelated. It becomes very difficult to first calculate the expected error of each network in the committee particularly *a priori* and to construct a population of committee of networks that are sufficiently uncorrelated. Evolving networks has been a subject of investigation for some time. Rajan and Mohan (2007) implemented an evolutionary programming method, which was based on simulated annealing to solve the unit commitment problem, whereas Basu (2004) used an evolutionary programming method to construct an interactive fuzzy satisfying method and used this to solve a multi-objective short-term hydro-thermal scheduling. Shi and Xu (2001) adopted a self-adaptive evolutionary programming method and applied this to optimize a multi-objective operation of power systems, whereas Cao, Jiang, and Wu (2000) implemented an evolutionary programming method for a mixed-variable optimization problem. In this chapter, genetic programming and Monte Carlo method are adopted to evolve a committee of networks (Metropolis et al., 1953).

This section presents a genetic programming procedure (Jacob, 2001; He, 2008; Zhao, Gao, & Hu, 2007) combined with the Metropolis method (Metropolis et al., 1953) to evolve an optimal committee of networks that was used for missing data estimation. The presented method works by firstly establishing the parameters in the committee that will be evolved. These characteristics are the attributes that define the architectures of the specific machine learning tools such as network type, number of hidden nodes, activation functions etc. Once these attributes are identified, they are numerically codified. Then these attributes are transformed into binary number space. They then undergo crossovers and mutations (Choi, 2002; Dong et al., 2007), and are then transformed back into floating number space. The resulting committee is then constructed and its fitness evaluated. The current committee structure is accepted or rejected using the Metropolis method. The process is repeated until the resulting posterior probability distribution has converged. Ultimately, the committee with the highest probability is chosen as the optimal technique. This process is illustrated in Figure 7.3.

In Figure 7.3, S_n is the nth state and $E_{Com:\,n}$ is the committee of networks error for the nth state. The transition between states is enabled using the mutation and crossover operators of genetic programming. The attributes of the members of the committee, therefore, undergo crossover and mutation. In essence,

Figure 7.3. Evolving a committee of networks using a Monte Carlo method and genetic programming

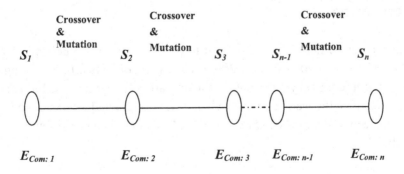

this process is aimed at sampling the posterior probability of the committee characteristics, given the observed data, $P(Com|[D])$, which can be written in Bayesian form as follows (Bishop, 1995):

$$P(Com|[D]) = \frac{P([D]|Com)P(Com)}{P([D])}$$

(7.30)

Here $P([D]|Com)$ is the likelihood function, $P(Com)$ is the prior probability of the committee of networks and $P([D])$ is the evidence. Formulating problems in the Bayesian framework is a powerful approach that facilitates solving mathematically intractable problems. Akkaladevi and Katangur (2007) used Bayesian formulation and a committee of networks for protein secondary structure prediction whereas Vasudevan, Bhaduri, and Raj (2007) used Bayesian neural networks for predicting ferrite numbers in stainless steel welds. If in equation 7.30, the prior probability is set to 1, then equation 7.30 may be rewritten in the following form:

$$P(Com|[D]) = \frac{1}{Z_{norm}} \exp(-E_{COM})$$

(7.31)

In equation 7.31, Z_{norm} is the normalization factor that ensures that the posterior probability function integrates to 1. In equation 7.31, the error of the committee is calculated by giving each network equal weights. Therefore, the weighting of the members of the committee will be inherently modeled in the evolved networks through the evolutionary process. As indicated before, the state that represents the structure of the committee is accepted or rejected using the Metropolis algorithm as follows:

If $P_{n+1} > P_n$ then accept the state S_{n+1}, else accept if $\frac{P_{n+1}}{P_n} > \zeta$ where ζ is uniformly sampled to fall in the interval [0, 1], else reject state S_{n+1} and return to state S_n. The evolutionary method described above is implemented as shown in Figure 7.4.

Basically, Figure 7.4 shows that the state (set of attributes) that has been visited the most is the evolved architecture.

NETWORKS USED IN THE COMMITTEE

In this chapter, neural networks are viewed as parameterized models that define interrelationships about the data.

Multi-Layer Perceptron

Multi-layer perceptron neural networks are trained to give a relationship between the input data to itself (Bishop, 1995; Marwala, 2007; Nabney, 2001). The MLP contains a linear or logistic activation function in the outer layer and the hyperbolic tangent activation function in the hidden layer. Training the neural network identifies the weights. In this chapter it is conducted as was done in Chapter II using the scaled conjugate gradient method (Møller, 1993). More details on multi-layer perceptrons were described in Chapter II.

Figure 7.4. Flow chart of the implementation evolutionary procedure

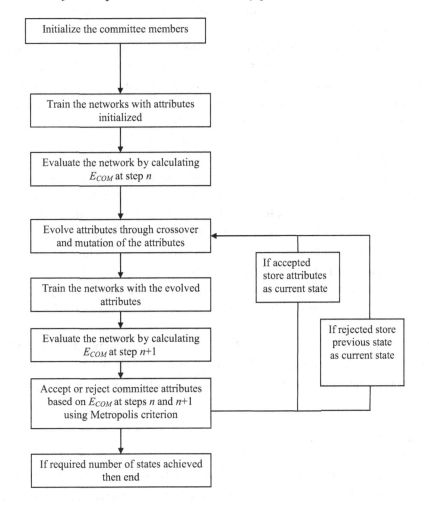

Multi-layer perceptrons have been successfully used for modeling complex systems. Niska et al. (2005) used a multi-layer perceptron model for forecasting urban airborne pollutant concentrations, while Frias-Martinez, Sanchez, and Velez (2006) established that support vector machines perform better than the multi-layer perceptrons for off-line signature recognition, whereas Rossi and Conan-Guez (2005) successfully used multi-layer perceptron for functional data analysis. Furthermore, Cigizoglu (2004) used multi-layer perceptrons for forecasting daily suspended sediment data whereas Pizzi, Somorjai, and Pedrycz (2006) successfully classified biomedical spectra using multi-layer perceptrons.

Further successful applications of multi-layer perceptrons were included in hypothesis testing of spatially dependent data (Walde, 2007), in voltage contingency ranking (Pandit, Srivastava, & Sharma, 2001) and in tool-state classification for metal turning (Dimla, 1999).

Radial Basis Functions

Radial Basis Function (RBF) neural networks are similar to the multi-layers perceptrons and are feed-forward networks, trained using a supervised training algorithm (Haykin, 1999). They are usually configured with one hidden layer of units, whose activation function is selected from a class of functions called the *basis function*. The activation of the hidden units in an RBF neural network is given by a non-linear function of the distance between the input vector and a prototype vector (Bishop, 1995).

Even though they are related to multi-layer perceptrons in many ways, radial basis function networks posses a number of advantages. Generally they are trained a great deal faster than multi-layer perceptron networks and are less prone to problems with non-stationary inputs due to the behavior of the radial basis function (Bishop, 1995). However, independent analysis performed as part of this chapter has found that the RBF becomes unstable during training particularly when the pseudo-inverse technique is used.

As stated in earlier chapters, the activation function in the hidden layers is chosen in this chapter to be either a Gaussian distribution or thin plate spline (Bishop, 1995). The radial basis function differs from the multi-layered perceptron in that it has weights in the outer layer only, while the hidden nodes have what are called the *centers*. Training the radial basis function network entails identifying two sets of parameters. These are the centers in the hidden nodes and the output weights. Even though the centers and network weights can be determined simultaneously, this chapter uses a two stage training process to identify the centers and subsequently the network weights. The first stage is to determine the centers. In this chapter, the Expectation Maximization (EM) algorithm (Dempster, Laird, & Rubin, 1977), which was described in Chapter IV, is used to determine the centers, as opposed to the *k-means* clustering method (Hartigan & Wong, 1979) which was described in detail in Chapter II.

Successful applications of radial basis functions include the analysis of the free vibration of composite material by Ferreira, Roque, and Jorge (2005), voltage stability monitoring by Chakrabarti and Jeyasurya (2007), image reconstruction by Magoulès, Diago, and Hagiwara (2007), fault diagnosis of electronic circuits by Catelani and Fort (2000) as well as for differential relaying by Moravej, Vishwakarma, and Singh (2003).

Support Vector Regression

The fundamental initiative that is behind a Support Vector Regression (SVR) is to map the input space to an output space that is the same as those of the multi-layer perceptron and radial basis functions. As explained in Chapter VI, SVR seeks to find a function that maps the training inputs to the training outputs by identifying the function that has at most a deviation ε from the actual training targets y_i. A number of types of functions $f(x)$ that are able to map the training inputs to training outputs can be found and these functions are known as *kernel functions*. A kernel function cannot be merely any function, however, it has to adhere to some mathematical criteria (Joachims, 1999).

To train an SVR, the process involves the use of an ε-insensitive loss function which is used to penalize certain training samples that are outside of the bound given by ε. This ε parameter is chosen by the user. There are various other loss functions such as the Huber loss function, which can also be, used. However, the most common one is the ε-insensitive loss function (Schölkopf, Burges, & Smola, 1999; Smola & Schölkopf, 2004). The value for the parameter called *capacity*, C, is used as the level to which deviations from the parameter ε are tolerated (Smola & Schölkopf, 2004). This parameter can also be viewed as a measure of over-fitting a function to its training points too well. If the value of C is set too high, then the function found, $f(x)$, will be too well fitted to the training data and will not predict very well on data that was not seen during the training stage. This means that points lying outside of the bounds given by the parameter ε are not penalized enough and this results in the function being too well fitted to the training samples (Smola & Schölkopf, 2004). Those samples that fall within the bounds of the function are not penalized at all and their corresponding slack variable values (ξ_i, ξ_i^*) are set to zero, and thus these samples play a foremost role in the determination of the function $f(x)$.

The optimization problem solved in the SVR training is then set up to be a quadratic programming problem by first finding the Lagrangian multiplier and applying the Karush-Kuhn Tucker (KKT) conditions (Schölkopf, 2003; Nocedal & Wright, 1999). Then the function can be determined so that the linear function needed to fit the training data can be explicitly found. In this chapter, the parameters that are evolved are the kernel functions. In this chapter, three kernel functions are used. These are the Gaussian radial basis function, sigmoid and the homogeneous polynomial (Schölkopf, 2003).

Successful implementations of support vector machines include the developing structural health monitoring system by Hagiwara and Mita (2002), analyzing stochastic mechanics by Hurtado (2003), segmentation of remote sensing images by Mitra, Shankar, and Pal (2004), pharmaceutical data analysis by Burbidge et al. (2001), bankruptcy prediction by Min and Lee (2005), web service quality control based on text mining by Lo (2008), prediction of protein structural classes by Cai et al. (2002), analyzing climate change by Tripathi, Srinivas, and Nanjundiah (2006), gene extraction in cancer diagnosis by Huang and Kecman (2005) as well as for automatic masking in image analysis.

HYBRID PARTICLE SWARM OPTIMIZATION AND GENETIC ALGORITHM

In this chapter, a Hybrid Genetic Algorithm and Particle Swarm Optimization (HGAPSO) is used for missing data estimation. The HGAPSO operates by using particle swarm optimization to come to a solution. That solution then undergoes crossover and mutation and then it is selected and the entire process is called an iteration step. Kao and Zahara (2008) used a hybrid genetic algorithm and particle swarm optimization for optimizing multimodal functions, which are functions that have more than one optimal solution. The results from the experimental studies using a group of 17 multimodal test functions showed that the HGAPSO method offers a better solution quality and convergence rates than other techniques. Fan, Liang, and Zahara (2006) implemented a hybrid genetic algorithm and a particle swarm optimizer and tested this on ten hard non-linear continuous functions, and compared the results with the best known heuristics in the literature. The results obtained indicate that the HGAPSO method is capable of reaching the global optimum solution with a comfortable computational expense. Shi et al. (2005) successfully implemented the HGAPSO algorithm for approximating a variety of radiation properties in a two-dimensional irregular medium when the temperatures were measured at only four points.

To understand what aspects of PSO are brought into the HGAPSO method, it is important to describe the PSO briefly. PSO was developed by Kennedy and Eberhart (1995). This procedure was inspired by algorithms that model the "flocking behavior" seen in birds. It is based on social-psychological principles inspired by swarm intelligence, offers an understanding of social behavior, and has contributed to engineering applications. Society enables an individual to maintain cognitive robustness through influence and learning. Individuals learn to tackle problems by communicating and interacting with other individuals and thereby develop a generally similar ways of tackling problems.

PSO models this type of social interaction within the context of evolutionary programming to tackle optimization problems by using a *fitness function* to describe a measure of the desired outcome, which in this chapter, was used to estimate missing variables. To reach an optimum solution, a social network representing a population of possible solutions is defined and randomly generated. The individuals within this social network are assigned neighbors with whom they will interact. Then, a process to update these particles is initiated by ensuring that each particle could remember the location where it had its best success as measured by the fitness function. The best solution for the particle is named the *local best* and each particle makes this information on the local best accessible to its neighbors who in turn also observe their neighbors' successes.

The second step is to use genetic algorithm (GA) to complete a single generation. GA is as a stochastic algorithm whose search method attempts to model the biological phenomena of genetic inheritance and natural selection (Ko et al., 1997; Cox, 2005). The basic concept of GA is designed to simulate processes in natural systems, necessary for evolution (Gen & Cheng, 1997), in particular, those that follow the principles of the survival of the fittest, first laid down by Charles Darwin. As such, they symbolize an intelligent exploitation of a random search within a defined search space to solve a problem. Like most evolutionary methods, the GA operation is based on a population of individuals that represent a number of possible solutions to the problem. Evolution of the population at each iteration step is accomplished using evolutionary operators such as selection, crossover and mutation that model the phenomena of natural selection, reproduction and genetic mutation, respectively (Ko et al., 1997; Gen & Cheng, 1997).

Genetic algorithm have been found to be successful in solving the following problems: designing mesh networks (Ko et al., 1997); molecular design (Devillers, 1996); optimizing pin jointed structures (Mahachi & Dundu, 2001); diagnosing cracks in shafts (Saridakis et al., 2006); silico drug design (Embrechts et al., 2003); antenna design (Linden, 2002); corrosion detection (Amaya, Ridha, & Aoki, 2003); structural design (Arora, 2004); nuclear reactor design (Pereira & Lapa, 2003) as well as in modeling urban drainage (Rauch & Harremoës, 1999).

EXPERIMENTAL INVESTIGATION OF THE COMMITTEE APPROACH

In this section, the committee methods proposed in this chapter are applied to the problem of the artificial taster (Marwala, 2005).

Artificial Taster

The variables used in the autoassociative neural networks are: *alcohol*; *present extract*; *real extract*; PE-LE, which stands for *Present Extract minus Limit Extract*; *pH*; *iron*; *acetaldehyde*; DMS, which stands for *Di-Methyl Sulfide*; *eythyl acetate*; *iso-amyl acetate*; *total higher alcohols*; *color*; *bitterness* and *amount of carbohydrates*.

The auto-associative multi-layer perceptron neural network, which is a component of the three member committee networks presented in this chapter, had 14 inputs, 10 hidden neurons and 14 outputs. This MLP was trained on the data obtained from a brewery in South Africa (Marwala, 2005). The multi-layer perceptron implemented in this chapter had hyperbolic tangent function in the hidden layer and a linear activation function in the output layer.

The auto-associative radial basis function neural network, which is a second component of the three-member committee of networks, had 14 inputs, 10 hidden neurons and 14 outputs. This RBF architecture had a Gaussian activation function in the hidden layer It was trained using an Expectation Maximization (EM) algorithm to determine the network centers and a pseudo-inverse procedure to determine the network weights (Campbell & Meyer, 1991; Bishop, 1995).

The auto-associative support-vector regression model with Gaussian radial basis function kernel was also constructed and trained with the artificial taster dataset. Missing values in a single record were examined to investigate the accuracy of the approximated values as the number of missing cases within a single record increased.

Upon the implementation of the genetic programming to evolve committees of networks, three sets of committees were constructed. These were a committee of 10 RBFs, 10 MLPs and 10 SVMs. For the committee of RBFs, the attributes that were evolved were the number of hidden nodes, which was made to be distributed between 8 and 15, and the activation functions which were Gaussian distributions or thin plate splines. For the committee of MLPs the attributes that were evolved were number of hidden nodes, which were distributed between 8 and 15 and the activation function in the outer layer, either linear or logistic. For the committee of SVMs the attributes that were evolved were kernel functions and these were the Gaussian radial basis function, the sigmoid and the homogeneous polynomial. To evolve the networks using genetic programming, single point crossover and non-uniform mutation were used.

To assess the accuracy of the values approximated using the presented missing data estimation models, standard errors were calculated for each missing case. Using the presented models for missing data, the results are shown in Table 7.1.

The results in Table 7.1 show that for the data under consideration, the committee of networks performs better than the individual methods. The traditional committee is found to perform the best, followed by the committee of evolutionary MLPs, then the committee of evolutionary RBFs and finally

Table 7.1. Standard error of the four methods for the test dataset

Method	Standard Errors
Multi-layer perceptron	13.98
Radial basis function	14.17
Support vector regression	16.33
Traditional MLP-RBF-SVM committee method	12.31
Evolutionary Committee: 10 RBFs	13.88
Evolutionary Committee: 10 MLPs	12.93
Evolutionary Committee: 10 SVM	15.08

Table 7.2. Standard error of the four methods of missing data estimation

Method	Standard Errors with 1 missing value	Standard Errors with 2 missing values
Multi-layer perceptron + HGAPSO	13.98	14.18
Radial basis function + HGAPSO	14.70	15.88
Support vector regression + HGAPSO	16.33	17.07
Traditional MLP-RBF-SVM Committee method + HGAPSO	12.31	13.13
Evolutionary MLP Committee method + HGAPSO	12.78	13.81
Evolutionary RBF Committee method + HGAPSO	13.77	14.63
Evolutionary SVM Committee method + HGAPSO	15.52	16.63

the committee of evolutionary SVMs. The genetic approach to an evolutionary committee is found to be substantially more computationally expensive than the traditional committee is.

To implement the missing data procedure illustrated in Figure 7.1, the HGAPSO algorithm was used for optimization, and the committee of networks was used for auto-associative regression. For the genetic algorithm implementation, a rank selection method, single point crossover and non-uniform mutation were used. The parameters for the genetic algorithm implementation were 70% probability of reproduction, 3% mutation rate and 60% crossover rate. When the method was used for missing data estimation, the results in Table 7.2 are obtained.

The results in Table 7.2 do indeed demonstrate that the committee approach is an improvement over the stand-alone networks for missing data estimation. Further, it can also be observed that the missing data estimation process generally decreases in accuracy as the number of missing values increases. The radial basis function is found to outperform the other two stand-alone techniques. Again, the traditional committee performs the best, followed by the evolutionary MLP, then the evolutionary RBF and the evolutionary SVM. The gains that are realized from the implementation of the committee approach, however, come at a huge computational expense because instead of requiring one regression model for missing data estimation, many models are required.

Modelling HIV

The second set of data that was used for testing the committee of network approach is the HIV data set from antenatal clinics from South Africa which was gathered by the Department of Health of South Africa. The data set included multiple input fields.

Data gathered from surveys and other data collection techniques usually contain outliers, which are data that do not fit the profile by which that particular piece of data set is historically understood. These outliers are normally removed from the dataset because they can potentially contaminate conclusions that are drawn from it. In this chapter, data samples that have outliers were removed. The data variables and their ranges are shown in Table 7.3. The dataset was normalized to be between 0 and 1, and the training data was then split into 3 partitions: 60% was used for training, 15% was used for validation purposes

Table 7.3. HIV data variables

Variable	Type	Range
HIV Status	Binary	[0, 1]
Education	Integer	0 – 13
Age Group	Integer	14 – 60
Age Gap	Integer	1 – 7
Gravidity	Integer	0 - 11
Parity	Integer	0 – 40
Race	Integer	1 – 5
Province	Integer	1 – 9
Region	Integer	1 – 36
RPR	Integer	0 – 2
WTREV	Continuous	0.64 – 1.27

and the last 25% was used for the testing. More details on this can be found in Marivate, Nelwamondo, and Marwala (2008). The multi-layer perceptrons, radial basis function and support vector regression networks were implemented as it was done in the artificial taster dataset. However, there were now 11 inputs and 11 outputs that were used to form the auto-associative regressors and for the radial basis function and the multi-layer perceptron there were now 8 hidden neurons.

When the auto-associative regression models were constructed and the committees of networks were constructed as in the artificial taster example, the results in Table 7.4 are obtained.

The results in Table 7.4 show that the traditional committee of networks gives the best results, followed the evolutionary committee of MLPs, then the evolutionary committee of RBFs, then the evolutionary committee of SVMs, followed by the multi-layer perceptron, which is marginally better than the radial basis function. In addition, support vector regression is found to perform the worst, as was the case for artificial taster data set. Therefore, these results are consistent with the conclusions that were drawn in the previous example.

As was done in the artificial taster data set, the procedure in Figure 7.1 was implemented for estimating the missing data and the HGAPSO method was used for optimization, while the committee of networks was used for auto-associative regression. Again, the HGAPSO method implemented rank selection, single point crossover and non-uniform mutation. The parameters for genetic algorithm implementation were 70% for the probability of reproduction, 3% for mutation rate and 60% for crossover rate. Table 7.5 shows the missing data estimation results.

The results in Table 7.5 show that the committee approach improves over the stand-alone network for missing data estimation even though the improvement witnessed here is marginal. Again, these improvements come at a huge computational expense because instead of requiring one regression model for missing data estimation, three models are required.

CONCLUSION

In this chapter, a committee of multi-layer perceptron, radial basis functions and support vector machines was introduced for missing data estimation. This procedure was tested on an artificial taster data set and HIV data set, and then compared to using a multi-layer perceptron, a radial basis function and a support vector machine individually. The results obtained show that the committees of networks give

Table 7.4. Standard error of the four methods for the test data set

Method	Standard Errors
Multi-layer perceptron	17.01
Radial basis function	17.57
Support vector regression	20.65
Traditional MLP-RBF-SVM Committee method	16.88
Evolutionary Committee method: 10 MLPs	16.01
Evolutionary Committee method: 10 RBFs	16.97
Evolutionary Committee method: 10 SVNs	18.99

better results than the three methods used individually, while a stand-alone multi-layer perceptron gives similar results to the stand-alone radial basis function. The traditional committee of networks is found to perform the best, followed by the evolutionary programming derived MLP committee, followed by the evolutionary RBF committee and finally the evolutionary SVM committee. Both the MLP and RBF perform better than a support vector machine. However, these committees of networks have a higher computational load than the individual approaches. In particular, the evolutionary committees are more computationally expensive than a traditional committee.

FURTHER RESEARCH

For future work, a formal study on how to measure the diversity of the committee of networks should be undertaken. Furthermore, the relationship between the diversity of the committee of networks and its performance should be identified. In this chapter, the evolutionary approach to constructing a committee of networks only used one machine learning methodology. In a future study, an effective manner of evolving a committee of networks with more than one machine learning technique should be implemented.

REFERENCES

Abdel-Aal, R. E. (2005a). Improving electric load forecasts using network committees. *Electric Power Systems Research, 74*(1), 83-94.

Abdel-Aal, R.E. (2005b). Improved classification of medical data using abductive network committees trained on different feature subsets. *Computer Methods and Programs in Biomedicine, 80*(2), 141-153.

Abdella, M. (2005). *The use of genetic algorithms and neural networks to approximate missing data in database.* Unpublished master's thesis, University of the Witwatersrand, Johannesburg.

Table 7.5. Standard error of the four methods on missing data estimation

Method	Standard Errors with 1 Missing Value	Standard Errors with 1 Missing Value
Multi-layer perceptron + HGAPSO	17.98	17.95
Radial basis function + HGAPSO	18.23	19.73
Support vector regression + HGAPSO	20.97	21.13
Traditional MLP-RBF-SVM Committee method + HGAPSO	17.23	17.07
Evolutionary MLP Committee method + HGAPSO	16.53	16.79
Evolutionary RBF Committee method + HGAPSO	17.39	17.83
Evolutionary SVM Committee method + HGAPSO	19.32	20.04

Abdella, M., & Marwala, T. (2005). Treatment of missing data using neural networks. In *Proceedings of the IEEE International Joint Conference on Neural Networks,* Montreal, Canada (pp. 598-603).

Abdella, M., & Marwala, T. (2006). The use of genetic algorithms and neural networks to approximate missing data in database. *Computing and Informatics, 24,* 1001-1013.

Akkaladevi, S., & Katangur, A. K. (2007). Protein secondary structure prediction using Bayesian inference method on decision fusion algorithms. In *Proceedings - 21ˢᵗ International Parallel and Distributed Processing Symposium, IPDPS 2007: Abstracts and CD-ROM,* art. no. 4228158.

Allison, P. (2000). Multiple imputation for missing data: A cautionary tale. *Sociological Methods and Research 28,* 301-309.

Amaya, K., Ridha, M., & Aoki, S. (2003). Corrosion pattern detection by multi-step genetic algorithm. *Inverse Problems in Engineering Mechanics IV* (pp. 213-219).

Anthony, M. (2007). On the generalization error of fixed combinations of classifiers. *Journal of Computer and System Sciences, 73*(5), 725-734.

Arora, J. S. (2004). *Genetic algorithms for optimum design: Introduction to optimum design.* New York: John Wiley & Sons.

Bacauskiene, M., & Verikas, A. (2004). Selecting salient features for classification based on neural network committees. *Pattern Recognition Letters, 25*(16), 1879-1891.

Basu, M. (2004). An interactive fuzzy satisfying method based on evolutionary programming technique for multiobjective short-term hydrothermal scheduling. *Electric Power Systems Research, 69*(2-3), 277-285.

Bishop, C. M. (1995). *Neural networks for pattern recognition.* Oxford: Oxford University Press.

Burbidge, R., Trotter, M., Buxton, B., & Holden, S. (2001). Drug design by machine learning: support vector machines for pharmaceutical data analysis. *Computers & Chemistry, 26*(1), *5-14.*

Cai, Y-D., Liu, X-J., Xu, X-b., & Chou, K-C. (2002). Prediction of protein structural classes by support vector machines *Computers & Chemistry, 26*(3), 293 296.

Campbell, S. L., & Meyer, C. D. (1991*). Generalized inverses of linear transformations.* New York: Dover Publications.

Cao, Y. J., Jiang, L., & Wu, Q. H. (2000). An evolutionary programming approach to mixed-variable optimization problems. *Applied Mathematical Modelling, 24*(12), 931-942.

Catelani, M., & Fort, A. (2000). Fault diagnosis of electronic analog circuits using a radial basis function network classifier *Measurement, 28*(3), 147-158.

Chakrabarti, S., & Jeyasurya, B. (2007). An enhanced radial basis function Network for voltage stability monitoring considering multiple contingencies. *Electric Power Systems Research, 77*(7), 780-787.

Chandramouli, V., Brion, G., Neelakantan, T. R., & Lingireddy, S. (2007). Backfilling missing microbial concentrations in a riverine database using artificial neural networks. *Water Research, 41*(1), 217-227.

Chen, C.-H., & Lin, Z.-S. (2006). A committee machine with empirical formulas for permeability prediction. *Computers & Geosciences, 32*(4), 485-496.

Choi, D.-H. (2002). Cooperative mutation based evolutionary programming for continuous function optimization. *Operations Research Letters, 30*(3), 195-201.

Cigizoglu, H. K. (2004). Estimation and forecasting of daily suspended sediment data by multi-layer perceptrons. *Advances in Water Resources, 27*(2), 185-195.

Cox, E. (2005). *Fundamental concepts of genetic algorithms fuzzy modeling and genetic algorithms for data mining and exploration.* San Fransisco, CA: Morgan Kaufmann Publishers.

Das, A., Reddy, N. P., & Narayanan, J. (2001). Hybrid fuzzy logic committee neural networks for recognition of swallow acceleration signals. *Computer Methods and Programs in Biomedicine, 64*(2), 87-99.

Dempster, A. P, Laird, N. M., & Rubin, D. B. (1977). Maximum likelihood for incomplete data via the EM algorithm. *Journal of the Royal Statistical Society, B39*, 1-38.

Devillers, J. (1996). Genetic algorithms in computer-aided molecular design. *Genetic Algorithms in Molecular Modeling* (pp. 1-34).

Dimla, Snr, D. E. (1999). Application of perceptron neural networks to tool-state classification in a metal-turning operation *Engineering Applications of Artificial Intelligence, 12*(4), 471-477.

Dong, H., He, J., Huang, H., & Hou, W. (2007). Evolutionary programming using a mixed mutation strategy. *Information Sciences, 177*(1), 312-327.

Du, J., Zhai, C., & Wan, Y. (2007). Radial basis probabilistic neural networks committee for palmprint recognition. *Lecture Notes in Computer Science, 4492(2)*, 819-824.

Embrechts, M. J., Ozdemir, M., Lockwood, L., Breneman, C., Bennett, K., Devogelaere, D., & Rijckaert, M. (2003). Feature selection methods based on genetic algorithms for in silico drug design. In G. Fogel and D. Corne (Eds.), *Evolutionary computation in bioinformatics* (pp. 317-339). San Fransisco, CA: Morgan Kaufmann Publishers.

Fan, S-K. S., Liang, Y-C., & Zahara, E. (2006). A genetic algorithm and a particle swarm optimizer hybridized with Nelder–Mead simplex search *Computers & Industrial Engineering, 50*(4), 401-425.

Fernandes, A. M., Utkin, A. B., Lavrov, A. V., & Vilar, R. M. (2004). Development of neural network committee machines for automatic forest fire detection using lidar. *Pattern Recognition, 37*(10), 2039-2047.

Ferreira, A. J. M., Roque, C. M. C., & Jorge, R. M. N. (2005). Free vibration analysis of symmetric laminated composite plates by FSDT and radial basis functions. *Computer Methods in Applied Mechanics and Engineering, 194*(39-41), 4265-4278.

Frias-Martinez, E., Sanchez, A., & Velez, J. (2006). Support vector machines versus multi-layer perceptrons for efficient off-line signature recognition. *Engineering Applications of Artificial Intelligence, 19*(6), 693-704.

Gen, M., & Cheng, R. (1997). *Genetic algorithms and engineering design.* New York: John Wiley & Sons.

Goldberg, D. E. (1989). *Genetic algorithms in search, optimization and machine learning.* Boston, MA: Kluwer Academic Publishers.

Hagiwara, H., & Mita, A. (2002). Structural health monitoring system using support vector machine. In M. Anson, J.M. Ko, and E.S.S. Lam *(Eds.), Advances in building technology* (pp. 481-488). New York: Elsevier.

Hartigan, J. A., & Wong, M. A. (1979). A K-Means clustering algorithm. *Applied Statistics, 28*(1), 100-108.

Haykin, S. (1999). *Neural networks.* Reading, New Jersey: Prentice-Hall.

He, J. (2008) (In Press). An experimental study on the self-adaption mechanism used by evolutionary programming. *Progress in Natural Science.*

Huang, T. M., & Kecman, V. (2005). Gene extraction for cancer diagnosis by support vector machines—An improvement. *Artificial Intelligence in Medicine, 35*(1-2), 185-194.

Hurtado, J. E. (2003). Relevance of support vector machines for stochastic mechanics. *Computational Fluid and Solid Mechanics* (pp. 2298-2301).

Jacob, C. (2001). *Illustrating evolutionary computation with Mathematica.* San Fransisco, CA: Morgan Kaufmann Publishers.

Joachims, J. (1999). Making large-scale SVM learning practical. In B. Scholkopf, C.J.C. Burges and A.J. Smola (Eds.), *Advances in Kernel Methods-Support Vector Learning* (pp. 169-184). Cambridge, MA: MIT Press.

Junninen, H., Niska, H., Tuppurainen, K., Ruuskanen, J., & Kolehmainen, M. (2004). Methods for imputation of missing values in air quality data sets. *Atmospheric Environment, 38*(18), 2895-2907.

Kao, Y-T., & Zahara, E. (2008). A hybrid genetic algorithm and particle swarm optimization for multimodal functions. *Applied Soft Computing, 8*(2), 849-857.

Kennedy, J., & Eberhart, R. (1995). Particle swarm optimization. In *Proceedings of the IEEE International Conference on Neural Networks,* Perth, Australia (pp. 1942-1948).

Ko, K. T., Tang, K. S., Chan, C. Y., Man, K. F., & Kwong, S. (1997). Using genetic algorithms to design mesh networks. *Computer, 30*(8), 56–61.

Linden, D. S. (2002). Antenna design using genetic algorithms. In *Proceedings of the 2002 Genetic and Evolutionary Computation Conference* (pp. 1133 – 1140).

Lo, S. (2008). Web service quality control based on text mining using support vector machine. *Expert Systems with Applications, 34*(1), 603-610.

Magoulès, F., Diago, L. A., & Hagiwara, I. (2007). Efficient preconditioning for image reconstruction with radial basis functions. *Advances in Engineering Software, 38*(5), 320-327.

Mahachi, J., & Dundu, M. (2001). Genetic algorithm operators for optimisation of pin jointed structures. In A. Zingoni (Ed.), *Proceedings of the international conference on structural engineering, mechanics and computation* (pp. 1155-1162). Cape Town: Elsevier.

Marivate, V. N., Nelwamondo, F. V., & Marwala, T. (2008) Investigation into the use of autoencoder neural networks, principal component analysis and support vector regression in estimating missing HIV data. In *Proceedings of the World Congress of the 2008 IFAC,* Seoul, South Korea (pp. 682-689).

Marwala, T. (2000). On damage identification using a committee of neural networks. *American Society of Civil Engineers, Journal of Engineering Mechanics, 126,* 43-50.

Marwala, T. (2001a). Probabilistic fault identification using a committee of neural networks and vibration data. *Journal of Aircraft, 38,* 138-146.

Marwala, T. (2001b). *Fault identification using of neural networks and vibration data.* Unpublished doctoral dissertation, University of Cambridge, Cambridge.

Marwala, T. (2005). The artificial beer taster. *Electricity+Control, 3,* 22-23.

Marwala, T. (2007). *Computational intelligence for modelling complex systems.* New Delhi: Research India Publications.

Marwala, T., de Wilde, P., Correia, L., Mariano, P., Ribeiro, R., Abramov, V., Szirbik, N., & Goossenaerts, J. (2001). *Scalability and optimisation of a committee of agents using genetic algorithm.* Paper presented at the International Symposia on Soft Computing and Intelligent Systems for Industry, Parsley, Scotland.

Marwala, T., & Hunt, H. E. M. (1999). Fault identification using finite element models and neural networks. *Mechanical Systems and Signal Processing, 13*(3), 475–490.

Metropolis, N., Rosenbluth, A.W., Rosenbluth, M.N., Teller, A.H., and Teller, E. (1953). Equations of state calculations by fast computing machines. *Journal of Chemical Physics, 21,* 1087-1092.

Min, J. H., & Lee, Y-C. (2005). Bankruptcy prediction using support vector machine with optimal choice of kernel function parameters. *Expert Systems with Applications, 28*(4), 603-614.

Mitra, P., Shankar, B. U., & Pal, S. K. (2004). Segmentation of multispectral remote sensing images using active support vector machines. *Pattern Recognition Letters, 25*(9), 1067-1074.

Møller, A. F. (1993). A scaled conjugate gradient algorithm for fast supervised learning. *Neural Networks, 6,* 525-533.

Moravej, Z., Vishwakarma, D. N., & Singh, S. P. (2003). Application of radial basis function neural network for differential relaying of a power transformer. *Computers & Electrical Engineering, 29*(3), 421-434.

Nabney, I. T. (2001). *Netlab: Algorithms for pattern recognition.* Heidelberg: Springer-Verlag.

Nelwamondo, F. V. (2008). *Computational intelligence techniques for missing data imputation.* Unpublished doctoral dissertation, University of the Witwatersrand, Johannesburg.

Niska, H., Rantamäki, M., Hiltunen, T., Karppinen, A., Kukkonen, J., Ruuskanen, J., & Kolehmainen, M. (2005). Evaluation of an integrated modelling system containing a multi-layer perceptron model and the numerical weather prediction model HIRLAM for the forecasting of urban airborne pollutant concentrations. *Atmospheric Environment, 39*(35), 6524-6536.

Nocedal, J., & Wright, S. (1999). *Numerical optimization.* Heidelberg: Springer-Verlag.

Pandit, M., Srivastava, L., & Sharma, J. (2001). Voltage contingency ranking using fuzzified multilayer perceptron. *Electric Power Systems Research, 59*(1), 65-73.

Pelckmans, K., De Brabanter, J., Suykens, J. A. K., & De Moor, B. (2005). Handling missing values in support vector machine classifiers. *Neural Networks, 18*(5-6), 684-692.

Pereira, C. M. N., & Lapa, C. M. F. (2003). Coarse-grained parallel genetic algorithm applied to a nuclear reactor core design optimization problem. *Annals of Nuclear Energy, 30*(5), 555-565.

Perrone, M. P., & Cooper, L. N. (Ed.). (1993). When networks disagree: Ensemble methods for hybrid neural networks. In RJ Mammone (Ed.), *Artificial neural networks for speech and vision* (pp. 126–142). London: Chapman and Hall.

Pizzi, N. J., Somorjai, R. L., & Pedrycz, W. (2006). Classifying biomedical spectra using stochastic feature selection and parallelized multi-layer perceptrons. In B. Bouchon-Meunier, Giulianella Coletti, and Ronald Yager (Eds.), *Modern information processing* (pp. 383-393). London: Elesevier.

Rajan, C. C. A., & Mohan, M. R. (2007). An evolutionary programming based simulated annealing method for solving the unit commitment problem. *International Journal of Electrical Power & Energy Systems, 29*(7), 540-550.

Rauch, W., & Harremoës, P. (1999). On the potential of genetic algorithms in urban drainage modeling *Urban Water, 1*(1), 79-89.

Reddy, N. P., & Buch, O. A. (2003). Speaker verification using committee neural networks. *Computer Methods and Programs in Biomedicine, 72*(2), 109-115.

Rossi, F., & Conan-Guez, B. (2005). Functional multi-layer perceptron: a non-linear tool for functional data analysis *Neural Networks, 18*(1), 45-60.

Saridakis, K. M., Chasalevris, A. C., Dentsoras, A. J., & Papadopoulos, C. A. (2006). *Fusing neural networks, genetic algorithms and fuzzy logic for diagnosis of cracks in shafts.* Paper presented at the Virtual International Conference on Intelligent Production Machines and Systems.

Schölkopf, B. (2003). An introduction to support vector machines. In M.G. Akritas and D.N. Politis (Eds.), *Recent advances and trends in nonparametric statistics* (pp. 3-17). New York: JAI Press.

Schölkopf, B., Burges, C. J. C., & Smola, A. J. (1999). *Advances in Kernel Methods: Support Vector Learning.* Cambridge, MA: MIT Press

Sheikh-Ahmad, J., Twomey, J., Kalla, D., & Lodhia, P. (2007). Multiple regression and committee neural network force prediction models in milling FRP. *Machining Science and Technology, 11*(3), 391-412.

Shi, X. H., Liang, Y. C., Lee, H. P., Lu, C., & Wang, L. M. (2005). An improved GA and a novel HGAPSO-based hybrid algorithm. *Information Processing Letters, 93*(5), 255-261.

Shi, L., & Xu, G. (2001). Self-adaptive evolutionary programming and its application to multi-objective optimal operation of power systems. *Electric Power Systems Research, 57*(3), 181-187.

Smola, A. J., & Schölkopf, B. (2004). A tutorial on support vector regression. *Journal of Statistics and Computing, 14*(3), 1573-1375.

Tripathi, S., Srinivas, V. V., & Nanjundiah, R. S. (2006). Downscaling of precipitation for climate change scenarios: A support vector machine approach. *Journal of Hydrology, 330*(3-4), 621-640.

Vasudevan, M., Bhaduri, A. K., & Raj, B. (2007). Prediction of ferrite number in stainless steel welds using Bayesian Neural Network model. *Welding in the World, 51*(7-8), 15-28.

Walde, J. F. (2007). Valid hypothesis testing in face of spatially dependent data using multi-layer perceptrons and sub-sampling techniques. *Computational Statistics & Data Analysis, 51*(5), 2701-2719.

Zhao, X., Gao, X-S., & Hu, Z-C. (2007) Evolutionary programming based on non-uniform mutation. *Applied Mathematics and Computation, 192*(1), 1-11.

Zhao, Z-Q., Huang, D-S., & Sun, B-Y. (2004). Human face recognition based on multi-features using neural networks committee *Pattern Recognition Letters, 25*(12), 1351-1358.

Zhong, M., Lingras, P., & Sharma, S. (2004). Estimation of missing traffic counts using factor, genetic, neural, and regression techniques. *Transportation Research Part C: Emerging Technologies, 12*(2), 139-166.

Chapter VIII
Online Approaches to Missing Data Estimation

ABSTRACT

The use of inferential sensors is a common task for online fault detection in various control applications. A problem arises when sensors fail when the system is designed to make a decision based on the data from those sensors. Various techniques to handle missing data are discussed in this chapter. First, a novel algorithm that classifies and regresses in the presence of missing data online is presented. The algorithm was tested for using both classification and regression problems and was compared to an off-line trained method that used auto-associative networks as well as a Hybrid Genetic Algorithm (HGA) method and a Fast Simulated Annealing (FSA) technique. The results showed that the presented methods performed well for online missing data estimation. Second, an online estimation algorithm that uses an ensemble of multi-layer perceptron regressors, HGA and FSA and genetic programming is presented for missing data estimation and compared with a similar procedure that was trained off-line.

INTRODUCTION

Fault detection is one of the most active research areas with several applications. There are a number of challenges faced by online detection and identification systems (Marwala, 2001; Benitez-Perez, Garcia-Nocetti, & Thompson, 2007; Vilakazi & Marwala, 2006; Vilakazi & Marwala, 2007a&b; Marwala, 2007; Hulley & Marwala, 2007). Marwala and Hunt (2000) proposed pseudo modal energies for fault detection in structures. The problem with the method they proposed is that for its successful implementation it had to be continuously retrained and this process required human intervention (Marwala, 2001). One of the biggest problems hindering the performance of online condition monitoring is in dealing with missing data (Abdella & Marwala, 2005). Missing data problems often arise due to sensor failure in online condition monitoring systems. For example, Dhlamini, Nelwamondo, and Marwala (2006) implemented sensor failure compensation techniques for high voltage bushing monitoring using evolutionary computing. This chapter investigates a problem of condition monitoring where fault detection is confounded by missing data. The biggest challenge is that standard neural networks cannot process input data with missing values and hence, cannot perform classification or regression when some input data are missing. More often, there is a limited time between the readings depending on how frequently the sensor is sampled. As a result, missing data problem becomes a huge obstacle in deciding the condition of the machine being monitored from the limited time between readings. For both classification and regression, all decisions concerning how to proceed must be taken during this finite period.

Three general ways have been used to deal with the problem of missing data (Little and Rubin, 1987). The simplest method is known as *listwise deletion*. This method simply deletes instances with missing values (Little & Rubin, 1987). Even though this approach to the problem can be valid for surveys, the major disadvantage of this method is the dramatic loss of information in datasets (Kim & Curry, 1997). Another disadvantage is that it assumes that the observation with missing values is unimportant and can be ignored. For engineering problems where the observations are required for automated decision-making, this option is not viable. The second common technique *imputes* the data by finding estimates of the values and replaces missing entries with these estimates. Various estimates have been used and these estimates include zeros, means and other statistical calculations. These estimates are then used as if they were the observed values. This procedure is valid for engineering problems but the problem is that it results in bad decisions. Another common technique assumes some model for the prediction of the missing values and uses the maximum likelihood approach to estimate the missing values (Little & Rubin, 1987; Nelwamondo, Mohamed, & Marwala, 2007).

Much research has been done to find new ways of approximating missing values. Among others, Abdella and Marwala (2006) as well as Mohamed and Marwala (2005) used neural networks together with Genetic Algorithms (GA) to approximate missing data. Qiao, Gao, and Harley (2005) used neural networks and Particle Swam Optimization (PSO) to keep track of the dynamics of a power plant in the presence of missing data. In the aforementioned examples, auto-associative neural networks were used together with GA or PSO to predict the missing values and to optimize the prediction to be as accurate as possible. On the other hand, Yu and Kobayashi (2003) used semi-hidden Markov models to predict missing data in mobility tracking whereas Huang and Zhu (2002) used a pseudo-nearest-neighbor approach for missing data recovery of Gaussian random data sets. Nauck and Kruse (1999) and Gabrys (2002) have also used neuro fuzzy techniques in the presence of missing data. A different approach was taken by Wang (2005) who replaced incomplete patterns with fuzzy patterns. Along with fuzzy patterns, the patterns without missing values were used to train the neural network. In Wang's model, the neural network learns to classify without actually predicting the missing data.

This chapter introduces techniques to handle missing data in an online system. Online systems require continuous learning. Many techniques have been proposed to deal with this problem. Lunga and Marwala (2006) proposed an ensemble of classifiers for online classification of a time series. Mohamed, Marwala, and Rubin (2007) applied online learning for adaptive protein classification. An approach where no attempt was made to recover the missing values was also presented. In this approach, both classification and regression techniques were considered. For classification, an approach that uses an ensemble of classifiers, also called a *committee* as in Chapter VII, to classify even in the presence of missing data was presented. Unlike Wang (2005), it did not attempt to replace missing patterns with anything. Instead, the network was trained with various subsets of available data. The algorithm was further extended to a regression application where Multi-Layer Perceptrons (MLP) were used in an attempt to get the correct output with limited input variables.

The last part of this chapter extends the technique used by Dhlamini, Nelwamondo, and Marwala (2006) for the classification of faults in transformer bushings in the presence of missing data. In Chapter VII, the committee of networks approach is presented for missing data estimation. In this chapter, an ensemble (committee) of neural networks to approximate missing variables is used to address the problem of online missing data estimation. In the techniques presented in this chapter, hybrid GA and Fast Simulated Annealing (FSA) are used to compensate for missing data online. In addition, the fuzzy ARTMAP is also presented to deal with online missing data estimation. Fuzzy ARTMAP and neural networks have been successfully used in the past for dealing with online missing data estimation (Nelwamondo & Marwala, 2007&2008). There are two crucial aspects of online missing data estimation. These are the speed and adaptation. This chapter addresses both these factors.

MISSING DATA

In this chapter, a number of methods are presented to deal with missing data in systems that operate online. One way of achieving online learning is to use the Fuzzy ARTMAP, which is known for its online learning capability. Another way is to use Bayesian neural networks as described in Chapter III, updating the prior distribution to facilitate online learning. One method trains networks with different input variables thereby assuming that one or two variables are missing. To achieve this, it is important to know which networks are likely to be missing. Finally, a committee or ensemble of networks is constructed to estimate missing data estimation and the online capability is facilitated using the evolutionary programming method described in detail in Chapter VII.

It is difficult to tell which sensor will fail first or what will cause the failure. However, sensor manufacturers often state specifications such as Mean-Time Between-Failures (MTBF) and Mean-Time-To-Failure (MTTF), which can help in detecting which sensors are more likely to fail than other sensors. MTTF is used in cases where a sensor is replaced after a failure, whereas MTBF denotes the time between failures, in cases where the sensor is repaired. There is, nevertheless, no guarantee that failures will follow manufacturers' specifications. Missing values from sensors will generally follow a structured failure process, where only values from one or two sensors will be missing. An example of this kind of missing data is shown in Table 8.1 and will be considered throughout this chapter.

BACKGROUND: NEURAL NETWORKS

Approaches presented in this chapter make use of neural networks. In this chapter, neural networks are viewed as systems that learn the complex input-output relationship from the data. The training

Table 8.1. Sample missing values from sensors

F_1	F_2	...	F_{n-1}	F_n
3.6	2.05	...	9.6	0.03
4.5	6.59	...	0.03	?
3.9	4.57	...	0.02	?
1.8	?	...	0.02	?
2.0	?	...	0.1	0.03
6.8	?	...	0.9	0.02

process of neural networks involves presenting the network with inputs and corresponding outputs and this process is termed *supervised learning*. There are various types of neural networks. In this chapter, only the Multi-Layer Perceptron (MLP) and the Fuzzy ARTMAP will be discussed in preparation for the rest of this chapter.

The Fuzzy ARTMAP

The *Fuzzy ARTMAP* is a neural network architecture developed by Carpenter et al. (1992) which has the structure shown in Figure 8.1. This architecture is based on Adaptive Resonance Theory (ART) that is capable of supervised learning of an arbitrary mapping between clusters of the input space and their associated class labels. Fuzzy ARTMAP has been successfully used by Lopes, Minussi, and Lotufo (2005) for electricity load forecasting while Tan, Rao, and Lim (2008) used fuzzy ARTMAP for conflict resolution. The Fuzzy ARTMAP has been used in condition monitoring by Javadpour and Knapp (2003). However, their application was not online. The Fuzzy ARTMAP architecture used in this chapter is composed of two ART modules (ART$_a$ and ART$_b$) that create stable recognition categories in response to sequences of the input pattern. The two ART modules are inter-connected by a series of weighted connections between the F_2 layers as shown in Figure 8.1.

The connection between the two ART modules forms the Map Field (Carpenter et al., 1992). Various parameters need to be set for the training process and this is achieved in the ART Network. The vigilance parameter, $\rho[0, 1]$ is the only user-specified parameter. The vigilance parameter controls the *network resonance*. The second parameter that is adjusted is the *training rate $\beta[0, 1]$* which controls the *adaptation speed*, where 0 implies a slow speed and 1 is the fastest. This adaptation process facilitates

Figure 8.1. The Fuzzy ARTMAP architecture

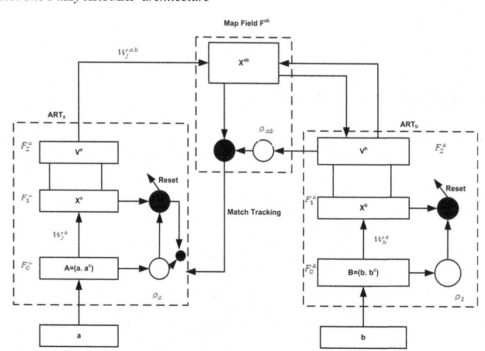

online learning. The variable α acts as a parameter to decide the category class and is always $0 < \alpha < 1$. During supervised learning, an analog signal vector $X=(a, a^c)$ is input to the ART_a input layer F_0^a in the form of a complement code. Both the original input a, and its complement a^c are presented to the fuzzy ARTMAP network as explained by Javadpour and Knapp (2003). Each component in the input vector corresponds to a single node in F_1^a. The main function of this F_1^a block is to compare the hypothesized class propagated from the F_2^a block to the input signal. Simultaneously, a vector b is presented to the ART_b module. This vector contains the desired outputs corresponding to the vector a in the ART_a module. The network then uses some hypothesis testing to deduce to which category the input pattern should belong. Mapping is done in two steps. First, the ART_a module allows data to be clustered into categories that are mapped to a class in the ART_b side of the module. The Fuzzy ARTMAP map-field then maps the data cluster in the 'A-side' to the label cluster in the 'B-side'. During the learning process, each template from the 'A-side' is mapped to one template on the 'B-Side', ensuring a many-to-one mapping. The weights w_{jk}^{AB} are used to control the association between the F_2 nodes on both sides. When vectors a and b are presented to ART_a and ART_b, respectively, both models soon enter resonance.

When the vigilance criterion is respected, the map field learns the association between vectors a and b by modifying its weights following the initial weight. A fuzzy ARTMAP has an internal controller that ensures autonomous system operation in real time, thereby facilitating online learning. The inter-ART module has a self-regulatory mechanism named *match tracking*, whose objective is to maximize the generalization and minimize the network error. A complete description of the Fuzzy ARTMAP is provided by Carpenter et al. (1992). The *vigilance criterion* is given by:

$$\frac{\left| y^b \wedge w_{JK}^{ab} \right|}{y_b} \geq \rho_{ab} \tag{8.1}$$

where ρ_{ab} is the *vigilance parameter*, and w_{JK}^{ab} represents the model's weight as presented in Figure 8.1. The vigilance criterion is reached by making a small increment on the vigilance parameter such that a certain category is excluded. This is done until the instant that the active category corresponds to the desired output. After the input has completed the resonance state by the vigilance criterion, the weight adaptation is accomplished. The adaptation of the ART_a and ART_b module weights is given by:

$$w_j^{new} \beta (I \wedge w_j^{old}) + (1-\beta) w_j^{old} \tag{8.2}$$

In this chapter, the Fuzzy ARTMAP is used to predict a class given inputs that are not complete. This is because the Fuzzy ARTMAP is trained using different configuration of inputs.

Multilayered Perceptron (MLP)

As described before, the MLP is a feed-forward neural network with an architecture consisting of the input layer and the output layer. Compared to the Fuzzy ARTMAP, the MLP is a relatively simple machine. On its own, it does not have any online learning capabilities. Each layer is formed from smaller units known as *neurons*. Neurons in the input layer receive the input signals and distribute them forward to the network. In the next layers, each neuron receives a signal, which is a weighted sum of the outputs of the nodes in the layer below. Inside each neuron, an *activation function* is used to control the output thereof. Such a network determines a nonlinear mapping from an input vector to the output vector, parameterized by a set of network weights, which are referred to as the *vector of weig*hts. Details of the MLP were given in Chapter II.

The first step in approximating the weight parameters of the model is to find the approximate architecture of the MLP, where the architecture is characterized by the number of hidden units, the type of activation function, as well as the number of input and output variables. The second step estimates the weight parameters using the training set (Japkowicz, 2002; Mi & Takeda, 2007). The training process estimates the network weights vector to ensure that the output is as close as possible to the target vector. The problem of identifying the weights in the hidden layers is solved in this chapter by maximizing the probability of the weight parameter given the measured data using Bayes' rule (Marwala, 2007) as in Chapter III. More information on neural networks can be found in (Bishop, 1995). The training of the neural network using the Bayesian approach is conducted in this chapter by using the hybrid Monte Carlo method, also described in detail in Chapter III. The training of the neural network using the Bayesian framework is advantageous particularly in this chapter where the question of online missing data estimation had to be dealt with. This is because the Bayesian framework can facilitate the online functionality by continuously adjusting the prior probability distribution. As new information is observed in an online fashion, the prior probability is adjusted and a number of Monte Carlo simulation steps are taken to simulate a number of new states that incorporate the new information and an equal number of states from the initial states are removed. This process of continuously adjusting the prior and time windowing is therefore able to facilitate online learning.

Auto-Associative Neural Networks

As described in earlier chapters, auto-associative neural networks are simply neural networks that are trained to recall their own inputs (Thompson, Marks, & El-Sharkawi, 2003). To achieve this, auto-associative neural networks have a number of inputs equal to the number of outputs, and the number of hidden units is less than the number of input units. Auto-associative neural networks have been used in various applications, including in the treatment of missing data by a number of researchers (Frolov et al., 1995; Lu & Hsu, 2002). More details on auto-associative networks were given in the previous chapters. In this chapter, trained auto-associative networks are adjusted to be re-trained online through continuously adjusting the priors to facilitate online learning.

MISSING DATA ESTIMATION TECHNIQUES

As described in earlier chapters, a Genetic Algorithm (GA) is an iterative heuristic deriving its operation from biology. The GA mimics the natural selection of species in an environment where only the fittest species can survive. The GA creates a set of possible solutions that can be viewed as a population of organisms (Goldberg, 1989; Shtub, LeBlanc, & Cai, 1996; Davis, 1991; Holland, 1975).

Hybrid Genetic Algorithms (HGA)

In this chapter, the HGA is used and it combines the GA with the local gradient descent optimizer (Wang & Wang, 2005). The gradient descent method was described in Chapter II. The traditional GA is known to excel at finding the space near the global optimum, but not necessarily finding the global optimum.

In the case of the HGA, the gradient descent local optimizer took over when the GA converges on a solution (Davis, 1991) and this helps to fine-tune the solution (McGookin & Murray-Smith, 2006). In the traditional GA, a chromosome in the current generation is selected into the next generation with a certain probability. The best chromosomes of the current generation may be lost due to mutation, cross-over or selection during the evolving process, and subsequently cause difficulty in reaching convergence. Therefore, it may take more generations and hence more running time, to get quality solutions. To avoid this, an *elitism method* proposed by Tamaki, Kita, and Kobayahi (1996) which permits chromosomes with the best fitness to survive and to be carried into the next generation is used in this chapter. The Hybrid genetic algorithm has been successfully used in many applications such as flow-shop scheduling with limited buffers (Wang, Zhang, & Zheng, 2006), a vehicle routing problem (Jeon, Leep, & Shim, 2007) and in digital finite impulse response studies (Cen, 2007).

In this chapter, the auto-associative networks are continually adjusted to facilitate online learning by adjusting the priors in conjunction with a hybrid genetic algorithm for missing data estimation. On implementing HGA, simple crossover, binary mutation, roulette reproduction and conjugate gradient methods are used.

Fast Simulated Annealing (FSA)

In contrast to the HGA, *Simulated Annealing* (SA) does not maintain a population of trial solutions. Instead, it generates a trajectory through the search space by making incremental changes to a single set of parameters. SA is a stochastic relaxation technique that has its origins in statistical mechanics. It is shown in Figure 8.2.

The SA was first developed by Kirkpatrick, Gelatt, and Vecchi (1983) as a local search algorithm following the initial algorithm proposed by (Metropolis et al., 1953). The SA is a probabilistic, hill-climbing technique that is based on the annealing process of metals (Tamaki, Kita, & Kobayahi, 1996). The annealing process occurs after the heat source is removed from a molten metal. As a result, the temperature starts to decrease. The metal becomes more rigid with this decrease in temperature. This decrease in temperature continues until the temperature of the metal is equal to that of the surroundings. It is at this temperature that the metal is perfectly solid (Tamaki, Kita, & Kobayahi, 1996).

Like most hill-climbing search techniques, SA searches the space by piece-wise perturbations of parameters that are being optimized (Tamaki, Kita, & Kobayahi, 1996). These perturbations depend on the temperature T, which decreases with every iteration of the search. Due to this, the perturbations are larger at the beginning of the search and they become smaller towards the end of the search.

At each iteration step, the cost is evaluated. If the cost value is lower than the previous one, the previous parameter is replaced by the new parameter. Should the cost function be negative (down hill), the new parameter is accepted. However, if the new cost is higher than the previous one (up hill), the cost gets subjected to a probability check where the probability, P of the new parameters cost C_{new} relative to the previous best cost C_{prev} is calculated using the Boltzmann's equation as follows (Tamaki, Kita, & Kobayahi, 1996):

$$P = \exp\left(\frac{c_{prev} - c_{new}}{T}\right)$$

(8.3)

Figure 8.2. Flowchart showing the simulated annealing algorithm

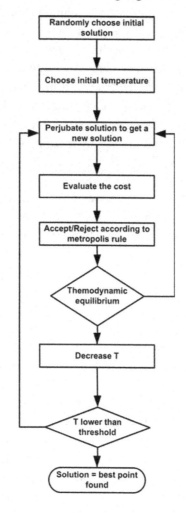

This probability is compared to a certain threshold that is randomly sampled to fall in the interval [0 1] and *P* is only accepted if it is above that threshold. This process is known as the Metropolis criterion and is used to control the acceptance probability of every step in the search process (Nascimento et al., 2001).

Recently, some techniques have been designed to improve the convergence rates of optimizations techniques. The FSA is simply an improvement to the popular SA algorithm. Many techniques of making this process faster have been proposed in the literature (Chen & Chang, 2006). The technique used in this chapter combine hill-climbing with random distribution and this improved the annealing speed as random moves are semi-local (Nascimento et al., 2001). Fast simulated annealing has been successfully used in the estimation of time delay in brainstem auditory evoked potentials (Cherrid, Naït-Ali, & Siarry, 2005) and in studying permanent prostate implants (Martin et al., 2007). In this chapter, the auto-associative networks are used in conjunction with the FSA for missing data estimation.

Proposed Approaches

First, this section presents a technique of dealing with missing data without actual prediction of the missing variables. A technique that uses an ensemble or committee of auto-associative networks to approximate the missing data is also presented.

Online Classification in the Presence of Missing Data

The algorithm presented in this chapter is motivated by the incremental learning approach proposed by Polikar et al. (2001) and improved by Hulley and Marwala (2007). In the algorithm presented here, an ensemble of Fuzzy ARTMAP classifiers is used because of its online abilities. Online training is performed and testing is done online. Training is achieved using a number of Fuzzy ARTMAP networks, each trained with a different combination of input features with one or two variables missing. Fuzzy ARTMAP classifiers are preferred for their incremental learning ability. The presented algorithm can work with any classifier.

For a monitoring system that contains n sensors, the user has to state how many features are most likely to be available at any given time. Such information can be deduced from the reliability of the sensors, which is quantified in terms of Mean Time Between Failures (MTBF). When the number of sensors or features that are most likely to be available has been determined, the number of all possible classifiers can be calculated using:

$$C = \binom{n}{n_{avail}} = \left(\frac{n!}{n_{avail}!(n-n_{avail})!} \right) \tag{8.4}$$

where C is the total number of all possible classifiers, n is the total number of features and n_{avail} is the number of features most likely to be available at any time. Each classifier is then trained with n_{avail} features. Although the number n_{avail} can be statistically calculated, it has an effect on the number of networks that can be available. Let us consider a simple example where the input space has 5 features, labeled, a, b, c, d and e and there are 3 features that are most likely to be available at any time. Using equation 8.4, C is found to be 10. These classifiers are trained with features [abc, abd, abe, acd, ace, ade, bcd, bce, bde, cde] as inputs. In a case where one variable is missing, say, a, only four networks can be used. These are the classifiers that do not use a in their training input sequence and are [bcd, bce, bde, cde]. If a situation where two variables are missing, say a and b, is encountered, then one classifier with input feature cde remains. Each network is trained with n_{avail} features. Therefore, for the above example, 10 networks are trained. The validation process is then conducted and the outcome is used to decide on the combination scheme. For example in the above problem if feature a is missing, then the estimated output is constructed by a committee of a weighted combination of networks with features [bcd, bce, bde]. The training process requires complete data to be available as initial training is done off-line. The available data set is divided into the 'training set' and the 'validation set'. Each network created is tested on the validation set and is assigned a relative weight according to its performance on the validation set. A diagrammatic illustration of the proposed ensemble approach is presented in Figure 8.3. After all possible networks have been trained, each classifier performance is determined based on the prediction on the validation set.

In Figure 8.3, different networks are trained with different sets of combination of inputs. Then each of these networks is used to contribute in a weighted fashion to the final output. The weightings are obtained by ensuring that each network is assigned a weight according to its performance on the validation set. The weight is assigned using the weighted majority scheme as given by (Merz, 1997; Opitz & Shavlik, 1996):

$$\alpha_i = \frac{1 - \varepsilon_i}{\sum_{j=1}^{N} (1 - \varepsilon_i)} \qquad (8.5)$$

where ε_i is the estimate of model i's error on the validation set. In Figure 8.3, if one of the inputs is not available, then the combined output is the weighted average of the networks, which are created without that particular input variable.

This kind of weight assignment intuitively ensures that model i gets more weight if its performance is higher than the performance of other models. The objective here is to have a set of models that are likely to have uncorrelated errors (Merz, 1997; Freund and Schapire 1995). This method has its roots in what is called *boosting* as explained in Merz (1997) and which is explained in detail in Chapter VII. Boosting is based on the fact that a set of networks which produce varying results can be combined to produce better results than each individual network that has been used in the set. The training pseudo-code is presented in Figure 8.4.

The next stage that follows the training is the testing process. It is at this stage that the parameters obtained in the training stage are used for the online detection or classification of defects. The pseudo-code in Figure 8.5 summarizes the testing phase. It should be noted that testing is done online and as a result, is done one instance at a time. In Figure 8.5, the first stage is to determine which networks are relevant. For example, in our earlier case, if feature a is missing then the relevant networks are [*bcd*, *bce*, *bde*]. Then each of these networks gives its predicted class. Based on its individual performance, calculated using equation 8.5, the networks are combined to form a joint class prediction.

In this section, the online aspect is facilitated by using a Fuzzy ARTMAP, which is ideally suited for this task.

Figure 8.3. The ensemble based approach for missing data

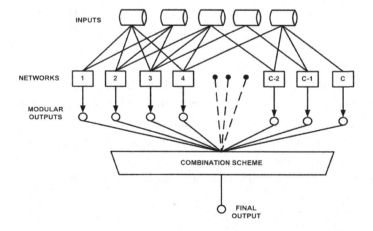

Figure 8.4. Pseudo-code for the training algorithm for classification

> **TRAINING PHASE**:
> User specified parameters: n_{avail}
> **BEGIN** Training 1.
>   1. Calculate the number of maximum networks C required using equation 8.4.
>   2. Create all networks each with n_{avail} inputs.
> **DO** for all classifiers:
>   3. Train network with a subset of a different combination of n_{avail} features.
>   4. Subject each classifier to a validation set as follows:
>   (a) Select only those features with which a particular network has been trained.
>   (b) Obtain a classification performance for each classifier on the validation set.
>   (c) Store the performance of each network for use in the ensemble creation.
> **END** loop
>   5. Assign weights to each classifier according to performance and using equation 8.5.
>   6. Store all weights and other relevant parameters for future use.
> **END** training

Figure 8.5. Pseudo-code for the testing algorithm for classification

> **TESTING PHASE**:
> **Parameters**: Load all parameters from the training phase.
> For each instance with missing data select relevant network(s)
>   1. Classify using the networks.
>   2. Use the performance of each classifier from equation 8.5 from the training phase to create a weighted combination of the networks.
>   3. Use the networks and the relative weights to come up with the class.

Regression in the Presence of Missing Data

The technique used in this section is very similar to the technique proposed above. The proposed training and testing algorithms are presented in Figures 8.6 and 8.7, respectively. This technique is designed for a regression problem where inputs are used to determine the output value of a system. A standard implementation of MLP is used to generate regressors that attempt to map inputs to the output. The objective here is to show that the proposed algorithm can work with any kind of network that is available.

All networks used here consist of an input layer, single hidden layer and an output layer. An ensemble of MLP regressors that sets the number of inputs to be equal to n_{avail} is created. In doing this, all networks are trained to give the same output value using different subsets of features. Each regressor is then tested on a validation set as is done in a classification condition, and the error is computed for each data set. It should be noted that when this is done, each regressor is presented only with those features with which it has been trained.

As in classification, each network is assigned a relative weight according to its performance on the validation set. A population of regression estimates used to estimate a function y needs to be combined and this was explained in detail in Chapter VII. It is desired in this work that a network that generally offers better accuracy must be given more significance and hence must be accounted for more that a weaker regressor. The final predicted value is computed as follows (Perrone & Cooper, 1993):

Figure 8.6. Pseudo-code for the training algorithm for regression MSE (Mean Square Error)

TRAINING PHASE:
User specified parameters: n_{avail}
BEGIN Training
   Calculate number of maximum networks C required using equation 8.4.
   Create all networks each with inputs n_{avail}
DO for all regressors:
   Train the network with a subset combination of n_{avail} features.
   Subject each regressor to a validation set as follows:
Select only those features with which a particular network has been trained.
Obtain a regression performance for each regressor using the MSE on the validation set.
Store the performance
END loop
   Assign α to each classifier, according to its performance, using equation 8.5.
   Store all weights and other parameters for future use.
END training.

Figure 8.7. The presented testing algorithm for regression

TESTING PHASE:
Parameters:
Load all parameters from the training phase.
For each instance with missing data:
   Get regression estimates from all networks.
   Use their weights, to compute the final value.
END testing.

$$f(x) = y = \sum_{i=1}^{N} \alpha_i f_i(x)$$

(8.6)

where α is the weight assigned during the validation stage and N is the total number of regressors. The algorithm used for this task takes a very similar approach to the one used for classification. The MLPs used in this section are trained as is in Chapter III, using a Bayesian framework. As it can be recalled, the Bayesian framework trains the network by estimating the posterior probability function of the network weights given the data from the likelihood function, the prior probability and the evidence. In this chapter, the online aspects are facilitated by updating the prior as new data come and performing a number of hybrid Monte Carlo simulation (Marwala, 2007) steps n, to factor into account recent data and thereby creating m additional states. Then the earliest m states are removed to retain the size of states that form a posterior probability distribution.

Ensemble of Networks (Committee) to Approximate Missing Data

The method presented in this section has also been derived from the adaptive boosting method (Freund & Schapire, 1995). This algorithm uses an ensemble of autoassociative networks to estimate the missing values and uses a technique of rewarding or penalizing a particular network using the known performance of the network. A similar approach was used by Perrone and Cooper (1993) as well as Mohamed and Marwala (2005) where they showed that a committee of networks could optimize the decision as described in Chapter VII. In this section, an ensemble of networks committee is used to estimate the values of the missing data. As the data are observed online, the committee is evolved as was done in Chapter VII. This evolutionary computing based committee method uses an ensemble of auto-associative networks in a configuration shown in Figure 8.8.

As the committee approach was used, fast simulated annealing and hybrid genetic algorithms are each used to obtain missing data estimates and their results are compared. Each optimizer (FSA and Hybrid GA) tries to minimize the missing data-estimation error equation, described in earlier chapters, which can be written as follows:

$$\varepsilon = \left\| \left(\begin{Bmatrix} \{x_k\} \\ \{x_u\} \end{Bmatrix} - f\left(\{W\}, \begin{Bmatrix} \{x_k\} \\ \{x_u\} \end{Bmatrix} \right) \right) \right\| \tag{8.7}$$

Here $\| \; \|$ is Euclidean norm, $\{x_k\}$ is the known (observed) data vector, $\{x_u\}$ is the unknown data vector to be estimated and f is the auto-associative network, which in this chapter is the multi-layer perceptron with mapping weight vector $\{w\}$.

Figure 8.8. An ensemble of auto-associative networks (also called auto-encoders) to approximate missing data through the use of a GA or an SA

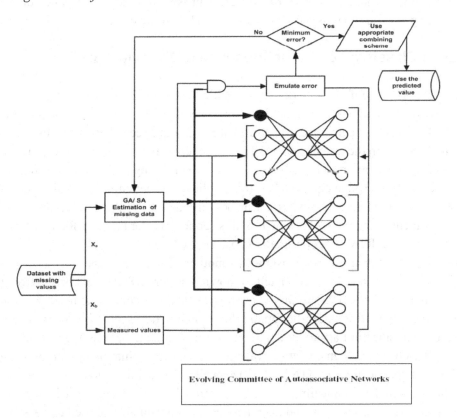

In this chapter, the auto-associative networks are evolved by adopting genetic programming and Monte Carlo methods (Metropolis et al., 1953) for missing data estimation. The presented method works by firstly establishing the parameters in the committee that will be evolved. These characteristics are the attributes that define the architectures of the specific machine learning tools such as number of hidden nodes, and activation functions. In this chapter, the parameter of the MLP to be evolved is the number of hidden nodes. This attribute is transformed into binary space and then it undergoes crossover and mutation and is then transformed back into floating number space. Finally the resulting committee is constructed and its fitness is evaluated. The current committee structure is either accepted or rejected using the Metropolis method. The Metropolis method works as follows: given the new data observation, the auto-associative networks that form a committee undergo crossover and mutation, resulting in the new network architectures. The new committee is either accepted or rejected using the Metropolis method. This procedure was described in detail in Chapter VII.

EXPERIMENTAL RESULTS AND DISCUSSION

Data set: The experiment conducted in this chapter is implemented using the methods discussed in the previous sections. A database on the monitoring of transformer bushings obtained from an electricity power company is used. In the power distribution network, bushings are used to connect lines to equipment such as breakers, transformers, isolators, ring main units, capacitor banks, voltage regulators and constant voltage transformers. Bushing failures are grouped into three categories namely chemical, electrical or mechanical. The objective of this experiment is to test the performance of the proposed method in detecting whether there is a fault in bushings when some of the information is missing.

Case Study 1: Classification in the Presence of Missing Data

The hypothesis here is to see if the state (faulty or healthy) could be determined when some of the data are missing. Training is conducted when no data is missing whereas testing is done under the missing data condition. The database contained 10 features for each instance. These features are all different gases that were dissolved in the transformer oil and were then used to determine if the bushing is faulty or not. Two specific inputs are considered more likely to be missing and as a result, 8 are considered most likely to be available. Using equation 8.4, this implies that there are 45 possible sets of networks that have different combinations of 8 inputs that can be constructed. Training and testing are done online with online adjustment of the networks, which in this section are the Fuzzy ARTMAP. The vigilance parameter for the fuzzy ARTMAP is optimally chosen to be 0.75.

The results obtained varied from one instance to another. However, a generalized performance was calculated after 4000 cases have been evaluated as shown in Figure 8.8. As can be seen from this figure, classification increased with an increase in the number of classifiers that were used. Although none of these classifiers were trained with all the inputs, their combination seemed to work better than with one network. The classification accuracy obtained for missing data went as high as 98.2%, which is very close to a 99% obtained when no datum was missing. The combination of auto-associative neural network and hybrid genetic algorithm (NN-HGA) as well as the combination of auto-associative neural network and fast simulated annealing (NN-FSA) approaches were trained off-line for missing data estimation, as presented in Chapter II, and then these estimated values were used to classify faults. The results obtained are tabulated in Table 8.2.

The results obtained show that the Fuzzy ARTMAP method performs better than the NN-FSA approaches. The reason why the online Fuzzy ARTMAP performs better than the NN-HGA and NN-FSA approaches is that the online approach forgets information that is no longer relevant, whereas the offline approach combines all the information from the entire time space, even though it may no longer be relevant. The results presented in Table 8.2 clearly show that the proposed algorithms can be used as a means of solving the missing data problem. The Fuzzy ARTMAP compares well with NN-HGA and NN-FSA approaches. The run-times for testing the performance of the method varies considerably. It can be noted from the table that for the NN-HGA and NN-FSA methods, the run-times increase with an increasing number of missing variables per instance. Contrary to the NN-HGA and NN-FSA, the Fuzzy ARTMAP offers run times that decrease with increasing number of inputs. The reason for this is that the number of Fuzzy-ARTMAP networks available reduces with an increasing number of inputs as mentioned earlier.

Case Study 2: Regression in the Presence of Missing Data

In this section, the algorithm implemented above is extended to a regression problem. Instead of using an ensemble of Fuzzy ARTMAP networks as in classification, the MLP networks are used. The reasons for this practice are two fold: firstly, MLPs are excellent regressors and secondly, to show that the proposed algorithm can be used with any architecture of neural networks.

Figure 8.9. Performance versus number of classifiers

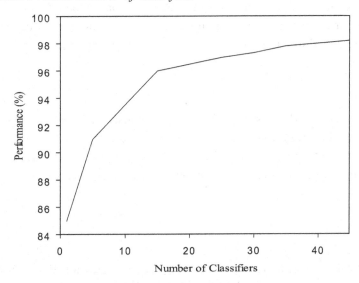

Table 8.2. Comparison between the Fuzzy ARTMAP, NN-HGA and NN-FSA

	Fuzzy ARTMAP		NN-HGA		NN-FSA	
Number of missing	1	2	1	2	1	2
Accuracy (%)	98.2	97.2	99	89.1	98	90.3
Run time (seconds)	0.86	0.77	0.67	1.33	0.48	1.07

Database

The data used in this experiment come from a model of a Steam Generator at the Abbort Power Plant in Urbana-Champaign in Illinois. This data set has four inputs, which are the *Fuel, Air, Reference Level* and the *Disturbance,* which is defined by the load level. This work attempts to regress in order to obtain two outputs, which are the drum pressure and the steam flow. More information on the inputs and the setup can be found in De Moor (2006).

Experimental Setup

Although Fuzzy-ARTMAP cannot be used for regression, the same approach as presented above is extended using MLP neural networks for the regression problem. In this section, regression is conducted to obtain two outputs, which are the *drum pressure* and the *steam flow*. It is assumed that n_{avl} = 2 is the case. As a result, only two inputs can be used. An ensemble of MLP networks is created, each with five hidden nodes and trained using only two of the inputs to obtain the output. Due to the limited features in the dataset, this work only considers a maximum of one sensor failure per instance. Each network is trained with 1200 training cycles, using the scaled conjugate gradient algorithm and a hyperbolic tangent activation function.

All these training parameters are again empirically determined. The MLPs used in this section are trained as in Chapter III using a Bayesian framework. The Bayesian framework trains the network by estimating the posterior probability function of the network weights given the data from the likelihood function, the prior probability and the evidence. In this section the online aspects is facilitated by updating the prior as new data come and by performing a number of hybrid Monte Carlo simulation (Marwala, 2007) steps, *n,* to factor recent data into account and thereby creating *m* additional states. Then the earliest *m* states are removed to retain the size of states that form a posterior probability distribution.

Results

In this chapter, a tolerance of 20% is used and is arbitrarily chosen. The results are summarized in Table 8.3. Since testing is done online, where one input arrives at a time, evaluation of the performance at each instance will not give a general view of how the algorithm works. The work, therefore, evaluates the general performance using the following equation only after N instances have been predicted.

$$Error = \frac{n_\tau}{N} x100\%$$

$$(8.8)$$

Performance indicates the accuracy in percentage and *time* indicates the running time in seconds. The results show that the ensemble method is well suited for the problem under investigation. The method performs better than the NN-HGA and NN-FSA in the regression problem under investigation. The ensemble approach presented here does not suffer from this problem, as there is no attempt to approximate the missing variables. It is also observed that the ensemble-based approach takes less time that the NN-HGA and NN-FSA methods. The reason is that HGA and FSA take longer to converge to reliable estimates of the missing values, depending on the objective function to be optimized. Although the prediction times are negligibly small, an ensemble-based technique takes longer to train since training involves many networks but it is more efficient online than the NN-HGA and NN-FSA methods.

Case Study 3: Evolving Ensemble of Machines to Approximate Missing Data

As was already mentioned, there are some cases where it is very important to get the missing value itself. This section investigates how close to the actual value the network can predict. The database used in the regression problem above is again used here. Both the hybrid GA and Fast Simulated Annealing are used and their results are compared. The ensemble is evolved online using genetic programming in a similar fashion to that described in Chapter VII. The parameter of the ensemble of auto-associative MLP networks in Figure 8.8 that is evolved is the number of hidden nodes. The ensemble of three auto-associative networks is evolved so that it is able to track the online changes in the dynamics of the system it is modeling. The ensemble approach is then compared to using individual networks alone. These individual networks sre tested online using a Bayesian approach as described in the previous section.

The results obtained are presented in Table 8.4. In this table, the results obtained by each network in the ensemble (committee) are compared to those obtained using the single MLP auto-associative networks.

It can be noted in Table 8.4 that the results obtained from the FSA and the HGA are closely related to each other. It can also be seen that the performance obtained using an ensemble of networks is even better than that obtained using the best network in the ensemble with the exception of the prediction of the *Disturbance*. As it can be seen in Table 8.4, the accuracy of the ensemble is lower than that obtained with the first network, labeled Network 1 in the table.

The average results obtained using the HGA and the FSA are very close to each other. This simply proves that the two optimization techniques used in this chapter are likely to converge to the same values

Table 8.3. Regression accuracy obtained without estimating the missing values

	Ensemble of MLPs		NN-HGA		NN-FSA	
Number missing	1	2	1	2	1	2
	Performance (%)	Time (seconds)	Performance (%)	Time (seconds)	Performance (%)	Time (seconds)
Drum Pressure	99.0	100.3	68	126	69	88
Steam Flow	85	0.92	84	98	87	100

Table 8.4. Comparison between hybrid GA and FSA for the presented approach. All values are in percentages

Inputs	Network 1		Network 2		Network 3		Evolved Ensemble	
	HGA%	FSA%	HGA	FSA	HGA	FSA	HGA	FSA
Fuel	84	87	67	60	94	95	100	100
Air	83	80	59	67	18	12	100	99
Level ref	73	74	76	76	89	93	94	97
Disturbance	91	91	58	51	46	48	79	84
Average	82.75%	83%	65%	63.5%	61.75%	62%	93.25%	95%

for the problem under investigation. Unfortunately, both techniques are found to require more time to produce accurate results. It can also be concluded that the bigger the ensemble, the longer it will take to produce accurate results. The biggest trade-off in this task is to decide whether more accurate results or results in a very short time are required. Obviously, for online condition monitoring, reliable results are required on time.

In this analysis, the statistical relationship between the predicted output and the actual target is further examined. This relationship is quantified in terms of correlation coefficients between the two quantities. Table 8.5 presents the results obtained using both the HGA and the FSA. As it can be seen from this table, the relationship is very close. The findings are in agreement with those obtained by Mohamed and Marwala (2005). Therefore, this shows that the method used to approximate the missing data does not add any bias to the data. Results obtained here are similar to those obtained when there is no missing data.

CONCLUSION

Treatment of missing data is a challenge in online condition monitoring systems. In this chapter, techniques for dealing with missing data for online condition monitoring problem are studied. An ensemble of neural networks is used in all attempts. Firstly, the problem of classifying in the presence of missing data is addressed. The problem domain is then extended to regression where no attempt to approximate the missing data is done and the results show that this practice can be used to obtain good results. The advantage of this technique is that it eliminates the need to find the best estimate of the data, and hence, saves time. An ensemble of auto-associative networks together with hybrid GA and fast simulated annealing is used to approximate missing data. The evolutionary committee is implemented for missing data estimation and is found to give good results. Several insights are deduced from the simulation results. It is observed that hybrid GA and FSA can converge to the same search space and to almost the same values but may differ in the duration they take. FSA is found to be faster than the hybrid GA but this can vary from one problem to another.

FURTHER RESEARCH

This chapter dealt with the issue of online data imputation by using ensemble models as well as a Fuzzy ARTMAP. For future research, the whole concept of online learning should be firmly embedded within

Table 8.5. Comparison between HGA and FSA

Inputs	HGA Correlation	MSE	FSA Correlation	MSE
Fuel	1	0.02	1	0.024
Air	0.92	0.3	0.96	0.2
Level ref	0.78	0.8	0.76	0.9
Disturbance	0.69	1.01	0.60	1.2

the field of evolutionary learning, where the concepts of incremental learning and reverse incremental learning can be fully explored. As future work, the techniques explored should be implemented in real time on many, more complex problems.

REFERENCES

Abdella, M., & Marwala, T. (2005). Treatment of missing data using neural networks. In *Proceedings of the IEEE International Joint Conference on Neural Networks*, Montreal, Canada (pp. 598-603).

Abdella, M. I., & Marwala, T. (2006). The use of genetic algorithms and neural networks to approximate missing data in databases. *Computing and Informatics*, *24*, 1001-1013.

Benitez-Perez, H., Garcia-Nocetti, F., & Thompson, H. (2007). Fault classification based upon self organizing feature maps and dynamic principal component analysis for inertial sensor drift. *International Journal of Innovative Computing, Information and Control*, *3*(2), 257–276.

Bishop, C. M. (1995). *Neural networks for pattern recognition*. Oxford: Oxford University Press.

Carpenter, G. A., Grossberg, S., Markuzon, N., Reynolds, J. H., & Rosen, D. B. (1992). Fuzzy ARTMAP: A neural network architecture for incremental supervised learning of analog multidimensional maps. *IEEE Transactions on Neural Networks*, *3*, 698-713.

Cen, L. (2007). A hybrid genetic algorithm for the design of FIR filters with SPoT coefficients. *Signal Processing*, *87*(3), 528-540.

Chen, T., & Chang, Y. (2006). Modern floor planning based on fast simulated annealing. In *Proceedings of the International Symposium on Physical Design*, California, USA (pp. 104-112).

Cherrid, N., Naït-Ali, A., & Siarry, P. (2005). Fast simulated annealing algorithm for BAEP time delay estimation using a reduced order dynamic model. *Medical Engineering & Physics*, *27*(8), 705-711.

Davis, L. (1991). *Handbook of genetic algorithms*. New York: Van Nostrand.

De Moor B.L.R. (ed.), (2006). *DaISy. Database for the identification of systems*. Department of Electrical Engineering, ESAT/SISTA, K.U.Leuven, Belgium, [Used dataset: Model of a steam generator at Abbott Power Plant, Process Industry Systems (pp. 98-003)]. Retrieved April 2, 2006, from http://www.esat.kuleuven.ac.be/sista/daisy/

Dhlamini, S. M., Nelwamondo, F. V., & Marwala, T. (2006). Condition monitoring of HV bushings in the presence of missing data using evolutionary computing. *WSEAS Transactions on Power Systems*, *1*, 296–302.

Freund, Y., & Schapire, R. E. (1995). A decision theoretic generalization of online learning and an application to boosting, In *Proceedings of the Second European Conference on Computational Learning Theory* (pp. 23–37).

Frolov, A., Kartashov, A., Goltsev, A., & Folk, R. (1995). Quality and efficiency of retrieval for Willshaw-like auto-associative networks. *Computation in Neural Systems*, *6*, 535–549.

Gabrys, B. (2002). Neuro-fuzzy approach to processing inputs with missing values in pattern recognition problems. *International Journal of Approximate Reasoning, 30*, 149–179.

Goldberg, D. (1989). *Genetic algorithms in search, optimization and machine learning.* Reading, MA: Addison-Wesley.

Holland, J. H. (1975). *Adaptation in natural and artificial systems.* Ann Arbor: University of Michigan Press.

Huang, X., & Zhu, Q. (2002). A pseudo-nearest-neighbor approach for missing data recovery on Gaussian random data sets. *Pattern Recognition Letters, 23*, 613–1622.

Hulley, G., & Marwala, T. (2007). Genetic algorithm based incremental learning for optimal weight and classifier selection. *Computational Models for Life Sciences. 952*, 258-267.

Japkowicz, N. (2002). Supervised learning with unsupervised output separation. In *Proceedings of the International Conference on Artificial Intelligence and Soft Computing* (pp. 321–325).

Javadpour, R. & Knapp, G. M. (2003). A fuzzy neural network approach to condition monitoring. *Computers and Industrial Engineering, 45*, 323–330.

Jeon, G., Leep, H. R., & Shim, J. Y. (2007). A vehicle routing problem solved by using a hybrid genetic algorithm. *Computers & Industrial Engineering, 53*(4), 680-692.

Kim, J. O., & Curry, J. (1997). The treatment of missing data in multivariate analysis. *Sociological Methods and Research, 6*, 215-241.

Kirkpatrick, S., Gelatt, C. D., & Vecchi, M. P. (1983). Optimization by simulated annealing. *Science, 220*, 671-680.

Little, R., & Rubin, D. (1987). *Statistical analysis with missing data.* New York: John Willey and Sons.

Lopes, M. L. M., Minussi, C. R., and Lotufo, A. D. P. (2005). Electric load forecasting using a fuzzy ART&ARTMAP neural network. *Applied Soft Computing, 5*(2), 235-244.

Lu, P., & Hsu, T. (2002). Application of auto-associative neural network on Gas-Path sensor sata validation. *Journal of Propulsion and Power, 18*(4), 879–888.

Lunga, D., & Marwala, T. (2006). Time series analysis using fractal theory and online ensemble classifiers. *Lectures Notes in Artificial Intelligence, 4304*, 312-321.

Martin, A. G., Roy, J., Beaulieu, L., Pouliot, J., Harel, F., & Vigneault, E. (2007). Permanent prostate implant using high activity seeds and inverse planning with fast simulated annealing algorithm: A 12-year Canadian experience. *International Journal of Radiation Oncology, Biology and Physics, 67*(2), 334-341.

Marwala, T. (2001). *Fault identification using neural networks and vibration data.* Unpublished doctoral dissertation, University of Cambridge, Cambridge.

Marwala, T. (2007). *Computational intelligence for modelling complex systems.* Delhi: Research India Publications.

Marwala, T., & Hunt, H. E. M. (2000). Fault detection using pseudo-modal-energies and modal properties. *International Journal of Engineering Simulation, 1*, 4-7.

McGookin, E. W., & Murray-Smith, D. J. (2006). Submarine maneuvering controllers optimisation using simulated annealing and genetic algorithms. *Control Engineering Practice 14*, 1-15.

Merz, C. J. (1997). Using correspondence analysis to combine classifiers. *Machine Learning, 36*(1-2), 1–26.

Metropolis, N., Rosenbluth, A. W., Rosenbluth, M. N., Teller, A., & Teller, E. (1953). Equation of state calculation using fast computing machines. *Journal of Chemical Physics, 21*, 1087-109.

Mi, L., & Takeda, F. (2007). Analysis on the robustness of the pressure-based Individual identification system based on neural networks. *International Journal of Innovative Computing, Information and Control, 3*(1), 1365–1380.

Mohamed, S., & Marwala, T. (2005). Neural network based techniques for estimating missing data in databases. In *Proceedings of the 16th Annual Symposium of the Pattern Recognition Association of South Africa*, Cape Town (pp. 27–32).

Mohamed, S., Marwala, T., & Rubin, D. M. (2007). Adaptive GPCR classification based on incremental learning. *SAIEE Africa Research Journal, 98*(3), 71-80.

Nascimento, V. B., de Carvalho, V. E., de Castilho, C. M. C., Costa, B. V., & Soares, E. A. (2001). The fast simulated algorithm applied to the search problem in LEED. *Surface Science, 487*, 15–27.

Nauck, D., & Kruse, R. (1999). Learning in neuro-fuzzy systems with symbolic attributes and missing values. In *Proceedings of the International Conference on Neural Information Processing*, Perth, Australia (pp. 142-147).

Nelwamondo, F. V. & Marwala, T. (2007). Fuzzy ARTMAP and neural network approach to online processing of inputs with missing values. *SAIEE Africa Research Journal, 98*(2), 45-51.

Nelwamondo, F. V. & Marwala, T. (2008), Techniques for handling missing data: applications to online condition monitoring. *International Journal of Innovative Computing, Information and Control, 4*(6), 1507-1526.

Nelwamondo, F. V., Mohamed, S., & Marwala, T. (2007). Missing data: A comparison of neural network and expectation maximisation techniques. *Current Science, 93*(11), 1514-1521.

Opitz, D., & Shavlik, J. W. (1996). Generating accurate and diverse members of a neural network ensemble. In D. Touretsky, M. Mozer, and M. Hasselmo, *Advances in neural information processing systems* (pp. 535–543). Denver, CO: MIT Press.

Perrone, P., & Cooper, L. N. (1993). When networks disagree: ensemble methods for hybrid neural networks. *Neural Networks for Speech and Image Processing* (pp. 126-142).

Polikar, R., Udpa, L., Udpa, S., & Honavar, V. (2001). Learn++: An incremental learning algorithm for supervised neural networks. *IEEE Transactions on System, Man and Cybernetics, Special Issue on Knowledge Management, 31*, 497–508.

Qiao, W., Gao, Z., & Harley, R. G. (2005). Continuous online identification of nonlinear plants in power systems with missing sensor measurements. In *Proceedings of the International Joint Conference on Neural Networks*, Montreal, Canada (pp. 1729–1734).

Shtub, A., LeBlanc, L. J., & Cai, Z., (1996). Theory and methodology scheduling programs with repetitive projects: A comparison of a simulated annealing, a genetic and a pair-wise swap algorithm. *European Journal of Operational Research*, *88*, 124–138.

Tamaki, H., Kita, H., & Kobayahi, S. (1996). Multi-objective optimization by genetic algorithms: A review. In T. Fukuda and T. Furuhashi (Eds.), *Proceedings of the 1996 International Conference on Evolutionary Computation (ICEC'96)*, (pp. 517-522). Nagoya, Japan: IEEE Press.

Tan, S. C., Rao, M. V. C., and Lim, C. P. (2008). Fuzzy ARTMAP dynamic decay adjustment: An improved fuzzy ARTMAP model with a conflict resolving facility. *Applied Soft Computing*, *8*(1), 543-554.

Thompson, B. B, Marks, R. J. & El-Sharkawi, M. A. (2003). On the contractive nature of autoencoders: Application to sensor restoration. In *Proceedings of the IEEE International Joint Conference on Neural Networks* (pp. 3011-3016).

Vilakazi, C. B., & Marwala, T. (2006). Application of feature selection and fuzzy ARTMAP to intrusion detection. In *Proceedings of the IEEE International Conference on Systems, Man and Cybernetics, Taiwan* (pp. 4880-4885).

Vilakazi, C. B., & Marwala, T. (2007a). Online incremental learning for high voltage bushing condition monitoring. In *Proceedings of the International Joint Conference on Neural Networks* (pp. 2521-2526).

Vilakazi, C. B., & Marwala, T. (2007b). Incremental learning and its application to bushing condition monitoring. *Lecture Notes in Computer Science*, *4491*(1), 1241-1250.

Wang, S. (2005). Classification with incomplete survey data: A Hopfield neural network approach. *Computers and Operations Research*, *32*, 2583-2594.

Wang, C., & Wang, Q. L. (2005). Shape optimization of HTS magnets using hybrid genetic algorithms. In *Proceedings of the Twentieth International Cryogenic Engineering Conference (ICEC20)* (pp. 713-716).

Wang, L., Zhang, L., & Zheng, D-Z. (2006). An effective hybrid genetic algorithm for flow shop scheduling with limited buffers. *Computers & Operations Research*, *33*(10), 2960-2971

Yu, S., & Kobayashi, H. (2003). A hidden semi-Markov model with missing data and multiple observation sequences for mobility tracking. *Signal Processing* (pp. 235–250).

Chapter IX
Missing Data Approaches to Classification

ABSTRACT

In this chapter, a classifier technique that is based on a missing data estimation framework that uses autoassociative multi-layer perceptron neural networks and genetic algorithms is proposed. The proposed method is tested on a set of demographic properties of individuals obtained from the South African antenatal survey and compared to conventional feed-forward neural networks. The missing data approach based on the autoassociative network model proposed gives an accuracy of 92%, when compared to the accuracy of 84% obtained from the conventional feed-forward neural network models. The area under the receiver operating characteristics curve for the proposed autoassociative network model is 0.86 compared to 0.80 for the conventional feed-forward neural network model. The autoassociative network model proposed in this chapter, therefore, outperforms the conventional feed-forward neural network models and is an improved classifier. The reasons for this are: (1) the propagation of errors in the autoassociative network model is more distributed while for a conventional feed-forward network is more concentrated; and (2) there is no causality between the demographic properties and the HIV and, therefore, the HIV status does change the demographic properties and vice versa. Therefore, it is better to treat the problem as a missing data problem rather than a feed-forward problem.

INTRODUCTION

This chapter proposes missing data approaches to classification. Problems in classification have been dealt with by a number of researchers from different areas such as engineering, mathematics and statistics. Classification entails categorizing objects into different classes. As an example, human beings can be classified into two classes, and these are male and female. Diseases can also be classified into classes, and in the case for tumors these can be either malignant or benign. Classification of objects into two classes is called binary classification. In computational intelligence techniques such as neural networks classifying data into two classes involves constructing neural network architecture with an output which is either 0 or 1, with 0 being one class and 1 being another class.

There are many classification techniques that have been developed thus far and these include support vector machines and neural networks. Habtemariam, Marwala, and Lagazio (2005) as well as Habtemariam (2005) have successfully used support vector machines to classify interstate conflict where a 0 represented expected peace and a 1 represented expected conflict. Tettey and Marwala (2006) as well as Tettey (2006) extended this work by using neuro-fuzzy systems to classify inter-state conflict into either peace or conflict. Cai and Chou (1998) used artificial neural network model to construct a model that predicts human immunodeficiency virus (HIV) protease cleavage sites in protein whereas Chamjangali, Beglari, and Bagherian (2007) trained an artificial neural network model using the Levenberg–Marquardt optimization algorithm and used this model for predicting cytotoxicity data (CC50) of anti-HIV 5-pheny-l-phenylamino-1H-imidazole derivatives. Vilakazi and Marwala (2006a) used extension neural networks to classify bushings, which are critical components of electrical transformers, into two classes faulty or healthy while Marwala, Mahola, and Nelwamondo (2006) successfully used hidden Markov models and Gaussian mixture models for detecting faults in mechanical systems where faults could be classified as either present or absent. There are many classification problems that have been tackled using computational intelligence techniques including mechanical fault detection (Mohamed, Rubin, & Marwala, 2006a&b; Mohamed, Tettey, & Marwala, 2006; Nelwamondo & Marwala, 2006; Marwala, Mahola, & Nelwamondo, 2006); intrusion detection system (Vilakazi & Marwala, 2006b); and multi-class protein sequence (Mohamed, Rubin, & Marwala, 2006a; Mohamed, Rubin, & Marwala, 2007). This chapter proposes a novel classification technique that is based on missing data approaches that have been discussed in earlier chapters. Therefore, the classification results are viewed in this chapter as missing data amongst other variables that have been measured.

The missing data approach adopted in this chapter is composed of the combination of the multi-layer autoassociative neural network and genetic algorithm as implemented earlier (Abdella, 2005; Abdella & Marwala, 2005a&b; Mohamed & Marwala, 2005; Leke & Marwala, 2006). This classification technique is then compared to a conventional classification technique that is based on a standard multi-layer perceptron and formulated using the Bayesian approach and the posterior probability estimated using the Gaussian approximation. This technique was proposed by MacKay (1992) and was successfully implemented for classifying inter-state conflict by Lagazio and Marwala (2005). These two classification methods are compared using the classification of HIV problem that has been addressed in the past on numerous occasions (Leke, Marwala, & Tettey, 2006; Marwala, Tettey, & Chakraverty, 2006; Leke, Marwala, & Tettey, 2007).

Acquired immunodeficiency syndrome (AIDS) was first defined (Root-Bernstein, 1993) in 1982 to describe the first cases of unusual immune system failure that was identified in the previous year. HIV was later identified as the cause of AIDS. Risk factor epidemiology examines the individual demographic and social characteristics and attempts to establish factors that position an individual at risk of acquiring a life-threatening disease (Poundstone, Strathdee, & Celectano, 2004). In this chapter, the demographic and social characteristics of the individuals and their behavior are utilized to establish the risk of HIV infection; known as biomedical individualism (Fee & Krieger, 1993). By identifying the individual risk factors that result in HIV infection, it is possible to modify social conditions, which give rise to the disease, and consequently design effective HIV prevention policies. In this chapter, a model is created and used to classify the HIV status of individuals based on demographic properties.

An artificial neural network is an inter-connected structure of processing elements. The artificial neural network structure employed in this chapter consists of three major components (Bishop, 1995) namely input layer, hidden nodes with one activation function and output nodes with one activation function. As indicated before, neural networks have been successfully used for medical informatics, for decision support and making, clinical diagnosis, prognosis, and prediction of outcomes (Tandon, Adak, & Kaye, 2006; Alkan, Koklukaya, & Subasi, 2005; Sawa & Ohno-Machado, 2003; Szpurek et al., 2005; Tan & Pan, 2005) and for classification. Marwala (2001) made use of a probabilistic committee of neural networks to classify faults in a population of nominally identical cylindrical shells and obtained an accuracy of 95% on classifying eight classes of fault cases. Ohno-Machado (1996) described the constraint on the accuracy of the neural network model due to the need for data balance and raised the accuracy by utilizing sequential neural networks.

This chapter is aimed at proposing neural network based methods to understand HIV. Lisboa (2002) evaluated the evidence of healthcare benefits using neural networks whereas Fernández and Caballero (2006) used neural networks to model the activity of cyclic urea HIV-1 protease inhibitors. They demonstrated that artificial neural networks are able to characterize the non-linearity in the HIV model. Lee and Park (2001) implemented neural networks to classify and predict the symptomatic status of HIV/AIDS patients based on publicly available HIV/AIDS data. Kim et al. (2008) successfully used rule based method for HIV-1 protease cleavage site analysis whereas Milac, Avram, and Petrescu (2006) used neural networks method for HIV-1 protease inhibitors and Tim (2006) used neural networks to predict HIV status from demographic factors.

An investigation was also conducted to predict the functional health status of HIV/AIDS patients defined as 'in good health' or 'not in good health', using neural networks (Sardari & Sardari, 2002). Laumann and Youm (1999) used the racial and ethnic group distinctions to model the incidence of HIV and succeeded in relating the demographic properties to the transmission of HIV. Poundstone, Strathdee, and Celectano (2004) identified relationships between demographic properties and the spread of HIV and their work justifies the utilization of such demographic properties in constructing and identifying a model to predict the HIV status of individuals, as is conducted in this chapter. The models mentioned above concluded that artificial neural networks are able to model HIV classification problems well.

The methodology presented in this chapter is intended for utilizing demographic and social factors to predict the HIV status of an individual, using a hybrid of a multi-layer autoassociative neural networks as well as genetic algorithm, and then compare this procedure to a feed-forward multi-layer perceptron than is trained using Bayesian approach with the posterior probability function being estimated by a Gaussian approximation method.

The most common neural network architecture is the multilayer perceptron (MLP). An alternative network is the radial basis function (RBF) (Bishop, 1995; Marwala, 2007). The employment of the MLP over the RBF can be attributed to the fact that the RBF usually necessitates the implementation of the pseudo-inverse of a matrix for training, which is often singular while the MLP uses conventional feed-forward optimization methods, which are stable (Bishop, 1995). In Chapter II, results showed that the MLP outperform the RBF. This can be attributed to the fact that the MLP networks, also known as universal approximators, are capable of modeling any complex relationship with one or two hidden layers and are thus most suited for this study. A great deal of successful implementation of neural networks have been conducted in biomedical engineering (Svozil, Kvasnicka, & Pospichal, 1997; Hudson & Cohen, 2000), prediction of breaking waves Deo and Jugdale (2003), in feed-back control (Narendra & Lewis, 2001) as well as in structures (Rafiq, Bugmann, & Easterbrook, 2001).

AUTOASSOCIATIVE MULTI-LAYER PERCEPTRON NETWORKS FOR HIV PREDICTION

In this chapter, missing data framework that is based on a multi-layered perceptron autoassociative networks is used to predict HIV status from demographic characteristics. As indicated before autoassociative neural networks are models that are trained to recall the inputs (Lu & Hsu, 2002). These networks in consequence predict the inputs as outputs, every time an input is presented and have been used in a number of applications. Lu and Hsu (2002) used autoassociative networks on gas path sensor validation while Frolov et al. (1995) studied and applied Willshaw-like autoassociative networks. Smauoi and Al-Yakoob (2003) studied the dynamics of cellular flames while Hines, Robert, and Wrest (1998) used autoassociative networks signal validation and Sohn, Worden, and Farrar (2001) studied mechanical systems using autoassociative neural networks.

An autoassociative neural network (or simply known as encoder) consists of an input and output layer with the same number of inputs and outputs, hence the name autoassociative, combined with a narrow hidden layer (Lu & Hsu, 2002). In this chapter, the autoassociative networks are trained using HIV/AIDS demographic data as well as HIV status. The hidden layer attempts to reconstruct the inputs to match the outputs, by minimizing the error between the inputs and the outputs when new data are presented. The narrow hidden layer, forces the network to reduce any redundancies, but still allows the network to detect non-redundant data. However, it must be noted that for missing data estimation it is absolutely crucial that the network be as accurate as possible and that this accuracy is not necessarily realized through few hidden nodes as is the case when these networks are used for data compression. It is, therefore, crucial that some process of identifying the optimal architecture be used. Genetic algorithm (GA) is used in this study to find the optimal autoassociative architecture by finding the global optimum solution (Holland, 1975). The autoassociative neural network architecture used in this study is shown in Figure 9.1.

The autoassociative network is constructed, in this chapter, using multi-layer perceptron (MLP) neural networks. This type neural network consists of a two-layer and as dictated by the universal approximation theory is quite capable of approximating any continuous function with arbitrary accuracy, provided the number of hidden neurons is sufficiently large (Bishop, 1995). If $\{x\}$ is the input vector to the multi-layer perceptron and $\{y\}$ is the output vector of the MLP, a relation mapping between the input and output may be written as follows (Bishop, 1995):

$$y = f_{output}\left(\sum_{j=1}^{M} w_j f_{hidden}\left(\sum_{i=1}^{N} w_{ij}x_i + b_0\right) + w_0\right)$$

(9.1)

In equation 9.1, N is the number of inputs units, M is the number of hidden neurons, x_i is the ith input unit, w_{ij} is the weight parameter between input i and hidden neuron j, and w_j is the weight parameter between hidden neuron j and the output neuron. If the activation function in the output layer f_{output} is sigmoid, as is the case in this chapter, then it can be written mathematically as follows:

$$f_{output} = \frac{1}{1+e^{-a}}$$

(9.2)

If the activation function in the hidden layer f_{hidden} is hyperbolic tangent function, as is also the case in this chapter, then it can be mathematically written as follows:

$$f_{hidden}(a) = \tanh(a) \tag{9.3}$$

CLASSIFICATION USING AUTOASSOCIATIVE NETWORK

In this chapter, the goal of our classification process is to develop an algorithm, which will assign an individual, represented by a vector $\{x\}$ describing the demographic, social and behavioral characteristics of that individual, to one of the HIV classes, C_1 or C_2. Here the parameters C_1 and C_2 represent the status of an individual, which may be positive or negative.

The data on which the model is based upon contain demographic characteristics of individuals, as well as the classes to which those individuals belong. The output of the classification system is assigned to the variable y. The autoassociative model is, therefore, required to map the input vector $\{x_1,...,x_d\}$ to the output vector $\{y\}$ as follows:

$$\{y\} = f(\{x\},\{w\}) \tag{9.4}$$

In equation 9.4 $\{w\}$ is the mapping weights vector and $\{x\}$ represents the demographic input parameters vector and $\{y\}$ represents the output vector. The network weights are obtained using the back-propagation method, which is described in detail in Chapter II. Essentially, an error is constructed that defines the difference between the model's estimated output (equation 9.4) and the target output. Then the gradient of this error is calculated as was done in Chapter II using the back-propagation method

Figure 9.1. Autoassociative neural network architecture

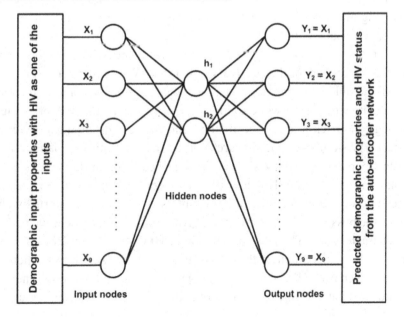

and then an optimization method is used to solve for the weight vector in equation 9.4. The optimization method that is chosen, in this chapter, is the scaled conjugate method because of its computational efficiency (Bishop, 1995).

To predict the HIV status of individuals, the HIV status input, in the input vector $\{x\}$ is assumed as an unknown input, while the demographic input properties are considered as the known inputs. When the input vector $\{x\}$ has unknown elements, the input vector set can be categorized into $\{x\}$ known represented by $\{x_k\}$ and $\{x\}$ unknown represented by $\{x_u\}$. The missing data estimation error equation that is based on an autoassociative network can thus be written as follows as described in Chapter II:

$$\varepsilon = \left\| \left(\left\{ \begin{matrix} \{x_u\} \\ \{x_k\} \end{matrix} \right\} - f \left(\left\{ \begin{matrix} \{x_u\} \\ \{x_k\} \end{matrix} \right\}, \{w\} \right) \right) \right\| \tag{9.5}$$

In equation 9.5, $\{x_u\}$ is the unknown vector, $\{x_k\}$ represents the demographic input parameters of the individuals, $\| \; \|$ is Euclidean norm, $\{w\}$ represents the weight vector that maps the autoassociative network, f, input vector $\{x\}$ to the same input vector $\{x\}$. In this chapter, there is one missing variable and this is the HIV status and thus equation 9.5 can thus be re-written as follows:

$$\varepsilon = \left\| \left(\left\{ \begin{matrix} HIV_{status} \\ \{x_k\} \end{matrix} \right\} - f \left(\left\{ \begin{matrix} HIV_{status} \\ \{x_k\} \end{matrix} \right\}, \{w\} \right) \right) \right\| \tag{9.6}$$

In equation 9.6 the unknown vector $\{x_k\}$, from equation 9.6, thus becomes the HIV status now represented as HIV_{status}. This estimated HIV status can thus be estimated by minimizing equation 9.6 using some optimization method. As stated in earlier chapters, the optimization method to be used must be able to identify a global optimum solution lest the estimated HIV status might otherwise be erroneous. In this chapter the optimization method chosen is genetic algorithm (GA) and it has a high probability of finding global optimum solutions than conventional gradient methods such as scaled conjugate methods which are used for training the autoassociative neural network. The minimization of equation 9.6 which represents the mapping in Figure 9.1 results in the propagation of errors in the demographic data that is distributed as there are many outputs.

In summary, GA is an evolutionary based method that was inspired by Darwin's theory of evolution, whose main tenant is the concept of the survival of the fittest. Within the context of this chapter, from the population of all possible HIV statuses given the measured demographic characteristics, $\{x_k\}$, and the known autoassociative neural network architecture, f, as well as the network weights, $\{w\}$, which are identified through the back-propagation method and the scaled conjugate gradient optimization technique, the fittest HIV statuses population are evolved and from this population the best HIV status is chosen. The fittest population of HIV statuses is identified through the drivers of evolution which are crossover, mutation and selection (Holland, 1975; Goldberg, 1989; Davis, 1991; Michalewicz, 1996) and these have been described in detail in earlier chapters. The procedure starts by choosing the size of the population of the possible HIV statuses for a given trained autoassociative neural networks and the demographic characteristics. Then this population of HIV statuses undergoes crossover, which is essentially the mixing of genes amongst the HIV statuses. Then a small proportion of possible HIV statuses are mutated. Then each HIV status in the population is given a fitness function by evaluating how each HIV status estimate performs in terms of achieving a lower estimation error from equation

9.6. Then based on this population of fitness, the HIV statuses that are fitter (here fitter means giving the low error from equation 9.6) are reproduced, undergo crossover, mutation and selection steps and then the process is repeated until convergence.

ARTIFICIAL NEURAL NETWORKS FOR HIV ESTIMATION

As indicated earlier, artificial neural networks are perceived as a generalized framework for representing non-linear mappings between multi-dimensional spaces, where the structure of the mapping process is administered by a set of adjustable parameters (Bishop, 1995). With supervised learning, the training data set consists of both the input to the artificial neural networks and an associated target output. In this chapter, the input is a set of features from the demographic parameters and the target output is 1 or 0, where 1 represents HIV positive and 0 represents HIV negative.

The Architecture

The Bayesian multi-layer perceptron neural network architecture trained with Bayesian approach and with Gaussian approximation to the posterior probability is selected for this part of this chapter (Bishop, 1995). The MLP architecture implemented in this section is illustrated in Figure 9.2 where the input vector consists of the demographic characteristics while the output is the HIV status. This architecture suggests that there is causality between demographic characteristics and HIV status. Furthermore, the errors in the demographic characteristics are propagated and summed in the prediction of the HIV status as there is one output.

Figure 9.2. Feed-forward MLP network architecture

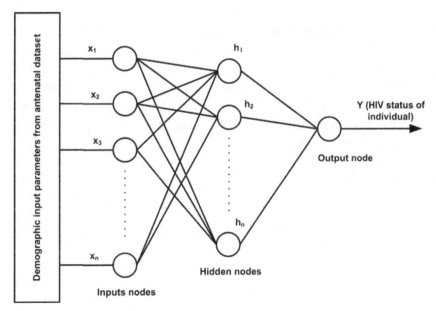

This architecture has been applied successfully to the epileptic SSW detection problem in the past (Mohamed, Rubin, & Marwala, 2006b). These researchers investigated the application of neural networks to the problem of detecting interictal epileptiform activity in the electroencephalogram (EEG). The proposed detector consisted of a segmentation, feature extraction and classification stages. For the feature extraction stage, coefficients of the Discrete Wavelet Transform (DWT), real and imaginary parts of the Fast Fourier Transform and raw EEG data were all found to be well suited for EEG classification. Principal component analysis was used to reduce the dimensionality of the features. For the classification stage, multi-layer perceptron neural networks were implemented in accordance to the maximum likelihood and Bayesian learning formulations. The latter was found to make better use of training data and consequently produced better trained neural networks.

Other examples of the use of Bayesian neural networks that were trained using the Gaussian approximation is the work by Marwala and Lagazio (2004) as well as Lagazio and Marwala (2005) who modeled the relationship between input parameters that include allies, contingency, distance, capability, dependency and major power, and the output parameter which was either peace or conflict

As indicated before the problem that is tackled in this chapter, the classification of HIV status is a two-class problem with 0 representing HIV negative and 1 representing HIV positive. For a two-class classification problem, Bishop (1999) showed that the error between the network output in Equation 9.1 and the target output t for all training patterns P is the cross-entropy error function given by:

$$E_{CEE} = -\sum_{p=1}^{P}\left[t_p \ln(y_p) + (1-t_p)\ln(1-y_p)\right]$$

(9.7)

For the duration of the training, the artificial neural networks weight parameter vector $\{w\}$ is iteratively adjusted in order to minimize the error function in equation 9.7 using an optimization method. In the initial stage, the weight parameters vector is set to random values. An optimization method known as the scaled conjugate gradient method (Møller, 1993) is used in this chapter to estimate the magnitude of the parameter updates. The scaled conjugate gradient method is selected instead of other optimization methods because of its fast convergence properties.

Bishop (1995) proved that if the cross-entropy error function is minimized during artificial neural network training and the activation function of an artificial neural network is given by a logistic function, which is shown in equation 9.2, the output of an artificial neural networks approximates the posterior probability of membership to a pattern class given the vector of inputs $\{x\}$, which in this chapter represents the demographic characteristics. In this chapter, therefore, the output approximates the posterior probability of the input demographic characteristics. If the HIV negative class is symbolized by C_1 and the pattern class HIV positive is represented by C_2, the relationship for the posterior probability of class membership can be written as (Bishop) follows:

$$P\left(C_1|\{x\}\right) = y$$

(9.8)

and

$$P\left(C_2|\{x\}\right) = 1 - y$$

(9.9)

The relationships in equations 9.8 to 9.9 provide a probabilistic interpretation to the artificial neural network output. Based on these relationships, it is understandable that an input vector has a high probability of belonging to class C_1 when y is close to 1, and C_2 when y is close to 0. If y is close to 0.5, there is uncertainty in the class membership of the input vector. A simple method to ensure that the classifier makes classification decisions only where there is a high degree of certainty is to apply an upper and lower rejection threshold to the artificial neural network output as put forward by Bishop (1995). This classification decision rule is defined as follows:

Decide C_1 if $y \geq \theta$; decide C_2 if $y < (1-\theta)$ (9.10)
    Otherwise do not classify $\{x\}$

The parameter θ sets the level of the rejection threshold. However, in this chapter θ is set to 0.5 but equation 9.10 is stated for generality's sake.

Bayesian Formulation

The maximum likelihood approach to neural network training assumes that weight parameters have fixed unknown values and estimates these values by minimizing a suitable error function and this procedure is pursued in the first part of this chapter to train the autoassociative neural networks. On the other hand, as the vector of weight parameters $\{w\}$ is approximated from a finite set of data, there is at all times some degree of uncertainty linked to the value of this weight vector. For instance, if a single training pattern of data is eliminated from the training set, the weight vector that is learnt during training will be different from the weight vector that is learnt when equivalent training pattern is there in the training set.

Bayesian learning takes into account the uncertainty in the weight vector by assigning a probability distribution to the weight vector that represents the relative degrees of belief in the values of the weight vector. For example, in this chapter a multi-layer perceptron model structure reflecting the number of hidden layer neurons is first defined as H, and a prior distribution $P(\{w\}|H)$ is assigned to the weight vector to reflect any initial beliefs regarding the distribution of the weight parameters. Once a set of training data, [D], is observed, the prior distribution can be converted to a posterior distribution of network weights using the Bayes theorem as follows (Bayes, 1793; Barnard, 1958):

$$P\big(\{w\}|[D],H\big) = \frac{P([D]|\{w\},H)P(\{w\}|H)}{P([D]|H)}$$ (9.11)

In equation 9.11 $P([D]|\{w\},H)$ is the *likelihood* function and $P([D]|H)$ is the *evidence* of the multi-layer perceptron model H. MacKay (1992) uses a Gaussian prior as:

$$p\big(\{w\}|H\big) = \frac{1}{(2\pi/\alpha)^{W/2}} \exp\left(-\frac{\alpha}{2}\sum_i w^2\right)$$ (9.12)

In equation 9.12, α is the regularization coefficient and $\{w\}$ is the number of weight parameters in the artificial neural networks. The distribution in equation 9.12 is Gaussian with variance $1/\alpha$. When the magnitudes of weight parameters w's are large, $p(\{w\}|H)$ will be small. In this approach, the prior distribution favors smaller values for weight parameters. By utilizing the prior probability distribution function in equation 9.12, the posterior probability of the weight vector given the training data in equation 9.11 may be derived as (MacKay, 1992) follows:

$$p(\{w\}|D,H) = \frac{1}{Z_S}\exp(-S(\{w\}))$$

(9.13)

where

$$S(\{w\}) = E_{CEE} + \frac{\alpha}{2}\sum_i w_i^2$$

(9.14)

In accordance to equation 9.13, the weight vector that corresponds to the maximum of the posterior distribution and $\{w_{MP}\}$ can be derived by minimizing the negative logarithm of $P(\{w\}|[D],H)$. This is the same as minimizing an error function in equation 9.7, which consists of the cross-entropy error in Equation 9.7 added to a weight penalty term called the *weight decay regularizer*. Evaluating the posterior probability in equation 9.2 is analytically intractable as it more often than not involves integration over a high dimensional weight space. One method of ensuring that these computations are more tractable is to bring in approximations that make simpler their computation and these are MacKay's (1992) evidence framework or Buntine and Weigend's (1991) approximation to the posterior distribution. These procedures assume that the posterior probability distribution, which is typically canonical, can be approximated with a Gaussian distribution centered around one of its modes at $\{w_{MP}\}$.

Penny and Roberts (1999) further investigated the evidence framework and found it to be a viable method for analyzing real-world data while Cawley and Talbot (2005) successfully applied the evidence framework to kernel logistic regression. Neal (1992) suggested a more general and exact approach where Markov Chain Monte Carlo methods are used to compute integrations over weight space and this is explained in Chapter II. The technique of MacKay (1993) is utilized in this chapter for the reason that it offers a reduced amount of computational requirements than Neal's method (1992) and because of the observations by Mackay (1993) that the evidence approximation should generate better-quality results than the Buntine and Weigend's technique when estimating regularization coefficients. MacKay's evidence framework provides principled methods for model selection, prediction and estimating regularization coefficients with no requirement for a validation set. This permits additional data to be employed for training the artificial neural networks, which is valuable when only inadequate amounts of data exist for training. The evidence framework approach entails the following steps:

Estimating Regularization Coefficients

The evidence of α is calculated by integrating over network weight parameters as (Sugiyama & Ogawa, 2002):

$$p(D|\alpha,H) = \int p(D|\{w\},H)p(\{w\}|\alpha)d\{w\}$$

(9.15)

By utilizing equation 9.15, MacKay (1992) developed an equation for the most probable regularization coefficient, α_{MP}, that maximize $p(D|\alpha, H)$ and this expression has the form of:

$$\alpha_{MP} = \frac{\gamma}{\sum w_i^2} \tag{9.16}$$

where:

$$\gamma = \sum_{i=1}^{W} \frac{\lambda_i}{\lambda_i + \alpha} \tag{9.17}$$

λ_i's are eigenvectors of the Hessian matrix [A] or second derivative of the error function in equation 9.17 with respect to weight parameters. The Hessian matrix is calculated using the Pearlmutter (1994) technique and on training, α is first initialized to a small random value. Training is conducted by adjusting the weight parameters by utilizing the scaled conjugate gradient algorithm in order to minimize equation 9.7. At a point when a minimum is achieved, α_{MP} is obtained by using equation 9.16 by holding {w} constant. The method stipulated above is then repeated over again until a self-consistent solution for both α and {w} is accomplished.

Model Order Selection

The evidence framework compares artificial neural network models of different complexity by evaluating the evidence of the model. For an artificial neural network model H_i, the evidence is given by:

$$p(D|H_i) = \int p(D|\alpha, H_i) p(\alpha|H_i) d\{w\} \tag{9.18}$$

MacKay (1993) derives an expression for the logarithm of equation 9.18 as follows:

$$\ln p(D|H_i) = -S(\{w\}) - \frac{1}{2}\ln \det([A]) + \frac{W}{2}\ln \alpha + \ln(2^N N!) + \frac{1}{2}\ln\left(\frac{4\pi}{\gamma}\right) \tag{9.19}$$

Studies have reported a correlation between the model evidence and generalization error of a neural network (MacKay, 1992). The expression in equation 9.19 permits the model with the highest evidence to be chosen without having to divide the data into a training and validation set.

Moderated Output

Artificial neural network estimate a mapping function by interpolating in a region of function space generated by a finite set of training data. The confidence of predictions within this space will be higher than predictions outside of this space. Moderated outputs were proposed by MacKay (1992) to adjust the artificial neural network output by an amount that is reflective of the uncertainty of the weight parameters. MacKay (1992) derived an equation for the moderated output by assuming that the activation function of the output neuron, a, is a locally linear function of artificial neural network weights. The expression can be written as follows:

$$P\big(C_1\big|\{x\}\big) = g\big(K(s)\alpha_{MP}\big) \tag{9.20}$$

where:

$$K(s) = \big(1 + \pi s^2 / 8\big)^{-1/2} \tag{9.21}$$

and:

$$s^2(\{x\}) = \{g\}^T[A]^{-1}\{g\} \tag{9.22}$$

Here, a_{MP} is the activation of the output neuron when $\{w\}=\{w_{MP}\}$ and g is the derivative of a with respect to $\{w\}$.

CLASSIFICATION OF HIV USING MISSING DATA APPROACHES AND NEURAL NETWORKS

The aim of this chapter is to examine the method which is based on a multi-layer perceptron autoassociative models combined with genetic algorithm to classify the HIV status of an individual based on the demographic properties and this method is characterized as the missing data estimation error method. This method is tested on the classification of the HIV status of individuals using a data set obtained from the South African antenatal sero-prevalence survey. The method is then compared to the feed-forward multi-layer perceptron neural network models that are trained using Bayesian approach where the posterior probability is approximated by a Gaussian distribution and also tested on the HIV modeling problem.

The literature review conducted, in this chapter, has shown that models for HIV prediction and classification have been developed using conventional feed-forward neural networks architectures and have worked well. This chapter thus focuses on a methodology for HIV classification from demographic properties using the autoassociative neural networks and genetic algorithm.

HIV Classification Using Autoassociative Networks

The NETLAB toolbox is used to create and train the autoassociative multi-layer perceptron architecture. This estimated value from the autoassociative network and genetic algorithm is a continuous value representing the HIV status. A threshold is thus required to convert the HIV output node value to a binary value, representative of the HIV class of the individual. Figure 9.3 shows the implementation of this proposed model in a flowchart.

HIV Classification Using Neural Networks

In this model, the NETLAB toolbox (Nabney, 2003) is applied to generate and train the multilayer perceptron neural network architecture. The network implemented consists of an input layer, representing different demographic inputs of an individual, mapped to an output layer representing the HIV status

Figure 9.3. Flow chart of the missing data estimation classification model

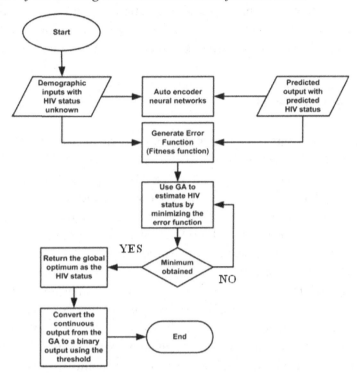

of an individual via the hidden layer. The network thus maps the demographic inputs of individuals to the HIV status. This network is shown in Figure 9.2. The neural network equation can be written as in equation 9.4. In this model, nonetheless, the output vector $\{y\}$ represents the HIV status of the individual as indicated in equation 9.6. The network is as a result trained to find the relationship between the HIV status of the individual and the individual's demographic input properties.

An error, however, exists between the individual's predicted HIV status which is the output vector $\{y\}$ in equation 9.1 and the individual's actual HIV status, target vector $\{t\}$ during training, which can be expressed as the difference between the target and output vector. For the neural network HIV classification, the mean square error function between the target output vector $\{t\}$ and the output vector $\{y\}$ is insufficient as a classification accuracy measure, as it only indicates the total number of correct classifications. A confusion matrix is thus constructed and the accuracy is obtained from the confusion matrix. The accuracy can be formulated as follows:

$$Accuracy = \frac{TN + TP}{TN + FN + TP + FP} \qquad (9.23)$$

Here *TN* indicates true negatives (where network predicts an HIV negative person as negative), *FP* indicates false positives (where network predicts an HIV negative person as positive), *FN* indicates false negatives (where network predicts an HIV positive person as negative) and *TP* indicates true positives (where network predicts an HIV positive person as positive).

The accuracy function is then used as the fitness function in the genetic algorithm to obtain the optimal neural network parameters. As mentioned before, genetic algorithm is used as it finds the maximum value of the fitness function, which is required in this case. Genetic algorithm is also used to obtain the threshold value to convert the continuous network output to a binary value representative of HIV.

RESULTS AND DISCUSSION

The demographic and medical data, used in this chapter are obtained from the South African antenatal sero-prevalence survey (Department of Health, 2001). This is a national survey, and pregnant women attending selected public health care clinics and who are participating for the first time in the survey are eligible to participate. The variables obtained are shown in Table 9.1. These include: age of mother, age of partner, educational level of mother, gravidity (number of complete or incomplete pregnancies), parity (number of pregnancies reaching viability), province of origin, race of mother, and region of origin. The qualitative variables such as the province of origin, race of mother and region of origin are encoded to integers. For example, the encoding scheme for race is shown in Table 9.2. The HIV status is encoded using an integer scheme, whereby a 1 represents a positive HIV status meanwhile a 0 represents a negative HIV status. The parameter distributions are also listed in Table 9.1. A total of 1986 training inputs are provided for the network. The genetic algorithm used for the autoassociative network model proposed in this chapter and the neural network model used arithmetic crossover, non-uniform mutation and normalized geometric selection. The probability of crossover is chosen to be 0.75 as proposed in Marwala and Chakraverty (2006). The probability of mutation is chosen to be 0.0333 as recommended by Marwala and Chakraverty (2006). Genetic algorithm has a population of 40 and is run for 150 generations.

Table 9.1. Summary of input and output variables

Variable	Type	Range
Input variables		
Education	Integer	0 – 13
Age Group	Integer	14 – 60
Age Gap	Integer	1 – 7
Gravidity	Integer	0 - 11
Parity	Integer	0 – 40
Race	Integer	1 – 5
Province	Integer	1 – 9
Region	Integer	1 – 36
RPR	Integer	0 – 2
WTREV	Continuous	0.64–1.27
Output variable		
HIV Status	Binary	[0, 1]

The first experiment investigates the use of autoassociative networks for HIV classification. An autoassociative network with 9 inputs and 9 outputs is constructed and several numbers of hidden units are investigated, using Matlab (Mathworks, 2004). A genetic algorithm method is implemented to achieve the optimum number of hidden units and yields an optimum number of hidden units of 2, hence the structure 9–2–9. Linear optimization using the mean square error versus hidden units is also investigated. As shown in Figure 9.4, the linear optimization yields 6 hidden units as the optimal network that gives the best prediction since the error does not change significantly from 6 units onwards (the difference in error is about 8.5% from 6 hidden units to 20 hidden units). It must be noted, however, that it is generally assumed that the best autoassociative network is the one that has the lowest possible number of hidden units (Kramer, 1991). A hidden unit of 2 is thus used as the optimal autoassociative network number of hidden units.

The performance analysis for the autoassociative network model is based on classification accuracy and the area under the receiver operating characteristics (ROC) curve (Heckerling, 2002). A receiver operating characteristic or merely ROC curve is a graphical representation of the sensitivity versus (1 - specificity) for a binary classifier system as its discrimination threshold is varied. The sensitivity is a statistical measure that measures the extent to which a binary classifier correctly identifies a class while specificity is a measure of the extent to which a binary classifier correctly identifies the negative cases.

Table 9.2. Example of an encoding scheme of a qualitative parameter (race)

Qualitative Parameters (race)	Integer Encoding
European	1
African	2
Mixed	3
Indian	4
Other	5

Figure 9.4. The prediction error versus the number of hidden nodes

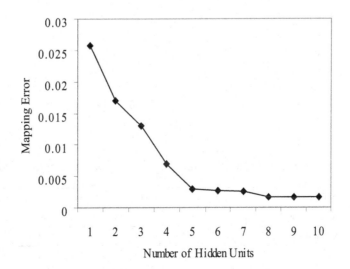

The ROC curve is a powerful tool that has been used in a wide range of applications. Lasko et al. (2005) used ROC in the classification of biomedical informatics whereas Lind et al. (2002) used the receiver operating characteristic curves to evaluate prediction methods for radiation-induced symptomatic lung injury. Mohamed (2006) used the receiver operating characteristics curves for dynamic protein classification whereas Yonelinas and Parks (2007) used the ROC curve in recognition memory. Further information on the ROC curve can be found in the work of Beiden, Wagner, and Campbell (2000); Dorfman, Berbaum, and Lenth (1995); Halpern et al. (1996) and Zou et al. (2003). The ROC can also be represented similarly by plotting the fraction of true positives (TPR = true positive rate) versus the fraction of false positives (FPR = false positive rate) and this is known as a confusion matrix. The proposed autoassociative network model obtained an HIV classification accuracy of 92%. The confusion matrix obtained for the above network is shown in Table 9.3. The ROC curve for this classification is shown in Figure 9.5 and the area under the curve is computed as 0.86, thus giving a very good classifier in accordance to ROC curves (Au, Eissa, & Jones, 2004; Metz, 2006; Brown & Davis, 2006).

The second experiment investigates the use of conventional feed-forward neural network MLP architecture to classify the HIV status of an individual using the demographic properties as inputs. The MLP is constructed with 9 inputs and 1 output. A genetic algorithm is then used to obtain the optimal structure and yields an optimal number of hidden units of 77, hence the structure is 9–77–1.

The performance analysis for this network model is also based on classification accuracy and the area under the ROC curve. This network gives an accuracy of 84%. The confusion matrix obtained for the above network is shown in Table 9.4. The ROC curve obtained for this classification is shown in Figure 9.6 and the area under this ROC curve obtained is 0.80, which in accordance to ROC curves benchmarking results is a very good classifier (Brown & Davis, 2006).

Figure 9.5. ROC curve for the autoassociative network classifier

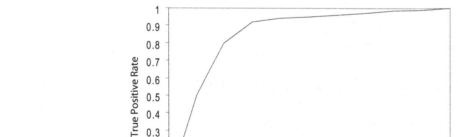

Table 9.3. Classifier confusion matrix from the missing data estimation autoencoder network classifier

Confusion Matrix	Predicted Positive	Predicted Negative
Actual Positive	899	94
Actual Negative	65	928

The rationale why the autoassociative network performs better than the conventional feed-forward neural network can be attributed to the fact there is a great deal of interdependencies in the demographical data. For example: Which causes what between HIV and geographic location? Does geographical location increase the risk of HIV? Or does the high HIV risk causes people to move to another geographic location? The difference in performance can also be attributed to the fact that in the autoassociative network, classification is done by choosing the best fitting model using probability distributions. The class of the network with the smallest reconstruction error is selected. Conventional feed-forward neural networks on the other hand just map an input vector to an output vector using scenario and encodes the classes directly. This plays a role because, for non-linear models such as in the HIV model, it is usually difficult to compute the derivatives for the scenarios since they require that all the possible representations that could have been used for each particular observed input vector be integrated. The distance measure in classification is thus better minimized in the autoassociative network than in the conventional feed-forward network model. Furthermore, in conventional feed-forward neural networks the errors from the input to the output are integrated and focused while in the autoassociative network the errors of the input are distributed.

CONCLUSION

A method based on autoassociative neural networks and genetic algorithms is proposed to classify the HIV status of an individual from demographic properties. This approach is proposed in order to investigate

Figure 9.6. ROC curve for the conventional feed-forward neural network classifier

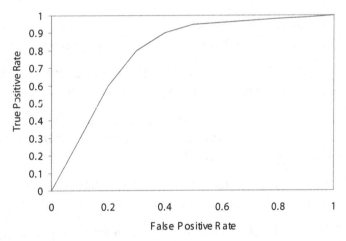

Table 9.4. Classifier confusion matrix from the multi-layer perceptron classifier

Confusion Matrix	Predicted Positive	Predicted Negative
Actual Positive	680	313
Actual Negative	0	993

whether using autoassociative networks improves on the accuracy of classification, of an individual's HIV status, from demographic properties. The proposed method is tested on an HIV data set obtained from the South African antenatal sero-prevalence survey of 2001. The method is then compared to a conventional feed-forward neural network model, implemented using the MLP architecture. A classification accuracy of 92% is obtained for the autoassociative network compared to 84% obtained for the conventional feed-forward neural network model implementation. The area under the ROC curve for the autoassociative network classifier is computed as 0.86 compared to 0.80 computed for the conventional feed-forward neural network classifier. The results thus suggest that autoassociative network models are more accurate and better classifiers for the HIV model than conventional feed-forward neural network models for the data used in this chapter.

FUTURE WORK

In this chapter a classification system based on the missing data estimation model is proposed and then compared to the Bayesian neural networks using the HIV data. For future research it is recommended that the proposed model be applied to other databases to discover the relationships between the nature of the data and the effectiveness of the proposed method. Furthermore, the missing data method should be tested using other algorithms such as particle swarm optimization as opposed to just genetic algorithm. On the Bayesian networks, the proposed method should be tested using Markov Chain Monte Carlo instead of assuming Gaussian approximation to simplify the calculations.

REFERENCES

Abdella, M. I. (2005). *The use of genetic algorithms and neural networks to approximate missing data in database.* Unpublished master's thesis, University of the Witwatersrand, Johannesburg.

Abdella, M., & Marwala, T. (2005a). Treatment of missing data using neural networks. In *Proceedings of the IEEE International Joint Conference on Neural Networks,* Montreal, Canada (pp. 598-603).

Abdella, M., & Marwala, T. (2005b). The use of genetic algorithms and neural networks to approximate missing data in database. In *Proceedings of the IEEE 3rd International Conference on Computational Cybernetics,* Mauritius (pp. 207-212).

Alkan, A., Koklukaya, E., & Subasi, A. (2005). Automatic seizure detection in EGG using logistic regression and artificial neural network. *Journal of Neuroscience Methods, 148,* 167–176.

Au Y. H. Eissa, J, S., & Jones B. E. (2004). Receiver operating characteristic analysis for the selection of threshold values for detection of capping in powder compression. *Ultrasonics, 42,* 149-153.

Barnard, G. A. (1958). Studies in the history of probability and statistics: IX. Thomas Bayes' essay towards solving a problem in the doctrine of chances. *Biometrika 45,* 293–295.

Bayes, T. (1763). *An essay towards solving a problem in the doctrine of chances.* Communicated by Mr. Price, in a letter to John Canton *Philosophical Transactions, Giving Some Account of the Present Undertakings, Studies and Labours of the Ingenious in Many Considerable Parts of the World.*

Beiden, S. V., Wagner, R. F., & Campbell, G. (2000). Components-of-variance models and multiple-bootstrap experiments: An alternative method for random-effects, receiver operating characteristic analysis. *Academic Radiology, 7,* 341-349.

Bishop, C. M. (1995). *Neural networks for pattern recognition. Oxford University Press,* Oxford.

Brown, C. D., & Davis, H. T. (2006). Receiver operating characteristics curves and related decision measures: A tutorial. *Chemometrics and Intelligent Laboratory Systems, 80,* 24-38.

Buntine, W. L., & Weigend, A. S. (1991). Bayesian back-propagation. *Complex Systems, 5,* 603-643.

Cai, Y-D. & Chou, K-C. (1998). Artificial neural network model for predicting HIV protease cleavage sites in protein. *Advances in Engineering Software, 29,* 119-128.

Cawley, G. C., & Talbot, N. L. C. (2005). The evidence framework applied to sparse kernel logistic regression. *Neurocomputing, 64,* 119-135

Chamjangali, M. A, Beglari, M., & Bagherian, G. (2007). Prediction of cytotoxicity data (CC50) of anti-HIV 5-pheny-l-phenylamino-1H-imidazole derivatives by artificial neural network trained with Levenberg–Marquardt algorithm. *Journal of Molecular Graphics and Modelling, 26,* 360-367.

Davis, L. (1991). *Handbook of genetic algorithms.* New York: Van Nostrand.

Deo, M. C., & Jagdale, S. S. (2003). Prediction of breaking waves with neural networks. *Journal of Ocean Engineering, 30,* 1163–1178.

Department of Health. (2001). *Republic of South Africa, HIV syphilis survey data,* Retrieved March 7, 2005, from http:// www.health.gov.za.

Dorfman, D. D., Berbaum, K. S., & Lenth, R. V. (1995). Multireader, multicase receiver operating characteristic methodology: A bootstrap analysis. *Academic Radiology, 2,* 626-633.

Fee, E., & Krieger, N. (1993). Understanding AIDS: Historical interpretations and limits of biomedical individualism. *American Journal of Public Health, 83,* 1477–1488.

Fernández, M., & Caballero, J. (2006). Modeling of activity of cyclic urea HIV-1 protease inhibitors using regularized-artificial neural networks. *Bioorganic & Medicinal Chemistry, 14,* 280-294.

Frolov, A., Kartashov, A., Goltsev, A., & Folk, R. (1995). Quality and efficiency of retrieval for Willshaw-like autoassociative networks. Part II: Recognition. *Network: Computational Neural Systems, 6,* 535–549.

Goldberg, D. E. (1989). *Genetic algorithms in search optimization and machine learning.* Addison-Wesley, Reading.

Habtemariam, E. (2005). *Artificial intelligence for conflict management.* Unpublished master's thesis, University of the Witwatersrand, Johannesburg.

Habtemariam, E., Marwala, T., & Lagazio, M. (2005). Artificial intelligence for conflict management. In *Proceedings of the IEEE International Joint Conference on Neural Networks,* Montreal, Canada (pp. 2583-2588).

Halpern, E. J., Albert, M., Krieger, A. M., Metz, C. E., & Maidment, A. D. (1996). Comparison of receiver operating characteristic curves on the basis of optimal operating points. *Academic Radiology, 3*, 245-253.

Heckerling, P. S. (2002). Parametric receiver operating characteristic curve analysis using mathematica. *Computer Methods and Programs in Biomedicine, 69*, 65-73.

Hines, J. W., Robert, E. U., & Wrest, D. J. (1998). Use of autoassociative neural networks for signal validation. *Journal of Intelligent Robotic Systems, 21*, 143–154.

Holland, J. (1975). *Adaptation in natural and artificial systems.* University of Michigan Press, Ann Arbor.

Hudson, D. L., & Cohen, M. E. (2000). *Neural networks and artificial intelligence for biomedical engineering.* New Jersey: IEEE Press.

Kim, H., Zhang, Y., Heo, Y-S, Oh, H-B., & Chen, S-S. (2008). Specificity rule discovery in HIV-1 protease cleavage site analysis. *Computational Biology and Chemistry, 32*, 72-79.

Kramer, M. A. (1991). Nonlinear principal component analysis using autoassociative neural Networks. *American Institute of Chemical Engineering Journal, 37*, 233–234.

Lasko, T. A., Bhagwat, J. G., Zou, K. H., & Ohno-Machado, L. (2005). The use of receiver operating characteristic curves in biomedical informatics. *Journal of Biomedical Informatics, 38*, 404-415.

Laumann, E. O., & Youm, Y. (1999). Racial/ethnic group differences in the prevalence of sexually transmitted diseases in the United States: A network explanation. *Sexually Transmitted Diseases, 26*, 250–261.

Lee, C., & Park, J. (2001). Assessment of HIV/AIDS-related health performance using an artificial neural network. *Journal of Information Management, 38*, 231–238.

Leke, B. B., & Marwala, T. (2006). Ant colony optimization for missing data estimation. In *Proceeding of the Pattern Recognition of South Africa* (pp. 183-188).

Leke, B. B., Marwala, T., & Tettey, T. (2006). Autoencoder networks for HIV classification. *Current Science, 91*, 1467-1473.

Leke, B. B., Marwala, T., & Tettey, T. (2007). Using inverse neural network for HIV adaptive control. *International Journal of Computational Intelligence Research, 3*, 11-15.

Lind, P. A., Marks, L. B., Hollis, D., Fan, M., Zhou, S-M., Munley, M. T., Shafman, T. D., Jaszczak, R. J., & Coleman, R. E. (2002). Receiver operating characteristic curves to assess predictors of radiation-induced symptomatic lung injury. *International Journal of Radiation Oncology, Biology and Physics, 54*, 340-347.

Lisboa, P. J. G. (2002). A review of evidence of health benefit from artificial neural networks in medical intervention. *Neural Networks, 15*, 11–39.

Lu, P. J., & Hsu, T. C. (2002). Application of autoassociative neural network on gas-path sensor data validation. *Journal of Propulsion and Power, 18*, 879–888.

MacKay, D. J. C. (1992). *Bayesian methods for adaptive models.* Unpublished doctoral dissertation, *California Institute of Technology*, Pasadena.

MacKay, D. J. C. (1993). Comparison of approximate methods for handling hyperparameters. *Neural Computation, 11*, 1035-1068.

Marwala, T. (2001). Probabilistic fault identification using a committee of neural networks and vibration data. *Journal of Aircraft, 38*, 138–146.

Marwala, T. (2007). *Computational intelligence for modelling complex systems.* Delhi: Research India Publications.

Marwala, T., & Chakraverty, S. (2006). Fault classification in structures with incomplete measured data using autoassociative neural networks and genetic algorithm. *Current Science, 90*, 542–549.

Marwala, T., & Lagazio, M. (2004). Modelling and controlling interstate conflict. In *Proceedings of the IEEE International Joint Conference on Neural Networks,* Budapest, Hungary (pp. 1233-1238).

Marwala, T., & Lagazio, M. (2005). Assessing different Bayesian neural network models for militarized Interstate Dispute. *Social Science Computer Review, 24*, 1-12.

Marwala, T., Mahola, U., Nelwamondo, F. V. (2006). Hidden Markov models and Gaussian mixture models for bearing fault detection using fractals. In *Proceedings of the IEEE International Joint Conference on Neural Networks,* British Columbia, Canada (pp. 5876-5881).

Marwala, T., Tettey, T., & Chakraverty, S. (2006). *Fault classification in structures using pseudomodal energies and neuro-fuzzy modelling.* Paper presented at the Asia-Pacific Workshop on Structural Health Monitoring, Yokohama, Japan.

Mathworks. (2004). *Matlab 7.1 manual.* New York: Mathworks Press.

Metz, C. E. (2006). Receiver operating characteristic analysis: A tool for the quantitative Evaluation of observer performance and imaging systems. *Journal of the American College of Radiology, 3*, 413-422.

Michalewicz, Z. (1996). *Genetic algorithms + Data structures = Evolution programs.* 3rd Edition. Berlin, Springer.

Milac, A-L, Avram, S., & Petrescu, A-J. (2006). Evaluation of a neural networks QSAR method based on ligand representation using substituent descriptors: Application to HIV-1 protease inhibitors. *Journal of Molecular Graphics and Modelling, 25*, 37-45.

Mohamed, S. (2006). Dynamic protein classification: adaptive models based on incremental learning strategies. Unpublished master's thesis, University of the Witwatersrand, Johannesburg.

Mohamed, S., & Marwala. T. (2005). Neural network based techniques for estimating missing data in databases. In *Proceedings of the 16th Annual Symposium of the Pattern Recognition Society of South Africa),* Langebaan, South Africa (pp. 27-32).

Mohamed, S., Rubin, D., & Marwala, T. (2006a). Multi-class protein sequence classification using fuzzy ARTMAP. In *Proceedings of the IEEE International Conference on Systems, Man and Cybernetics,* Taiwan (pp. 1676-1681).

Mohamed, N., Rubin, D.M., & Marwala, T. (2006b). Detection of epileptiform activity in human EEG signals using Bayesian neural networks. *Neural Information Processing - Letters and Reviews, 10*, 1-9.

Mohamed, S., Rubin, D., & Marwala, T. (2007). Incremental learning for classification of protein sequences. In *Proceedings of the International Joint Conference on Neural Networks*, BC, Canada (pp. 11-24).

Mohamed, S., Tettey, T., & Marwala, T. (2006). An extension neural network and genetic algorithm for bearing fault classification. In *Proceedings of the IEEE International Joint Conference on Neural Networks,* Orlando, Florida (pp. 7673-7679).

Møller, M. F. (1993). A scaled conjugate gradient algorithm for fast supervised learning. *Neural Networks, 6*, 525-533.

Nabney, I. T. (2003). *NETLAB: Algorithms for pattern recognition*. London: Springer–Verlag.

Narendra, K., & Lewis, F. (2001). Introduction to the special issue on neural network feedback control. *Automatica, 37*, 1147–1148.

Neal, R. M. (1992). *Bayesian training of backpropagation networks by the hybrid Monte Carlo method* (Tech. Rep. CRG-TR-92-1). Toronto, Canada: Department of Computer Science, University of Toronto.

Nelwamondo, F. & Marwala, T. (2006). Fault detection using Gaussian mixture models, mel-frequency ceptral coefficient and kurtosis. In *Proceedings of the IEEE International Conference on Systems, Man and Cybernetics* Taiwan (pp. 290-295).

Ohno-Machado, L. (1996). Sequential use of neural networks for survival prediction in AIDS. In *Proceedings American Medical Information Association Annual Fall Symposium,* Los Angeles, California (pp. 170–174).

Pearlmutter, B. A. (1994). Fast exact multiplication by the Hessian. *Neural Computation, 6*, 147-160.

Penny, W. D., & Roberts, S. J. (1999). Bayesian neural networks for classification: How useful is the evidence framework? *Neural Networks, 12*, 877-892.

Poundstone, K., Strathdee, S., & Celectano, D. (2004). The social epidemiology of human immunodeficiency virus/acquired immunodeficiency syndrome. *Epidemiologic Reviews, 26*, 22–35.

Rafiq, M. Y., Bugmann, G., & Easterbrook, D. J. (2001). Neural network design for engineering applications. *Computers and Structures, 79*, 1541–1552.

Root-Bernstein, R. (1993). *Rethinking AIDS*. New York: MacMillan.

Sardari, S., & Sardari, D. (2002). Applications of artificial neural network in AIDS research and therapy. *Current Pharmaceutical Design, 8*, 659–670.

Sawa, T., & Ohno-Machado, L. (2003). A neural network-based similarity index for clustering DNA microarray data. *Computational Biology and Medicine, 33*, 1–15.

Smauoi, N., & Al-Yakoob, S. (2003). Analyzing the dynamics of cellular flames using Karhunen–Loeve decomposition and autoassociative neural networks. *Society for Industrial and Applied Mathematics, 24,* 1790–1808.

Sohn, H., Worden, K. & Farrar, R. C. (2001). Novelty detection using auto-associative neural network. Paper presented at the Symposium on Identification of Mechanical Systems: International Mechanical Engineering Congress and Exposition, New York.

Sugiyama, M., & Ogawa, H. (2002). Optimal design of regularization term and regularization parameter by subspace information criterion. *Neural Networks, 15,* 349-361.

Svozil, D., Kvasnicka, V., & Pospichal, J. (1997). Introduction to multilayer feed-forward neural networks. *Chemometrics and Intelligent Laboratory Systems, 39,* 43–62.

Szpurek, D., Moszynski, R., Smolen, A., & Sajdak, S. (2005). Artificial neural network computer prediction of ovarian malignancy in women with adnexal masses. *International Journal of Gynaecology and Obstetrics, 89,* 108–113.

Tan, A-H., & Pan, H. (2005). Predictive neural network for gene expression data analysis. *Neural Networks, 18,* 297–306.

Tandon, R., Adak, S., & Kaye, J. (2006). A neural network for longitudinal studies in Alzheimer's disease. *Artificial Intelligence in Medicine, 36,* 245–255.

Tettey, T. (2006). A computational intelligence approach to modelling interstate conflict: Conflict and causal interpretations. Unpublished master's thesis, University of the Witwatersrand, Johannesburg.

Tettey, T, & Marwala, T. (2006). Controlling interstate conflict using neuro-fuzzy modeling and genetic algorithms. In *Proceedings of the 10ᵗʰ IEEE International Conference on Intelligent Engineering Systems,* London, UK (pp. 30-44).

Tim, T.H. (2006). *Predicting HIV status using neural networks and demographic factors.* Unpublished master's thesis, University of the Witwatersrand, Johannesburg.

Vilakazi, C. D., & Marwala, T. (2006a). Bushing fault detection and diagnosis using extension neural network. In *Proceedings of the 10ᵗʰ IEEE International Conference on Intelligent Engineering Systems,* London, UK (pp. 170-174).

Vilakazi, B. C. & Marwala, T. (2006b). Application of feature selection and fuzzy ARTMAP to intrusion detection. In *Proceedings of the IEEE International Conference on Systems, Man and Cybernetics,* Taiwan (pp. 4880-4885).

Yonelinas, A. P., & Parks, C. M. (2007). Receiver operating characteristics (ROCs) in recognition memory: A review. *Psychological Bulletin, 133,* 800-832.

Zou, K. H., Warfield, S. K., Fielding, J. R., Tempany, C. M. C., Wells, W. M., Kaus, M. R., Jolesz, F. A., & Kikinis, R. (2003). Statistical validation based on parametric receiver operating characteristic analysis of continuous classification data. *Academic Radiology, 10,* 1359-1368.

Chapter X
Optimization Methods for Estimation of Missing Data

ABSTRACT

This chapter presents various optimization methods to optimize the missing data error equation, which is made out of the autoassociative neural networks with missing values as design variables. The four optimization techniques that are used are: genetic algorithm, particle swarm optimization, hill climbing and simulated annealing. These optimization methods are tested on two datasets, namely, the beer taster dataset and the fault identification dataset. The results that are obtained are then compared. For these datasets, the results indicate that genetic algorithm approach produced the highest accuracy when compared to simulated annealing and particle swarm optimization. However, the results of these four optimization methods are the same order of magnitude while hill climbing produces the lowest accuracy.

INTRODUCTION

Missing data problem creates a variety of problems in many disciplines which depend on good access to accurate data and because of this reason, techniques to deal with missing data have been the subject of research in statistics, mathematics and other disciplines (Yuan, 2000; Allison, 2000; Rubin, 1978). In the engineering domain the missing data problem occurs because of a number of reasons including sensors failures, inaccessibility of measuring surfaces and so on. If the data component from the missing data is needed for decision-making purpose then it becomes very important that the missing values be estimated.

A number of methods have been proposed to deal with missing data (Little & Rubin, 2002). These include the use of Expectation-Maximization methods (Dempster, Liard, & Rubin, 1977), autoassociative neural networks (Abdella & Marwala, 2006), rough sets approach (Nelwamondo & Marwala, 2007) and many more. Wang (2008) successfully dealt with the problem of the probability density estimation in the existence of covariates when data were missing at random whereas Shih, Quan, and Chang (2008) devised a method for estimating the mean of the data which included non-ignorable missing values. Keesing et al. (2007) introduced missing data estimation procedure for 3D spiral CT image reconstruction while Wooldridge (2007) introduced inverse probability weighted estimation for general missing data problems. Kuhnen, Togneri, and Nordholm (2007) introduced the missing data estimation procedure for speech recognition that uses short-time-Fourier-transform while Mojirsheibani (2007) introduced a generalized empirical approach to non-parametric curve estimation with missing data.

In this chapter, the methods that use autoassociative neural networks for missing data estimation are investigated. These techniques essentially are made up of two main components: the autoassociative neural network component, which is a network that maps the input onto itself (Hines & Garvey, 2006) and the optimization component (Snyman, 2005). This chapter investigates a number of optimization methods for missing data estimation and then compares these methods. Spiller and Marwala (2007) used genetic algorithms, particle swarm optimization and simulated annealing for warp control point placement and found that particle swarm optimization is ideally suited for this problem than the other methods.

The optimization methods that are investigated for missing data estimation, in this chapter, are genetic algorithm (Garcia-Martinez et al. 2008; Tavakkoli-Moghaddam, Safari, & Sassani, 2008), simulated annealing (Güngör & Ünler, 2007; Chen & Su, 2002), particle swarm optimization (van den Bergh & Engelbrecht, 2006; Marwala, 2007a) and hill climbing (Tanaka, Toumiya, & Suzuki, 1997; Johnson & Jacobson, 2002). In earlier chapters, it is assumed that the optimization method to be used for missing data estimation to minimize the missing data error equation must have global properties. In this chapter, three global optimization methods genetic algorithm, simulated annealing and particle swarm optimization are compared to the local optimization method, hill climbing, with a sole purpose of answering the crucial question that states that: Is the missing data estimation error equation, which is derived in detail in Chapter II, best solved by global optimization method or by local optimization method?

MISSING DATA ESTIMATION APPROACHES

As explained earlier, the logical approach to handle missing data depends upon how data points become missing. As indicated in earlier chapters, Little and Rubin (1987) as well as Rubin (1978) have demonstrated that there are three types of missing data mechanisms and these are: Missing Completely at Random, Missing at Random and Missing Not At Random. Depending on the mechanism of missing data, currently various methods are used to treat missing data. More information with detailed discussions on the various missing data estimation methods used to handle missing data can be found in Allison (2000); Rubin (1978); Little and Rubin (1987); Mohamed and Marwala (2005); Leke, Marwala, and Tettey (2006); and Nelwamondo (2008). In this chapter, as in earlier chapters, the method for estimating missing data is able to estimate missing data irrespective of the missing data mechanism as long as the rules that describe inter-relationships in the data are known.

The missing data estimation algorithm considered in this chapter involves a neural network which is trained to recall itself and is, therefore, called an autoassociative neural network. Successful deployment of autoassociative neural network include that by Pomi and Olivera (2006) who developed a context-sensitive autoassociative memories and applied this for medical diagnosis, in object recognition (Caldara & Abdi, 2006; Yokoi et al., 2004), in nuclear engineering (Marseguerra, Zio, & Marcucci, 2006), in mechanical engineering (Marwala & Chakraverty, 2006), in fault detection of gearboxes (Del Rincon et al., 2005) and in spotting consonants in speech (Gangashetty, Sekhar, & Yegnanarayana, 2004). As described in earlier chapters, the missing data estimation error equation can be written as follows:

$$e = \left\| \left(\left\{ \begin{matrix} \{X_k\} \\ \{X_u\} \end{matrix} \right\} - f\left(\left\{ \begin{matrix} \{X_k\} \\ \{X_u\} \end{matrix} \right\}, \{w\} \right) \right) \right\|$$

(10.1)

In this equation $\{X_k\}$ is the known measurement, $\{X_u\}$ represents the missing values, $\| \ \|$ is Euclidean norm and f is the autoassociative model with mapping weight vector $\{w\}$. In this chapter, the autoassociative network function f is the multi-layered perceptron and has been found to be highly successful in modeling complex relationships in areas such as the prediction of nitrogen oxide and nitrogen dioxide concentrations (Juhos, Makra, & Tóth, 2008), in the evaluation of one dimensional tracer concentration profile in a small river (Piotrowski et al, 2007), on fault classification in cylinders (Marwala, Mahola, & Chakraverty, 2007) and in digital modulator classification (Ye & Wenbo, 2007).

To approximate the missing input values, equation 10.1 is minimized using an optimization method. In this chapter, genetic algorithm, particle swarm optimization, simulated annealing and hill climbing are all chosen, implemented and then compared. The details on these algorithms are outlined in the next sections.

OPTIMIZATION TECHNIQUES

Simple Genetic Algorithms (GA)

Genetic algorithm (GA) is a population based search method that is widely used due to its ease of implementation, intuitiveness and its ability to solve highly non-linear optimization problems. GA is a stochastic search procedure for combinatorial optimization problems based on the mechanism of natural selection (Malve & Uzsoy, 2007). Genetic algorithm is a particular class of evolutionary algorithms that uses techniques inspired by evolutionary biology such as inheritance, mutation, selection, and crossover (Youssef, Sait, & Adiche, 2001). The fitness function measures the quality of the represented solution, in this chapter, the solution to the missing data estimation error equation as described above. In this chapter, GA represents the missing variables with binary strings of 0's and 1's and these are referred to as chromosomes.

Kao, Zahara, and Kao (2008) hybridized genetic algorithm and particle swarm optimization (PSO) for global optimization of multi-modal functions. This hybrid system incorporates ideas from GA and PSO and generates individuals in a new generation by crossover and mutation operations which originate from GA as well as notions from PSO such as recalling the best in the group and the best gene that the individuals themselves have encountered. The experimental investigations utilizing a group of 17 multi-modal test functions shows that the hybrid GA-PSO method is superior to the other four search methods studied in terms of solution quality and convergence rates. The main shortcoming with this approach is that it does not make physical sense because there is no natural phenomenon that uses genetic evolution and swarm intelligence at the same time.

González and Pérez (2001) conducted an experimental investigation on the search method in a learning algorithm and compared hill-climbing methods with genetic algorithms while Polgár et al. (2000) compared genetic algorithms, simulated annealing and hill climbing algorithms for the evaluation of ellipsometric measurements random search. Yao and Chu (2008) used genetic algorithm to obtain optimal replenishment cycles that optimize warehouse space requirements. They used genetic algorithm to identify the optimal replenishment schedule and showed that GA was considerably better than the previously used heuristic methods.

Essafi, Mati, and Dauzere-Peres (2008) used a genetic local search algorithm for minimizing total weighted tardiness in job-shop scheduling. Genetic algorithm was hybridized with a local search method that utilized the longest path method on a disjunctive graph model and a design of experiments method was used to adjust the parameters and operators of the proposed algorithm. It was established that the effectiveness of genetic algorithm does not depend on the schedule builder when an iterated local search was used as opposed to previous observations.

Calvete, Gale, & Mateo (2008) solved linear bi-level problems, where two optimization problems have the constraint region of the first level problem inherently obtained by an additional optimization problem, through genetic algorithms. The results obtained demonstrated the effectiveness of the proposed procedure.

Defersha and Chen (2008) proposed genetic algorithm for solving an integrated cell formation and lot sizing problems while taking into account of the product quality. They developed a mathematical programming model that followed integrated method for cell arrangement and lot sizing within a dynamic manufacturing environment which took into account the effect of lot sizes on product quality. The results obtained demonstrated that the technique proposed was effective in searching for near optimal solutions.

Other applications of genetic algorithms include in ground water resource management by Sidiropoulos and Tolikas (2008), FIR filter design (Abu-Zitar, 2008), optimal batch plant design (Ponsich et al., 2008), diseases diagnosis (Yan et al., 2008), crack detection in mechanical systems (Vakil-Baghmisheh et al., 2008), scheduling problems (Valls, Ballestin, & Quintanilla, 2008), portfolio selection (Lin & Liu, 2008), finite element updating (Marwala, 2002), brewing control (Marwala, 2004), fault identification (Marwala & Chakraverty, 2006), HIV analysis (Crossingham & Marwala, 2007b), incremental learning (Hulley & Marwala, 2007), stock-market prediction (Marwala et al., 2001), genetic programming (Marwala, 2007a) interstate conflict (Tettey & Marwala, 2006) and bearing fault diagnosis (Mohamed, Tettey, & Marwala, 2006).

To perform optimization, GA employs three main operators to propagate its population from one generation to the next. The first operator is the selection. During each successive generation, a proportion of the population is selected to breed a new generation. This is performed on the basis of the "survival of the fittest" phenomenon. After each generation, the missing data estimation error equation, shown equation 10.1, is evaluated. This process continues until the termination criterion is reached. Several selection functions are available and, in this chapter, roulette wheel selection is used. Roulette wheel selection is a technique whereby members of the population of chromosomes are chosen in a way that is proportional to their fitness. The better the fitness of the chromosome, the greater the probability that it will be selected, however, it is not guaranteed that the fittest member goes to the next generation.

The second operator is the crossover function which mimics reproduction or mating in biological populations. The crossover technique used, in this chapter, is a uniform crossover. Uniform crossover first generates a random crossover mask (a binary string with the same size of chromosomes) and then exchanges relative genes between parents according to the mask. The parity of each bit in the mask determines, for each corresponding bit in an offspring, which parent it will receive that bit from (Lim, Yoon, & Kim, 2004).

The third operator used is mutation. This operation is a process whereby an arbitrary bit in a generic sequence or string are mutated or changed. The reason for implementing this operator is to promote diversity from one generation of chromosomes to another and this prevents genetic algorithm from getting stuck at a local optimum point but rather at a global optimum point. Boundary mutation is chosen in this chapter and this is an operation whereby a variable is randomly selected and is set to either the upper or lower bound depending on a randomly generated uniform number. The pseudo-code for the genetic algorithm is given below:

1. Initialize a population of chromosomes
2. Evaluate each chromosome (individual) in the population:
    a. Create new chromosomes by mating chromosomes in the current population (using crossover and mutation)
    b. Delete members of the existing population to make way for the new members
    c. Evaluate the new members and insert them into the population
3. Repeat stage 2 until some termination condition is reached
4. Return the best chromosome as the solution

In order to prevent premature convergence to a local optimum, the mutation diversification mechanism is implemented. Other mechanisms such as elitism can also be implemented in an attempt to improve the accuracy. The drawback of using the GA is that it is much slower than most traditional methods, i.e. a good initial guess will allow traditional optimization techniques to converge quickly towards the solution. Although GA will always approximate a solution, it must be noted that due to its stochastic nature, this approximation is only an estimate, whereas with traditional methods, if they can find an optimum solution, they will find it exactly (Goldberg, 1989).

Particle Swarm Optimization

Particle swarm optimization (PSO) is a stochastic, population-based optimization algorithm which was invented by Kennedy and Eberhart in the mid 1990's and is inspired by the simulation of a swarm (Kennedy & Eberhart, 1995; Hassan, Cohanim, & de Weck, 2005). PSO is based on the analogy of flocks of birds, schools of fish, and herds of animals and how they maintain an "evolutionary advantage" by either adapting to their environment, avoiding predators or finding rich sources of food, all by the process of "information sharing" (Hassan, Cohanim, & de Weck, 2005). A similar procedure to the PSO is the ant colony optimization which was used successfully by Lim, Rodrigues, and Zhang (2006a) for bandwidth minimization.

Guerra and Coelho (2008) used particle swarm optimization to train a radial basis neural network and utilized the radial basis network and clustering for non-linear identification of Lorenz's chaotic system. When the performance of this technique was compared to the existing radial basis neural networks, which was trained using the k-means and the Penrose-Moore pseudo-inverse, it was found to perform better in forecasting. Lian, Gu, and Jiao (2008) introduced a novel particle swarm optimization algorithm for flow-shop scheduling that minimized make-span. The results obtained showed that the proposed method performed better than genetic algorithm in the flow-shop scheduling problem aimed at minimizing the make-span. Shen, Shi, and Kong (2008) introduced the combination of the particle swarm optimization and tabu search method for choosing genes for tumor classification by utilizing gene expression data while Moradi and Fotuhi-Firuzabad (2008) applied particle swarm optimization algorithm for optimal switch assignment in distribution systems. Other successful applications of particle swarm optimization include in finite element analysis (Marwala, 2005a), in the design of higher-order digital differentiator (Chang & Chang, 2008), in the design of composite structures (Omkar et al., 2008), in parameter estimation in permanent magnet synchronous motors (Liu, Liu, & Cartes, 2008), in robust PID control (Kim, Maruta, & Sugie, 2008), in conceptual design (Liu, Liu, & Duan, 2007), in neural network training (Zhang et al., 2007), in inverse radiation problem (Qi et al., 2008), in scheduling problems (Jarboui et al., 2008) and in non-linear rational filter modeling (Lin, Chang, & Hsieh, 2008).

Particle swarm optimization initializes by creating a random population of solutions (particles), each solution's fitness is evaluated, according to a fitness function, and the fittest or best solution is noted and is often known as *pbest*. All the solutions in the problem space have their own set of co-ordinates and these co-ordinates are associated with the fittest solution found thus far. Another value noted is the best solution found in the neighborhood of the particular solutions or particles also often known as *lbest*. The particle swarm optimization concept consists of accelerating the velocities of each particle towards *pbest* and *lbest* locations, the acceleration for both *pbest* and *lbest* are separate randomly weighted values (Poli, Langdon, & Holland, 2005). The next iteration takes place after all particles have been moved. It is this behavior that mimics the swarm of birds.

The above described method can be represented by the concept of velocity. Velocity of a particle can be modified using the following equation (Kennedy & Eberhart, 1995; Naka et al., 2003):

$$v_i^{k+1} = wv_i^k + c_1 rand_1 \times \left(pbest_i - s_i^k \right) + c_2 rand_2 \times \left(lbest - s_i^k \right) \tag{10.2}$$

In equation 10.2:

v_i^k = Velocity of particle i at iteration k
w = Weighting function
c_j = Weighting factor
$rand$ = Random number between 0 and 1
s_i^k = Current position of particle i at iteration k
$pbest_i$ = *pbest* of particle i
$lbest$ = *lbest* of the group

The following weighting function is used in equation 10.2:

$$w = w_{max} - \frac{w_{max} - w_{min}}{iter_{max}} \times iter \tag{10.3}$$

In equation 10.3:

w_{max} = Initial weight
w_{min} = Final weight
$iter_{max}$ = Maximum iteration number
$iter$ = Current iteration number

By using equation 10.2, a velocity that gradually moves the current searching point close to *pbest* and *lbest* can then be calculated. The current position of the searching point can then be calculated as follows (Kennedy & Eberhart, 1995):

$$s_i^{k+1} = s_i^k + v_i^{k+1} \qquad (10.4)$$

The next iteration takes place after all particles have been moved. PSO has several similarities with other optimization techniques, such as GA these include a randomly generated initial population and a search for an optimum through a series of generations. However, unlike in GA the PSO does not have evolutionary operators such as mutation and crossover, and instead it has particles (or swarms) that travel through the solution space looking for optimal point.

Simulated Annealing (SA)

Simulated annealing is an algorithm that locates a good approximation to the global optimum of a given function. It originated as a generalization to the Monte Carlo method and relies on the Metropolis algorithm (Metropolis et al., 1953). As it is the case with genetic algorithm, SA continuously updates the solution until a termination criterion is reached (Christober, Rajan, & Mohan, 2007). Simulated annealing is a well established stochastic technique originally developed to model the natural process of crystallization and later adopted as an optimization technique. The SA algorithm replaces a current solution with a "nearby" random solution with a probability that depends on the difference between the corresponding function values and the temperature (T). The temperature T decreases throughout the process, and when T starts approaching zero, there is less random changes in the solution. As is the case of greedy search methods, simulated annealing keeps moving towards the best solution, except that it has the advantage of reversal in fitness. That means that it can move to a solution with worse fitness than it currently has, but the advantage of that is that it ensures that the solution is not found at a local optimum, but rather at a global optimum solution. This is the major advantage that SA has over most other methods, but once again its drawback is its computational time. Simulated annealing algorithm will find the global optimum if specified but it can approach infinite time in doing so.

Muppani and Adil (2008) used simulated annealing for an efficient formation of storage classes for warehouse storage location assignment. This was conducted by solving an integer programming model for class formation and storage assignment by considering product combinations, storage-space cost and order-picking cost. The results obtained showed that simulated annealing offers better results than dynamic programming algorithm which is a benchmark in this problem. On one hand, Zahrani et al. (2008) successfully applied genetic local search in multi-cast routing whereby the pre-processing part was conducted using simulated annealing with logarithmic cooling schedule. On the other hand, Wu, Chang, and Chung (2008) applied simulated annealing for manufacturing cell formation problems. The results obtained showed that SA improved the grouping effectiveness for 72% of the test data. Rodriguez-Tello, Hao, and Torres-Jimenez (2008) used simulated annealing algorithm for bandwidth minimization problems and tested SA on a set of 113 cases and then compared simulated annealing to a number of algorithms. It was found that SA improved the results over other methods. Sante-Riveira et al. (2008) used simulated annealing for land-use allocation and applied this technique on the Terra Chá district of Galicia (N.W. Spain). It was found that including compactness in the simulated annealing objective function circumvent the exceedingly disperse allocations distinctive of optimizations that pay no attention to this sub-objective.

Other successful implementations of simulated annealing include in lung module registration (Sun et al., 2008), in music playlist generation (Pauws, Verhaegh, & Vossen, 2008), in electric power distribution (Ahmed & Sheta, 2008), in temperature prediction and control (Lee, Wang, & Chen, 2008; Azizi & Zolfaghari, 2004), in wireless ATM network (Din, 2007), in finite element analysis (Marwala, 2005b), in smart design assembly (Shen & Subic, 2006), in robot path planning (Tsuzuki, Martins, & Takase, 2006), in vehicle routing problems (Tavakkoli-Moghaddam, Safaei, & Gholipour, 2006) and in batch distillation process (Hanke & Li, 2000).

Simulated annealing operates by sampling through the parameter space, and in this chapter the missing data space given the selected temperature, by accepting or rejecting the parameters using the Metropolis algorithm (Metropolis et al., 1953). The probability of accepting the parameters is given by Boltzmann's equation (Bryan, Cunningham, & Bolshkova, 2006):

$$P(\Delta E) \propto e^{-\frac{\Delta E}{T}} \qquad (10.5)$$

In equation 10.5, the parameter ΔE is the difference in energy (or fitness function fitness given by equation 10.1) between the old and new states, and T is the temperature of the system. The rate at which the temperature decreases depends on the cooling schedule chosen. The following cooling model is used (Bryan, Cunningham, & Bolshkova, 2006):

$$T(k) = \frac{T(k-1)}{1+\sigma} \qquad (10.6)$$

where $T(k)$ is the current temperature, $T(k-1)$ is the previous temperature, and σ dictates the cooling rate. It must be noted that the precision of the numbers used in the implementation of SAs can have a significant effect on the outcome. A method to improve the computational time of simulated annealing is to implement either very fast simulated re-annealing or adaptive simulated annealing (Li, Koike, & Pathmathevan, 2004; Salazar & Toral, 2006).

Hill Climbing (HC)

Hill climbing is an optimization technique that belongs to a group of local search algorithms, meaning the algorithm moves from a solution to a better solution in the search space until an optimal solution is found. The algorithm tries to maximize the fitness function or the negative of equation 10.1 by iteratively comparing two solutions, and then it adopts the best solution and continues with the comparison (i.e. move further up the hill). This iteration terminates when there are no better solutions on either side of the current solution (i.e. it has reached the peak). There are several variants or methods of hill climbing, the first and most basic form is simple hill climbing. In simple hill climbing the first closest node is chosen for evaluation. A second variant is called steepest ascent hill climbing for the maximization process, in this method all successors are compared and the closest to the solution is chosen.

Other variants that can be investigated include next-ascent hill climbing and zero-temperature Monte Carlo hill climbing (Mitchell, Holland, & Forest, 1994). In this chapter, steepest ascent hill climbing is implemented. A major disadvantage of both simple and steepest ascent hill climbing is that they only find the local optimum. Hernandez, Gras, and Appel (2008) applied neighborhood functions and hill-climbing strategies to study the generalized un-gapped local multiple alignment which is generally applied to bioinformatics. Iclanzan and Dumitrescu (2007) proposed an HC method that operates in building block space that can solve hierarchical problems. The proposed Building block hill-climber utilized hill-climb search knowledge to learn the problem structure and the neighborhood structure was adapted every time the latest knowledge on the building block structure was included into the search. This permitted the technique to climb the hierarchical structure by solving successively the hierarchical levels. The results they obtained showed that this technique scaled approximately linearly with the size of the problem and, therefore, outperformed population based re-combinative techniques such as GA, PSO and SA. Other successful implementations of HC include in studying the role of DNA transposition in enduring artificial evolution (Khor, 2007), in the time-delay systems (Santos & Sanchez-Diaz, 2007), in the creedal networks (Cano et al., 2007), in decentralized job scheduling (Wang, Gao, & Liu, 2006), in simultaneous multi-threaded processors (Wang, Gao, & Liu, 2006), in traveling tournament problem (Lim, Rodrigues, & Zhang, 2006a&b), on evolving robot gaits (Garder & Hovin, 2006) and on modeling HIV (Crossingham & Marwala, 2007a&b).

The variant of HC methods used in this chapter is the gradient descent (GD) method, which is also called gradient ascent for maximization problems. GD is based on the examination that if the real-valued function $F(\{x\})$ is defined and differentiable in a neighborhood of a point $\{x\}$, then $F(\{x\})$ reduces the quickest if the simulation proceed from $\{x\}$ in the direction of the negative gradient of F at $\{x\}$. This can, therefore, be written in the following form (Avriel, 2003; Snyman, 2005):

$$\{x\}_{n+1} = \{x\}_n - \eta \frac{\partial F}{\partial \{x\}}\left(\{x\}_n\right)$$

(10.7)

In equation 10.7, if the learning rate ($\eta \approx 0$) is a small enough number, then $F\left(\{x\}_{n+1}\right) < F\left(\{x\}_n\right)$. By repeating this process, the series converges to the desired local minimum. It should be noted that the value of the step size η is permitted to alter every step (iteration).

Hill climbing uses the missing data equation denoted as equation 10.1 as the objective/fitness function, and because it is a "greedy" algorithm, other methods or variations of hill climbing should be investigated. A "greedy" algorithm is an algorithm that looks at the next solution, and if the next solution is a better solution (fitter), this algorithm will take on the position of the next solution. This is disadvantageous as it does not look at the entire search space, and if the next solution (as it is climbing the hill) is worse than the present one, it will return stating it has found the best solution, this may not be the case because of the problem of local and global maxima. In this chapter since HC is operating in continuous space and it is called gradient ascent method.

EXPERIMENTAL INVESTIGATION

The optimization methods described above are tested on a mechanical system and an artificial taster that is described below and the performances of these optimization methods are compared.

Mechanical System

In this section the procedure proposed is experimentally validated. The experiment is performed on a population of cylinders, which are supported by inserting a sponge rested on a *bubble-wrap*, to simulate a 'free-free' environment (see Figure 10.1) and the details of this may be found in Marwala (2000). The sponge is inserted inside the cylinders to control boundary conditions. This will be further discussed below.

Conventionally, a 'free–free' environment is achieved by suspending a structure usually with light elastic bands. A 'free–free' environment is implemented so that rigid body modes, which do not exhibit bending or flexing, can be identified. These modes occur at frequency of 0 Hz and they can be used to calculate the mass and inertia properties. The present chapter does not consider the rigid body modes. Here, a 'free–free' environment is approximated using a bubble-wrap. Testing the cylinders suspended is approximately the same as testing it while resting on a bubble-wrap, because the frequency of cylinder-on-wrap is below 100 Hz. The first natural frequency of cylinders being analyzed is over 300 Hz and this value is several orders of magnitudes above the natural frequency of a cylinder on a bubble-wrap. Therefore, the cylinder on the wrap is effectively decoupled from the ground. It should be noted that the bubble-wrap adds some damping to the structure but the damping added is found to be small enough for the modes to be easily identified. When the damping ratios are estimated, it is observed that the structure is lightly damped and, therefore, damping does not play significant role in this paper.

The impulse hammer test is performed on each of the 20 steel seam-welded cylindrical shells (1.75 ± 0.02 mm thickness, 101.86 ± 0.29 mm diameter and of height 101.50 ± 0.20 mm). The impulse is applied at 19 different locations as indicated in Figures 10.1: 9 on the upper half of the cylinder and 10 on the lower half of the cylinder. The sponge is inserted inside the cylinder to control boundary conditions and by rotating it every time a measurement is taken. The bubble wrap simulates the free–free

Figure 10.1. Illustration of a cylindrical shell showing the positions of the impulse, accelerometer, sub-structures, fault position and supporting sponge

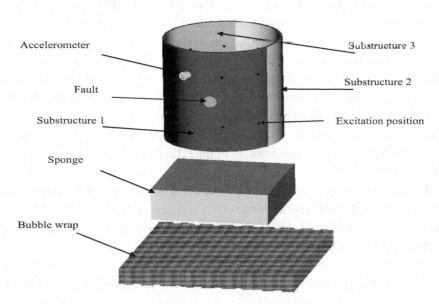

environment. The top impulse positions are located 25 mm from the top edge and the bottom impulse positions are located 25 mm from the bottom edge of the cylinder. The angle between two adjacent impulse positions is 36°.

Problems encountered during impulse testing include difficulty of exciting the structure at an exact position especially for an ensemble of structures and in a repeatable direction. Each cylinder is divided into three equal sub-structures and holes of 10–15 mm in diameter are introduced at the centers of the sub-structures to simulate faults. For one cylinder the first type of fault is a zero-fault scenario. This type of fault is given the identity [000], indicating that there are no faults in any of the three sub-structures. The second type of fault is a one-fault scenario, where a hole may be located in any of the three sub-structures. Three possible one-fault scenarios are [100], [010], and [001] indicating one hole in sub-structures 1, 2 or 3 respectively. The third type of fault is a two-fault scenario, where a hole is located in two of the three sub-structures. Three possible two-fault scenarios are [110], [101], and [011]. The final type of fault is a three-fault scenario, where a hole is located in all three sub-structures, and the identity of this fault is [111]. There are 8 different types of fault-cases considered (including [000]).

Because the zero-fault scenarios and the three-fault scenarios are over-represented, twelve cylinders are picked at random and additional one- and two-fault cases are measured after increasing the magnitude of the holes. This is done before the next fault case is introduced to the cylinders. The reason why zero-fault and three-fault scenarios are over-represented is because all cylinders tested give these fault-cases, whereas not all cylinders tested give all 3 one-fault and 3 two-fault cases. Only a few fault-cases are selected because of the limited computational storage space available. For each fault-case, acceleration and impulse measurements are taken. The types of faults that are introduced (i.e. drilled holes) do not influence damping.

Each cylinder is measured three times under different directions by changing the orientation of a rectangular sponge inserted inside the cylinder. The number of sets of measurements taken for undamaged population is 60 (20 cylinders times 3 for different directions). All the possible fault types and their respective number of occurrences are listed in Table 10.1. In Table 10.1 it should be noted that the numbers of one- and two-fault cases are each 72. This is because as mentioned above, increasing the sizes of holes in the sub-structures and taking vibration measurements generated additional one- and two-fault cases. The impulse and response data are processed using the Fast Fourier Transform to convert the time domain impulse history and response data into the frequency domain.

The data in the frequency domain are used to calculate the FRFs. The sample FRF results from an ensemble of 20 undamaged cylinders are shown in Figure 10.2. This figure indicates that the measurements are generally repeatable at low frequencies and are not repeatable at high frequencies. Axisymmetric structures such as cylinders have repeated modes due to their symmetry. In this chapter, the presence of an accelerometer and the imperfection of cylinders destroy the axisymmetry of the structures. Therefore, the problem of repeated natural frequencies is neatly avoided, thereby, making the process of modal analysis easier to perform. The problem of uncertainty of high frequencies is avoided by only using frequencies under 4000 Hz.

From the data measured, 10 parameters are selected. The multi-layer perceptron autoassociative network with 10 inputs and 10 outputs is constructed and several numbers of hidden units are tested and 8 hidden units are found to be the most optimal. The autoassociative network is, therefore, trained to understand the rules that define the inter-relationships among the measured variables.

After the autoassociative network is trained then the experimentation with 1, 2 and 3 missing variables is conducted and the estimated results are then compared to the true values and the results

Table 10.1. Number of different types of fault cases generated

Fault	[000]	[100]	[010]	[001]	[110]	[101]	[011]	[111]
Number	60	24	24	24	24	24	24	60

Figure 10.2. Measured frequency response functions from a population of cylinders

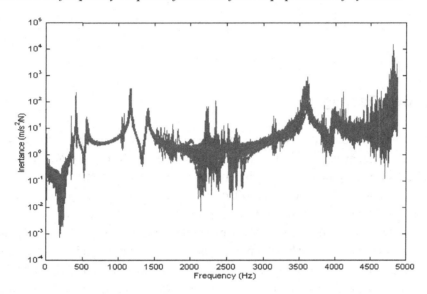

obtained are shown in Table 10.2. On the implementation of the missing data estimation procedure genetic algorithm is run with roulette wheel selection, a boundary mutation and uniform crossover, an initial population of 20 individuals and the termination function of 100 generations. On the other hand, particle swarm optimization is implemented with the maximum number of generations of 30, and the initial number of particles also set to 30 while steepest descent is implemented with 20 initial starting points. Furthermore, simulated annealing is implemented with the cooling model as given in equation 10.6. It is run with a random generator with the bounds of the maximum and minimum input values, an initial temperature of 1, a stopping temperature of 1e ^{8}, a maximum number of consecutive rejection of 200 and a maximum number of successes within one temperature set to t0.

Table 10.2 shows that genetic algorithms, particle swarm optimization and simulated annealing give similar results even though simulated annealing is the best of the four. Also it is observed that it is much easier to estimate one missing variable when compared to estimating two missing variables. Steepest descent method is found to perform worse than the three methods. This is because steepest descent method is a local optimization procedure and the missing data problem requires the identification of global optimal points. In general, genetic algorithm is found to the most computationally efficient followed very closely by simulated annealing and then steepest descent method. Simulated annealing is found to be the most computationally expensive method and this is mainly due to its Monte Carlo simulation nature which is generally computationally expensive. The results obtained verify the assumption made in earlier chapters that states that on the implementation of the missing data estimator using the missing data estimation error equation, global optimization method should be used.

Table 10.2. Accuracy in missing data estimation using GA, PSO, SA and SD optimization methods

Method	Accuracy: 1 Missing Variable(%)	Accuracy: 2 Missing Variable (%)	Accuracy: 3 Missing Variable (%)
GA	93.5	90.8	73.7
PSO	93.8	90.1	75.7
SA	93.7	91.3	76.0
SD	85.8	83.0	65.3

Modelling of Beer Tasting

As described in Chapter II, the autoassociative multi-layer perceptron neural network with 10 hidden neurons, 14 inputs, 10 hidden neurons and 14 outputs are trained on the data obtained from a Brewery in South Africa. The dataset contains information that is used to construct an artificial beer taster and more information on this is described by Marwala (2005c; 2007b). Generally the factors that human beings are sensitive to when tasting beer are: (1) color of beer; (2) smell; and (3) chemical components. In human beings, the color of objects is captured using the eye; smell is captured using the nose and taste using the tongue. As in Chapter II, the artificial beer taster that is under consideration in this chapter contains the following variables: *alcohol*; *present extract*; *real extract*; *present extract minus limit extract* (*PE-LE*); *pH*; *iron*; *acetaldehyde*; *dimethyl sulfide* (*DMS*); *eythyl acetate*; *iso-amyl acetate*; *total higher alcohols*; *color*; *bitterness* and amount of *carbohydrates*. These variables capture the color, smell and chemical components of the beer and are normally used to predict the beer taste score, which is conventionally obtained from a panel of tasters, using some regression analysis such as neural networks. All these parameters are measured and sometimes one or more of these measurements may not be available due to problems such as instrumentation failure. In such a situation, it is important to estimate these missing values because, otherwise, it is impossible to predict the taste score. More information on this artificial taster can be found in Marwala (2005c).

These beer characteristics variables are then used to train the neural networks with the architectures described earlier. These neural networks essentially mapped all the 14 variables onto themselves, in an autoassociative manner as described earlier in the chapter, using the multi-layer perceptron autoassociative neural network.

Cases of 1, 2, 3, 4 and 5 missing values in a single record are examined to investigate the accuracy of the approximated values as the number of missing cases within a single record increases. To assess the accuracy of the values approximated using the model the standard error is calculated for each missing case (Draper & Smith, 1998). More details on standard error are described in Chapter II. The standard error (S_e) estimates the capability of the model to predict the known dataset (Kolarik & Rudorfer, 1994). The higher the value of "S_e", the less reliable the approximations and vice-versa. The results of the standard error measures obtained from the experiment are given in Table 10.3.

In Table 10.3 genetic algorithms and particle swarm optimization are found to give similar results followed closely by simulated annealing. Hill climbing method is found to perform the worst. This is mainly due to the fact that the missing data estimation problem is a problem that requires a global optimization method and, therefore, global methods GA, PSO and SA would perform better than lo-

Table 10.3. Standard error

	Number of Missing Variables				
	1	2	3	4	5
GA	16.62	16.77	16.87	16.31	16.40
PSO	16.80	16.29	16.10	15.92	15.92
SA	15.21	14.87	15.63	15.33	15.62
HC	12.55	11.47	10.94	11.48	10.67

cal methods such as steepest gradient method and this again confirms the assumptions made in earlier chapters. In general, the accuracies of the results are found not to deteriorate as the number of missing values decrease. In general, genetic algorithm is found to be the most computationally efficient followed very closely by simulated annealing and then steepest descent method. Simulated annealing is found to be the most computationally expensive method and this is mainly due to its Monte Carlo simulation nature, which is generally computationally expensive.

CONCLUSION

In this chapter, four optimization methods genetic algorithm, particle swarm optimization, hill climbing and simulated annealing are implemented and compared in the missing data problem. These methods are tested on two sets of data and these are mechanical system and artificial taster. For the mechanical system case, the results obtained demonstrate that genetic algorithms, particle swarm optimization and simulated annealing give comparable results even though simulated annealing is the best of the four. Furthermore, it is observed that it is much easier to estimate one missing variables when compared to estimating two or three missing variables. Steepest descent method is found to perform worse than the three methods. This is because steepest descent method is a local optimization procedure and the missing data problem requires the identification of global optimal points. In general, genetic algorithm is found to the most computationally efficient followed very closely by simulated annealing and then steepest descent method. Simulated annealing is found to be the most computationally expensive method and this is mainly due to its Monte Carlo simulation nature, which is generally computationally expensive. For the artificial taster method, genetic algorithms and particle swarm optimization are found to give similar results followed closely by simulated annealing. Steepest descent method is found to perform the worst. This is mainly due to the fact that the missing data estimation problem is a problem that requires a global optimization method and, therefore, global methods GA, PSO and SA perform better than local methods such steepest gradient method. In general the accuracies of the results are found not to deteriorate as the number of missing values decrease. In general, genetic algorithm is found to be the most computationally efficient followed very closely by simulated annealing and then steepest descent method. Simulated annealing is found to be the most computationally expensive method and this is mainly due to its Monte Carlo simulation nature which is generally computationally expensive.

FUTURE WORK

In this chapter, four optimization methods, genetic algorithm, particle swarm optimization, hill climbing and simulated annealing are implemented and compared in the missing data problem. For future work hybrid optimization methods must be tested to evaluate their effectiveness and then compared. Furthermore, a study that would relate the characteristics of the data and the optimization method that is correspondingly most ideal should be undertaken.

REFERENCES

Abdella, M., & Marwala, T. (2006). The use of genetic algorithms and neural networks to approximate missing data in database. *Computing and Informatics, 24,* 1001-1013.

Abu-Zitar, R. (2008). The Ising genetic algorithm with Gibbs distribution sampling: Application to FIR filter design. *Applied Soft Computing Journal, 8*(2), 1085-1092.

Ahmed, W., & Sheta, A. F. (2008). Optimization of electric power distribution using hybrid simulated annealing approach. *American Journal of Applied Sciences, 5*(5), 559-564.

Allison, P. (2000). Multiple imputation for missing data: A cautionary tale. *Sociological Methods and Research 28,* 301-309.

Avriel, M. (2003). *Nonlinear programming: Analysis and methods.* New York: Dover Publishing.

Azizi, N., & Zolfaghari, S. (2004). Adaptive temperature control for simulated annealing: A comparative study. *Computers & Operations Research, 31*(14), 2439-2451.

Bryan, K., Cunningham, P., & Bolshkova, N. (2006). Application of simulated annealing to the biclustering of gene expression data. *IEEE Transactions on Information Technology in Biomedicine, 10,* 519–525.

Caldara, R., & Abdi, H. (2006). Simulating the 'other-race' effect with autoassociative neural networks: Further evidence in favor of the face-space model. *Perception, 35*(5), 659-670.

Calvete, H. I., Gale, C., & Mateo, P. M. (2008). A new approach for solving linear bilevel problems using genetic algorithms. *European Journal of Operational Research, 188*(1), 14-28.

Cano, A., Gomez, M., Moral, S., & Abellan, J. (2007). Hill-climbing and branch-and-bound algorithms for exact and approximate inference in credal networks. *International Journal of Approximate Reasoning, 44(3),* 261-280.

Chang, W.-D., & Chang, D.-M. (2008). Design of a higher-order digital differentiator using a particle swarm optimization approach. *Mechanical Systems and Signal Processing, 22*(1), 233-247.

Chen, T-Y., & Su, J-J. (2002). Efficiency improvement of simulated annealing in optimal structural designs. *Advances in Engineering Software, 33*(7-10), 675-680.

Christober, C., Rajan, A., & Mohan, M. R. (2007). An evolutionary programming based simulated annealing method for solving the unit commitment problem. *International Journal of Electrical Power & Energy Systems, 29*(7), 540-550.

Crossingham, B., & Marwala, T. (2007a). Using genetic algorithms to optimise rough set partition sizes for HIV data analysis. *Studies in Computational Intelligence, 78,*245-250.

Crossingham, B., & Marwala, T. (2007b). Using optimisation techniques to granulise rough set partitions. *Computational Models for Life Sciences, 952,* 248-257

Defersha, F. M., & Chen, M. (2008). A linear programming embedded genetic algorithm for an integrated cell formation and lot sizing considering product quality. *European Journal of Operational Research, 187*(1), 46-69.

Del Rincon, A. F., Rueda, F. V., Fenandez, P. G., & Herrera, R. S. (2005). Vibratory data fusion for gearbox fault detection using autoassociative neural networks. In *Proceedings of the ASME International Design Engineering Technical Conferences and Computers and Information in Engineering Conference - DETC2005* (pp. 635-642).

Dempster, A., Liard, N., & Rubin, D. (1977). Maximum likelihood from incomplete data via the EM algorithm (with discussion). *Journal of the Royal Statistical Society, B39,* 1-38.

Din, D.-R. (2007). Simulated annealing algorithm for solving network expanded problem in wireless ATM network. *Lecture Notes in Computer Science, 4570,* 1138-1147.

Draper, N., & Smith, H. (1998). *Applied regression analysis.* 3rd Edition. New York: J. Wiley.

Essafi, I., Mati Y., & Dauzere-Peres, S. (2008). A genetic local search algorithm for minimizing total weighted tardiness in the job-shop scheduling problem. *Computers and Operations Research, 35*(8), 2599-2616.

Gangashetty, S. V., Sekhar, C. C., & Yegnanarayana, B. (2004). Spotting consonant-vowel units in continuous speech using autoassociative neural networks and support vector machines. *Machine Learning for Signal Processing XIV - Proceedings of the 2004 IEEE Signal Processing Society Workshop* (pp. 401-410).

Garcia-Martinez, C., Lozano, M., Herrera, F., Molina, D., & Sanchez, A. M. (2008). Global and local real-coded genetic algorithms based on parent-centric cross-over operators. *European Journal of Operational Research, 185*(3), 1088-1113.

Garder, L. M., & Hovin, M. E. (2006). Robot gaits evolved by combining genetic algorithms and binary hill climbing. In *Proceedings of the Genetic and Evolutionary Computation Conference* (pp. 1165-1170).

Goldberg, D. (1989). *Genetic algorithms in search, optimization and machine learning.* Reading, MA: Addison-Wesley.

González, A., & Pérez, R. (2001). An experimental study about the search mechanism in the SLAVE learning algorithm: Hill-climbing methods versus genetic algorithms. *Information Sciences, 136,* 159-174.

Guerra, F. A., & Coelho, L. d. S. (2008). Multi-step ahead nonlinear identification of Lorenz's chaotic system using radial basis neural network with learning by clustering and particle swarm optimization. *Chaos, Solitons and Fractals, 35*(5), 967-979.

Güngör, Z., & Ünler, A. (2007). K-harmonic means data clustering with simulated annealing heuristic *Applied Mathematics and Computation, 184,* 199-209.

Hanke, M., & Li, P. (2000). Simulated annealing for the optimization of batch distillation processes. *Computers & Chemical Engineering, 24*(1), 1-8.

Hassan, R., Cohanim, B., & de Weck, O. (2005). *A comparison of particle swarm optimization and the genetic algorithm.* Paper presented at the 46th American Institute of Aeronautics and Astronautics, Austin, Texas.

Hernandez, D., Gras R., & Appel, R. (2008). Neighborhood functions and hill-climbing strategies dedicated to the generalized ungapped local multiple alignment. *European Journal of Operational Research, 185*(3), 1276-1284.

Hines, J. W., & Garvey, D. (2006). Sensor fault detectability measures for autoassociative empirical models. In *Proceedings of the 16th Annual Joint ISA POWID/EPRI Controls and Instrumentation Conference and 49th Annual ISA Power Industry Division, POWID Symposium* (pp. 835-847).

Hulley, G., & Marwala, T. (2007). Genetic algorithm based incremental learning for optimal weight and classifier selection. *Computational Models for Life Sciences, 952,* 258-267.

Iclanzan, D., & Dumitrescu, D. (2007). Overcoming hierarchical difficulty by hill-climbing the building block structure. In *Proceedings of GECCO 2007: Genetic and Evolutionary Computation Conference* (pp. 1256-1263).

Jarboui, B., Damak N., Siarry P., & Rebai, A. (2008). A combinatorial particle swarm optimization for solving multi-mode resource-constrained project scheduling problems. *Applied Mathematics and Computation, 195*(1), 299-308.

Johnson, A. W., & Jacobson, S. H. (2002). On the convergence of generalized hill climbing algorithms. *Discrete Applied Mathematics, 119*(1-2), 37-57.

Juhos, I., Makra, L., & Tóth, B. (2008) (in press). The behaviour of the multi-layer perceptron and the support vector regression learning methods in the prediction of NO and NO_2 concentrations in Szeged, Hungary. *Neural Computing and Applications.*

Kao, Y-T, Zahara, E., & Kao, I-W. (2008). A hybridized approach to data clustering. *Expert Sysytems with Applications: An International Journal, 34*(3), 1754-1762.

Keesing, D. B., O'Sullivan J. A., Politte, D. G., Whiting, B. R., & Snyder, D. L. (2007). Missing data estimation for fully 3D spiral CT image reconstruction. *Progress in Biomedical Optics and Imaging - Proceedings of SPIE, 6510,* Paper 65105V.

Kennedy, J., & Eberhart, R. (1995). Particle swarm optimization. In *Proceedings of the IEEE International Conference on Neural Networks* (pp. 1942-1948).

Khor, S. (2007). Hill climbing on discrete HIFF: Exploring the role of DNA transposition in long-term artificial evolution. In *Proceedings of GECCO 2007: Genetic and Evolutionary Computation Conference* (pp. 277-284).

Kim, T.-H., Maruta, I., & Sugie, T. (2008) (in press). Robust PID controller tuning based on the constrained particle swarm optimization. *Automatica.*

Kolarik, T., & Rudorfer, G. (1994). Time series forecasting using neural networks. In *Proceedings of the International Conference on APL: The Language and its Applications* (pp. 86-94).

Kuhnen, M., Togneri, R., & Nordholm, S. (2007). MEL-spectrographic mask estimation for missing data speech recognition using short-time-Fourier-transform ratio estimators. In *Proceedings of the IEEE International Conference on Acoustics, Speech and Signal Processing* (pp. 405-408).

Lee, L.-W., Wang, L.-H., & Chen, S.-M. (2008). Temperature prediction and TAIFEX forecasting based on high-order fuzzy logical relationships and genetic simulated annealing techniques. *Expert Systems with Applications, 34*(1), 328-336.

Leke, B., Marwala, T., & Tettey, T. (2007). Using inverse neural network for HIV adaptive control. *International Journal of Computational Intelligence Research, 3*(1), 11-15.

Li, X., Koike, T., & Pathmathevan, M. (2004). A very fast simulated re-annealing (VFSA) approach for land data assimilation. *Computers & Geosciences, 30*(3), 239-248.

Lian, Z., Gu, X., & Jiao, B. (2008). A novel particle swarm optimization algorithm for permutation flow-shop scheduling to minimize makespan. *Chaos, Solitons and Fractals, 35*(5), 851-861.

Lim, A., Lin, J., Rodrigues, B., & Xiao, F. (2006a). Ant colony optimization with hill climbing for the bandwidth minimization problem. *Applied Soft Computing, 6*(2), 180-188.

Lim, A., Rodrigues, B., & Zhang, X. (2006b). A simulated annealing and hill-climbing algorithm for the traveling tournament problem. *European Journal of Operational Research, 174*(3), 1459-1478.

Lim, C. H., Yoon, Y. S., & Kim, J. H. (2004). Genetic algorithm in mix proportioning of high-performance concrete. *Cement and Concrete Research, 34*, 409-420.

Lin, Y.-L., Chang, W.-D., & Hsieh, J.-G. (2008). A particle swarm optimization approach to nonlinear rational filter modeling. *Expert Systems with Applications, 34*(2), 1194-1199.

Lin, C.-C., & Liu, Y.-T. (2008). Genetic algorithms for portfolio selection problems with minimum transaction lots. *European Journal of Operational Research, 185*(1), 393-404.

Little, R., & Rubin, D. (1987). *Statistical analysis with missing data.* 1st Edition, New York: John Wiley and Sons.

Little, R., & Rubin, D. (2002). *Statistical analysis with missing data.* 2nd Edition, John Wiley and Sons.

Liu, X., Liu, H., & Duan, H. (2007). Particle swarm optimization based on dynamic niche technology with applications to conceptual design. *Advances in Engineering Software, 38*, 668-676.

Liu, L., Liu W., & Cartes D. A. (2008) (in press). Particle swarm optimization-based parameter identification applied to permanent magnet synchronous motors. *Engineering Applications of Artificial Intelligence.*

Malve, S., & Uzsoy, R. (2007). A genetic algorithm for minimizing maximum lateness on parallel identical batch processing machines with dynamic job arrivals and incompatible job families. *Computers and Operations Research, 34,* 3016–3028.

Marseguerra, M., Zio, E., & Marcucci, F. (2006). Continuous monitoring and calibration of UTSG process sensors by autoassociative artificial neural network. *Nuclear Technology, 154*(2), 224-236.

Marwala, T. (2000). *Fault identification using neural networks and vibration data.* Unpublished doctoral dissertation, University of Cambridge, Cambridge.

Marwala, T. (2002). Finite element updating using wavelet data and genetic algorithm. *American Institute of Aeronautics and Astronautics, Journal of Aircraft, 39,* 709-711.

Marwala, T. (2004). Control of complex systems using Bayesian neural networks and genetic algorithm. *International Journal of Engineering Simulation, 5*(2), 28-37.

Marwala, T. (2005a). Finite element model updating using particle swarm optimization. *International Journal of Engineering Simulation, 6*(2), 25-30.

Marwala, T. (2005b). Evolutionary optimization methods in finite element model updating. In *Proceedings of the International Modal Analysis Conference,* Orlando, Florida.

Marwala, T. (2005c). The artificial beer taster. *Electricity+Control, 3,* 22-23.

Marwala, T. (2007a). *Computational intelligence for modelling complex systems.* Delhi: Research India Publications.

Marwala, T. (2007b). Bayesian training of neural network using genetic programming. *Pattern Recognition Letters. 28,* 1452–1458

Marwala, T., & Chakraverty, S. (2006). Fault classification in structures with incomplete measured data using autoassociative neural networks and genetic algorithm. *Current Science, 90*(4), 542-548.

Marwala, T., de Wilde, P., Correia, L., Mariano, P., Ribeiro, R., Abramov, V., Szirbik, N., & Goossenaerts, J. (2001). *Scalability and optimisation of a committee of agents using genetic algorithm.* Paper presented at the International Symposia on Soft Computing and Intelligent Systems for Industry, Parsley, Scotland.

Marwala, T., Mahola, U., & Chakraverty, S. (2007). Fault classification in cylinders using multi-layer perceptrons, support vector machines and Gaussian mixture models. *Computer Assisted Mechanics and Engineering Sciences, 14*(2), 307-316.

Metropolis, N., Rosenbluth, A., Rosenbluth, M., Teller, A., & Teller, E. (1953). Equation of state calculations by fast computing machines. *The Journal of Chemical Physics, 21,* 1087–1092.

Mitchell, M., Holland, J., & Forest, S. (1994). When will a genetic algorithm outperform hill climbing? *Advances in Neural Information Processing Systems 6,* 51–58.

Mohamed, S., & Marwala, T. (2005). Neural network based techniques for estimating missing data in databases. In *Proceedings of the 16th Annual Symposium of the Pattern Recognition Society of South Africa,* Langebaan, South Africa (pp. 27-32).

Mohamed, S., Tettey, T., & Marwala, T. (2006). An extension neural network and genetic algorithm for bearing fault classification. In *Proceedings of the IEEE International Joint Conference on Neural Networks,* British Columbia, Canada, (pp. 7673-7679).

Mojirsheibani, M. (2007). Nonparametric curve estimation with missing data: A general empirical process approach. *Journal of Statistical Planning and Inference, 137*(9), 2733-2758.

Moradi, A., & Fotuhi-Firuzabad, M. (2008). Optimal switch placement in distribution systems using trinary particle swarm optimization algorithm. *IEEE Transactions on Power Delivery, 23*(1), 271-279.

Muppani (Muppant), V. R., & Adil, G. K. (2008). Efficient formation of storage classes for warehouse storage location assignment: A simulated annealing approach. *Omega, 36*(4), 609-618.

Naka, S. Genji, T., Yura, T., & Fukuyama, Y. (2003). Hybrid particle swarm optimization for distribution state estimation. *IEEE Transactions on Power Systems, 18,* 60–68.

Nelwamondo, F. V. (2008). *Computational intelligence techniques for missing data imputation.* Unpublished doctoral dissertation, University of the Witwatersrand, Johannesburg.

Nelwamondo, F. V., & Marwala, T. (2007). Rough set theory for the treatment of incomplete data. In *Proceedings of the IEEE Conference on Fuzzy Systems,* London, UK (pp. 338-343).

Omkar, S. N., Mudigere, D., Naik, G. N., & Gopalakrishnan, S. (2008). Vector evaluated particle swarm optimization (VEPSO) for multi-objective design optimization of composite structures. *Computers and Structures, 86*(1-2), 1-14.

Pauws, S., Verhaegh, W., & Vossen, M. (2008). Music playlist generation by adapted simulated annealing. *Information Sciences, 178*(3), 647-662.

Piotrowski, A., Wallis, S. G., Napiórkowski, J. J., & Rowinski, P. M. (2007). Evaluation of 1-D tracer concentration profile in a small river by means of multi-layer perceptron neural networks. *Hydrology and Earth System Sciences Discussions, 4*(4), 2739-2768.

Polgár, O., Fried, M., Lohner, T., & Bársony, I. (2000). Comparison of algorithms used for evaluation of ellipsometric measurements random search, genetic algorithms, simulated annealing and hill climbing graph-searches. *Surface Science, 457,* 157-177.

Poli, R., Langdon, W. B., & Holland, O. (2005). Extending particle swarm optimization via genetic programming. *Lecture Notes in Computer Science, 3447,* 291-300.

Pomi, A., & Olivera, F. (2006). Context-sensitive autoassociative memories as expert systems in medical diagnosis. *BMC Medical Informatics and Decision Making, 6,* Paper 39.

Ponsich, A., Azzaro-Pantel, C., Domenech, S., & Pibouleau, L. (2008). Constraint handling strategies in genetic algorithms application to optimal batch plant design. *Chemical Engineering and Processing: Process Intensification, 47*(3), 420-434.

Qi, H., Ruan, L. M., Shi, M., An, W., & Tan, H. P. (2008). Application of multi-phase particle swarm optimization technique to inverse radiation problem. *Journal of Quantitative Spectroscopy and Radiative Transfer, 109*(3), 476-493.

Rodriguez-Tello, E., Hao, J.-K., & Torres-Jimenez, J. (2008). An improved simulated annealing algorithm for bandwidth minimization. *European Journal of Operational Research, 185*(3), 1319-1335.

Rubin, D. B. (1978). Multiple imputations in sample surveys - A phenomenological Bayesian approach to non-response. In *Proceedings of the Survey Research Methods Section of the American Statistical Association* (pp. 20-34).

Salazar, R., & Toral, R. (2006). Simulated annealing using hybrid Monte Carlo. *arXiv:cond-mat/9706051.*

Sante-Riveira, I., Boullon-Magan, M., Crecente-Maseda, R., & Miranda-Barros, D. (2008). Algorithm based on simulated annealing for land-use allocation. *Computers and Geosciences, 34*(3), 259-268.

Santos, O., & Sanchez-Diaz, G. (2007). Suboptimal control based on hill-climbing method for time delay systems. *IET Control Theory and Applications, 1*(5), 1441-1450.

Shen, H., & Subic, A. (2006). Smart design for assembly using the simulated annealing approach. In *Proceedings of the Virtual Conference on Intelligent Production Machines and Systems* (pp. 419-424).

Shen, Q., Shi, W.-M., & Kong, W. (2008). Hybrid particle swarm optimization and tabu search approach for selecting genes for tumor classification using gene expression data. *Computational Biology and Chemistry, 32*(1), 52-59.

Shih, W. J., Quan, H., & Chang, M. N. (2008). Estimation of the mean when data contain non-ignorable missing values from a random effects model. *Statistics and Probability Letters, 19,* 249-252.

Sidiropoulos, E., & Tolikas, P. (2008). Genetic algorithms and cellular automata in aquifer management. *Applied Mathematical Modelling, 32*(4), 617-640.

Snyman, J. A. (2005). *Practical mathematical optimization: An introduction to basic optimization theory and classical and new gradient-based algorithms.* Heidelberg: Springer Publishing.

Spiller, J. M., & Marwala, T. (2007). Evolutionary algorithms for warp control point placement. In *Proceeding of the 2nd International Symposium on Intelligence Computation and Applications (ISICA 2007) Wuhan, China* (pp. 327-331).

Sun S., Zhuge F., Rosenberg J., Steiner R. M., Rubin G. D., & Napel S. (2008). Learning-enhanced simulated annealing: Method, evaluation, and application to lung nodule registration. *Applied Intelligence, 28*(1), 83-99.

Tanaka, T., Toumiya, T., & Suzuki, T. (1997). Output control by hill-climbing method for a small scale wind power generating system. *Renewable Energy, 12*(4), 387-400.

Tavakkoli-Moghaddam, R., Safaei, N., & Gholipour, Y. (2006). A hybrid simulated annealing for capacitated vehicle routing problems with the independent route length. *Applied Mathematics and Computation, 176*(2), 445-454.

Tavakkoli-Moghaddam, R., Safari, J., & Sassani, F. (2008). Reliability optimization of series-parallel systems with a choice of redundancy strategies using a genetic algorithm. *Reliability Engineering and System Safety, 93*(4), 550-556.

Tettey, T., & Marwala, T. (2006). Controlling interstate conflict using neuro-fuzzy modeling and genetic algorithms. In *Proceedings of the 10th IEEE International Conference on Intelligent Engineering Systems,* London, UK (pp. 30-44).

Tsuzuki, M. d. S., Martins, G. T. d. C., & Takase, F. K. (2006). Robot path planning using simulated annealing. In A. Dolgui, G. Morel and C. Pereira (Eds.), *Information control problems in manufacturing* (pp. 173-178). Ecole des Mines, Saint Etienne, France: IFAC Press.

Vakil-Baghmisheh, M.-T., Peimani, M., Sadeghi, M. H., & Ettefagh, M. M. (2008). Crack detection in beam-like structures using genetic algorithms. *Applied Soft Computing Journal, 8*(2), 1150-1160.

Valls, V., Ballestin, F., & Quintanilla, S. (2008). A hybrid genetic algorithm for the resource-constrained project scheduling problem. *European Journal of Operational Research, 185*(2), 495-508.

van den Bergh, F., & Engelbrecht, A. P. (2006). Study of particle swarm optimization particle trajectories. *Information Sciences, 176*(8), 937-971.

Wang, Q. (2008). Probability density estimation with data missing at random when covariables are present. *Journal of Statistical Planning and Inference, 138*(3), 568-587.

Wang, Q., Gao, Y., & Liu, P. (2006). Hill climbing-based decentralized job scheduling on computational grids. In *Proceedings of the First International Multi- Symposiums on Computer and Computational Sciences* (pp. 705-708).

Wooldridge, J. M. (2007). Inverse probability weighted estimation for general missing data problems. *Journal of Econometrics, 141*(2), 1281-1301.

Wu, T.-H., Chang, C.-C., & Chung, S.-H. (2008). A simulated annealing algorithm for manufacturing cell formation problems. *Expert Systems with Applications, 34*(3), 1609-1617.

Yan, H., Zheng, J., Jiang, Y., Peng, C., Xiao, S. (2008). Selecting critical clinical features for heart diseases diagnosis with a real-coded genetic algorithm. *Applied Soft Computing Journal, 8*(2), 1105-1111.

Yao, M.-J., & Chu, W.-M. (2008). A genetic algorithm for determining optimal replenishment cycles to minimize maximum warehouse space requirements. *Omega, 36*(4), 619-631.

Ye, Y., & Wenbo, M. (2007). Digital modulation classification using multi-layer perceptron and time-frequency features. *Journal of Systems Engineering and Electronics, 18*(2), 249-254.

Yokoi, T., Ohyama, W., Wakabayashi, T., & Kimura, F. (2004). Eigenspace method by autoassociative networks for object recognition. *Lecture Notes in Computer Science, 3138,* 95-103.

Youssef, H., Sait, S. M., & Adiche, H. (2001). Evolutionary algorithms, simulated annealing and tabu search: A comparative study. *Engineering Applications of Artificial Intelligence, 14*(2), 167-181.

Yuan, Y., (2000). Multiple imputation for missing data: Concepts and new development. *SUGI Paper* 267-25, Retrieved August 25, 2008, from http://support.sas.com/rnd/app/papers/multipleimputation.pdf.

Zahrani, M. S., Loomes, M. J., Malcolm, J. A., Dayem Ullah, A. Z. M., Steinhofel, K., & Albrecht, A. A. (2008). Genetic local search for multicast routing with pre-processing by logarithmic simulated annealing. *Computers and Operations Research, 35*(6), 2049-2070.

Zhang, J-R., Zhang, J., Lok, T-M., & Lyu, M. R. (2007). A hybrid particle swarm optimization–back-propagation algorithm for feedforward neural network training. *Applied Mathematics and Computation, 185*(2), 1026-1037.

Chapter XI
Estimation of Missing Data Using Neural Networks and Decision Trees

ABSTRACT

This chapter introduces a novel paradigm to impute missing data that combines a decision tree, autoassociative neural network (AANN) model and a principal component analysis-neural network (PCA-NN) based model. These models are designed to answer the crucial question of whether the optimization bounds actually matter. For each model, the decision tree is used to predict search bounds for a hybrid simulated annealing and genetic algorithm method that minimizes an error function derived from the respective model. The models' ability to impute missing data is tested and then compared using HIV sero-prevalance data. Results indicate an average increase in accuracy of 13% with the AANN based model's average accuracy increasing from 75.8% to 86.3% while that of the PCA-NN based model increasing from 66.1% to 81.6%.

INTRODUCTION

Missing data is a widely recognized problem affecting large databases that creates problems in many applications that depend on access to complete data records such as data visualization and reporting tools. This problem also limits data analysts interested in making policy decisions based on statistical inference from the data and thus estimating missing data is often invaluable as it preserves information and produces better, less biased estimates than simple techniques (Fogarty, 2006; Abdella, 2005; Nelwamondo, 2006) such as listwise deletion and mean-value substitution (Yansaneh, Wallace, & Marker, 1998; Allison, 2000).

Inferences made from available data for a certain applications depend on the completeness and quality of the data being used in the analysis. Thus, inferences made from a complete data are most likely to be more accurate than those made from incomplete data. Moreover, there are time critical applications, which require us to estimate or approximate the values of some missing variables that have to be supplied in relation to the values of other corresponding variables. Such situations may arise in system that uses a number of instruments, and in some cases one or more of the sensors used in the system fail. In such a situation, the value of the missing sensor has to be estimated within a short time and with great precision, and by taking into account of the values of the other sensors in the system. Approximation of the missing values in such situations requires the estimation of the missing values taking into account of the inter-relationships that exists amongst the values of other corresponding variables.

The neural network approach (Freeman and Skapura, 1991; Haykin, 1999), such as the one adopted by Abdella and Marwala (2005), involves an optimization process for the estimation of missing data. In earlier chapters, methods such as genetic algorithms, simulated annealing and particle swarm optimization are used without much regard to the optimization bounds (Michalewicz, 1996; Forrest, 1996; Banzhaf et al., 1998). Therefore, what is missing in this book is how to identify and incorporate optimization bounds in the estimation of missing data problem. This question necessarily requires an answer to the questions: What is the best method for identifying optimization bounds in the missing data problem? Is the incorporation of optimization bounds in the missing data problem significant? This chapter seeks to answer these questions.

Yu (2007) introduced a new non-monotone line search technique and combined it with the spectral projected gradient method for solving the bound constrained optimization problems. The results obtained showed that the identified global convergence and numerical tests are efficient. Yamada, Tanino, and Inuiguchi (2001) studied an optimization problem for minimizing a convex function over the weakly efficient set of a multi-objective programming problem.

Sun, Chen, and Li (2007) used C4.5 decision tree which was trained to diagnose faults in rotating machinery. The method was validated using six kinds of running states (normal or without any defect, unbalance, rotor radial rub, oil whirl, shaft crack and a simultaneous state of unbalance and radial rub), as simulated on a Bently Rotor Kit RK4. The results obtained showed that C4.5 has higher accuracy and needs less training time than back-propagation network. Kirchner, Tölle, and Krieter (2006) optimized a decision tree technique and successfully applied this to simulated sow herd datasets. Based on these successes, this chapter estimates optimization bounds using decision trees.

BACKGROUND

In this chapter, antenatal database is used to answer the question on whether optimization bounds are significant in the problem of missing data estimation. Before this is done, the dataset used is described within the context of missing data estimation with pre-defined bounds. As mentioned before, acquired immunodeficiency syndrome (AIDS) is a collection of symptoms and infections resulting from the specific damage to the immune system caused by the human immunodeficiency virus (HIV) in humans (Marx, 1982). The world has seen an increase in HIV infection rates in recent years as well as having the highest number of people living with the virus. This results from the high prevalence rate as well as resulting deaths from AIDS (South African Department of Health, Anonymous, 2000).

Research into the field is ongoing and is aimed at identifying ways of dealing with the virus. Demographic data are often used to classify people living with HIV/AIDS and how they are affected. Proper data collection needs to be done so as to understand where and how the virus is spreading. This will give more insight into ways in which education and awareness can be used to equip populations. Through being able to identify factors that deem certain people or populations in higher risk, governments can then be able to deploy strategies and plans within those areas that would help the people.

The problem with data collection in surveys is that is suffers from information loss. This can result from incorrect data entry or an unfilled field in a survey. This investigation explores the field of data imputation. The approach taken is to use regression models to model the inter-relationships between data variables and then undertake a controlled and planned approximation of data using a combination of regression and optimisation models (Draper & Smith, 1998; Houck, Joines, & Kay, 1995). Data imputation using autoassociative neural networks as a regression model has been carried out by Abdella and Marwala (2005) and others (Leke, Marwala, Tettey, 2006) while other variations are available in the literature including Expectation Maximization (Hu, Savucci, & Choen, 1998; Nelwamondo, Mohamed, & Marwala, 2007), rough sets (Nelwamondo and Marwala, 2007a) and decision trees (Barcena & Tussel, 2002). The dataset used in this chapter is the antenatal demographic data collected from clinics around South Africa.

THE MISSING DATA PROBLEM

Real time processing applications that are highly dependent on the data often suffer from the problem of missing input variables. Various methods of missing data imputation such as mean substitution and hot-deck imputation also depend on the knowledge of how data samples become missing. There are several reasons why the data may be missing, and as a result, missing data may follow an observable pattern. Exploring the pattern is important and may lead to the possibility of identifying cases and variables that affect the missing data. Having identified the variables that predict the pattern, a proper estimation method can be selected. In the context of missing data in surveys, the problem has been studied extensively (Hawkins & Merriam, 1991; Roth 1994; Huisman, 2000; Yuan, 2000; Allison, 2002; Schafer & Graham, 2002; Little & Rubin, 2002; Fogarty, 2006; Kim, 2005; He, 2006) and can arise from non-response by the interviewee or poorly designed questionnaires. Industrial databases also experience this problem especially in cases where important information, for example information on the equipment's operational environment, has to be entered manually (Lakshminarayan, Harp, & Samad, 1999) or due to instrumentation failures in on-line applications. These problems have been dealt with in the previous chapters.

Some of the missing data imputation methods that have been implemented are: regression techniques that estimate a missing value using a regression equation-based model derived from previously observed complete cases (Roth, 1994); as well as likelihood approaches like Expectation Maximization (EM) (Dempster, Laird, & Rubin, 1977; Nelwamondo & Marwala, 2007b) and variants of it such as the raw maximum likelihood that fit probabilistic statistical models of the data. The EM algorithm uses the following strategy: first, impute the missing data values; secondly, estimate the data model parameters using these imputed values; next, re-estimates the missing data values using these estimated model parameters and repeat until convergence (Pearson, 2006).

The above single imputation strategies possess a disadvantage in that they tend to artificially reduce the variability in the estimated data. This provides the motivation for using multiple imputation techniques where several imputed datasets are generated and subjected to the same analysis to give a set of results from which variability estimates (e.g. standard deviation) and other typical characterizations (e.g. mean) can be computed (Pearson, 2006). This chapter focuses on the imputation of missing data, concentrating mainly on continuous variables, by combining model-based and non-model based procedures as discussed below.

DECISION-TREE, NEURAL NETWORK AND HYBRID SIMULATED ANNEALING AND GENETIC ALGORITHM APPROACH

This section describes the combination of two machine-learning systems: a decision-tree based classifier algorithm and back-propagation for neural networks, along with the hybrid simulated annealing and genetic algorithm (HSAGA) optimization routine to impute missing data. Decision trees are used because they are able to offer optimization bounds. Within the ante-natal dataset that is used in this chapter, there are various ways in which bounds can be estimated. The first way is when bounds are estimated using what is termed in this chapter, "absolute bounds". For a given variable the absolute bounds are defined as the maximum and minimum values that have ever been encountered. Unfortunately, these absolute bounds are still wide and can be narrowed down. Narrowing down bounds, effectively, reduces the search space, thereby, reducing the number of possible optimum solutions, which in the context of missing data estimation is the reduction in the number of possible estimated values. Another way in which the search space can be narrowed is for each variable to define bounds as the average less standard deviation for the lower bound and average plus standard deviation for the upper bound. In this chapter, as already indicated, decision trees are used because of their simplicity for missing data estimation.

Decision Trees

A decision tree is basically a classifier that shows all possible outcomes and the paths leading to those outcomes in the form of a tree structure. A schematic diagram of decision trees is shown in Figure 11.1 for an example on whether to swim or not on a given day based on the weather.

Various algorithms for inducing a decision tree are described in existing literature, for example Classification and Regression Trees (CART) (Breiman et al., 1984), Oblique Classifier 1 (OC1) (Murthy et al., 1993), Iterative Dichotomiser 3 (ID3), rough set based decision trees (Wei et al., 2007), phonetic decision trees (Liu & Yan, 2004), ternary classification trees (Siciliano & Mola, 1998) and C4.5 (Quinlan, 1993). Decision trees have been applied in diverse fields such as land cover classification (Pal & Mather, 2003)

CART decision trees create classification and regression trees for predicting continuous dependent variables and categorical variables. Predicting continuous dependent variables is termed regression while predicting categorical variables is termed classification. Waheed et al. (2006) investigated the prospect of hyper-spectral remote sensing data to improve crop management information for use in precision farming. The ability of the CART decision tree algorithm to classify hyper-spectral data of experimental corn plots into categories was investigated and was able to successfully classify 12 treatment combinations with 75–100% accuracy even though the most excellent validation results were obtained at an early growth stage.

Spurgeon et al. (2006) used CART decision trees in prostate cancer screening. They collected data from 1,563 referred men with serum PSA 10 ng/ml or less who experienced an early prostate biopsy. The results observed demonstrated that cancer was detected in 26.1% of the men, greater cancer was detected in 8.3% of the men while the sensitivity and specificity of CART for detecting men with ag-

Figure 11.1. A decision tree for whether to swim or not

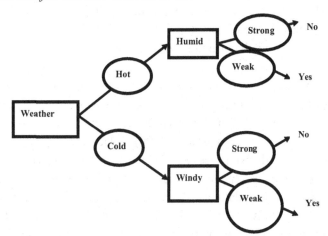

gressive cancer were found to be 100% and 31.8% for data used in model building, and 91.5% and 33.5% for the validation dataset, respectively.

OC1 is a decision tree type that is intended for applications where the problem has continuous values. OC1 creates decision trees that include linear combinations of one or more attributes at each internal node. OC1 has been utilized for classification of data into different subjects such as astronomy, bioinformatics and many others. Salzberg et al. (1995) successfully used OC1 for automated identification of cosmic ray hits in Hubble space telescope images.

ID3 is an algorithm that that is constructed using smaller decision trees, which use simple theories, as opposed to large ones. In spite of this, this approach does not guarantee the smallest tree. The ID3 is implemented by following these steps:

1. Take all unused attributes and count their entropy
2. Choose attribute for which the entropy is minimum
3. Make a node containing that particular attribute

Shao et al. (2001) proposed the application of an ID3 algorithm in knowledge discovery in tolerance design of injection-molded parts. This was achieved by calculating the information gain measure, representing entropy and creating the decision tree. Kinney and Murphy (1987) compared ID3 to discriminant analysis by applying this to cardiology and found that both methods produce similar results. Other successful applications of ID3 include: hand-written digit recognition (Chi, Suters, & Yan, 1996); tolerance design (Shao et al., 2001), breast cancer patients (Ture, Tokatli, & Kurt, 2007) and for incremental learning (Ichihashi et al., 1996).

A decision tree is, therefore, created recursively by dividing the training dataset into successively reduced sub-sets. A brief description of the basic decision tree algorithm is as follows (Salzberg, 1995; Tan, Steinbach, & Kumar, 2005):

For a given set of training records, S, associated with a node t, let C_i for $i = [1,2,...,m]$ be the class labels:

1. Split S_t into smaller subsets using a test on one or more attributes.
2. Check the split results. If all subsets are pure (all records in S_t belong to the same class C_l), label the leaf node with the class name C_t and stop.
3. Recursively split any partitions that are not pure.

The splitting criterion typically determines the difference between trees created with different algorithms. The C4.5 algorithm is used, in this chapter, and it measures the entropy of the initial set and sub-sets produced after splitting, and then it chooses attributes with the most information gain based on information theory (Quinlan, 1993; Han & Kamber, 2000). Polat and Güneş (2007) introduced a hybrid classification system based on C4.5 decision tree classifier and one-against-all method to classify problems in the areas of dermatology, image segmentation, and lymphography datasets taken from University of California Irvine machine learning database. The method proposed was tested using classification accuracy, sensitivity-specificity analysis, and 10-fold cross validation. When the C4.5 decision tree was tested in all dataset classification accuracies of 84.48%, 88.79%, and 80.11% were obtained for dermatology, image segmentation, and lymphography datasets using 10-fold cross validation, respectively.

If S contains s_i tuples of class C_i then the Information (entropy) required to classify any given tuple is given by (Han & Kamber, 2000):

$$I(s_{1j},...,s_{mj}) = -\sum_{i=1}^{m} \frac{s_i}{s} \log_2 (s_i / 2)$$

(11.1)

Assuming that an attribute A with v values is selected as a candidate root of a given tree, then S will be partitioned into sets $\{S_1, S_2,...,S_v\}$. The expected information needed to complete the tree with A as the root is:

$$E(A) = \sum_{j=1}^{v} \frac{s_{1i} + ... + s_{mj}}{s} I(s_{1j},...,s_{mj})$$

(11.2)

And thus the information gained by branching on attribute A is given by:

$$Gain(A) = I(s_1, s_2,...,s_m) - E(A)$$

(11.3)

Under C4.5, the attribute with the highest information gain is chosen to branch the given tree. A more detailed explanation on how C4.5 builds and prunes decision trees can be found in (Quinlan, 1993). Using C4.5 decision trees, optimization bounds are then calculated.

There are crucial issues that need to be taken into account when implementing decision trees to facilitate missing data estimation. One of these is on-line learning. For decision trees to be implemented on-line to facilitate on-line missing data estimation procedure, methods of evolving them need to be found. Aitkenhead (2008) proposed a method for evolving attributes that define decision trees and successfully applied this to classification problems. The problem that is solved in this chapter, missing data estimation in HIV database, contains both categorical and continuous variables. Therefore, a method that mixes the two should be devised.

Arentze and Timmermans (2007) combined continuous attribute variables into rule-based models for discrete choice, which is similar to a problem of HIV classification, which is the ultimate aim of missing data estimation in the database considered in this chapter. Other types of decision trees are fuzzy decision trees that have been used for path planning and audit fees (Hamzei & Mulvaney, 1999; Beynon, Peel, & Tang, 2004; Liu & Pedrycz, 2007), polynomial-fuzzy decision trees for medical applications (Mugambi et al., 2004) and scalable decision trees for intrusion detection (Li, 2005).

Neural Networks

As described in earlier chapters, neural networks are computational models that have the ability to learn and model systems (Yoon & Peterson, 1990; Mohamad, 1995). Neural networks have the ability to model non-linear systems (Bishop, 1995; Abdi, 1994; Rumelhart, Hinton, & Williams, 1986). They have been successfully used in many applications such as to investigate the Baldwin effect (Jones & Konstam, 1999) and in time series forecasting (Kolarik & Rudorfer, 1994). The neural network architecture used in this chapter is a multilayer perceptron network (MLP) (Bishop, 1995), which is found to be better than the radial basis function in Chapter II. The MLP is described in detail in earlier chapters. This has two layers of weights which connect the input layer to the output layer. The middle of the network is made up of a hidden layer. This layer can be made up of a different number of hidden nodes. This number has to be optimized so that the network can model systems better. An increase in hidden nodes translates into an increase in the complexity of the system. The output and the hidden nodes also have activation functions (Bishop, 1995). The general equation of a MLP neural network is shown as (Bishop, 1995):

$$y_k = f_{outer}(\sum_{j=1}^{M} w_{kj}^{(2)} f_{inner}(\sum_{i=1}^{d} w_{ji}^{(1)} + w_{j0}^{(1)}) + w_{k0}^{(2)})$$

(11.4)

The outer activation function (f_{outer}) chosen in this chapter is linear. The inner activation (f_{inner}) function chosen is the hyperbolic tangent function. These choices are made through trial-and-error and are thus found to increase the accuracy of the regression process. Thus the relation as implemented in this chapter thus becomes:

$$y_k = \sum_{j=1}^{M} w_{kj}^{(2)} \tanh(\sum_{i=1}^{d} w_{ji}^{(1)} + w_{j0}^{(1)}) + w_{k0}^{(2)}$$

(11.5)

In this chapter the Netlab (Nabney, 2001) MATLAB toolbox is utilized to implement the neural networks.

Autoassociative Networks

As indicated in earlier chapters, autoassociative neural networks are neural networks that are trained to recall their inputs. Thus the number of inputs is equal to the number of outputs. Autoassociative neural networks have a bottleneck that results from the structure of the hidden nodes (Thompson, Marks, & Choi, 2002). There are less hidden nodes than input nodes. This results in a butterfly structure. The autoassociative network is preferred in recall applications as it can map linear and non-linear relationships between all the inputs. The autoassociative structure results in the compression of data into a smaller dimension and then decompressing into the output space.

Autoassociative networks have been used in a number of applications including in missing data estimation and this is described in earlier chapters (Abdella & Marwala, 2005; Nelwamondo & Marwala, 2007a). In this investigation an autoassociative networks are constructed using the MLP structure discussed in the previous sub-section. The demographic and HIV data are fed into the network and the network is trained to recall the inputs. Thus the structure is as shown in Figure 11.2 results.

As it can be recalled from earlier chapters, the autoassociative networks are used in conjunction with some optimization method for missing data estimation. As it is observed in Chapter X, the results obtained from the choice of an optimization method offer more than one viable optimization method. Two optimization methods that are found in Chapter X to be successful are simulated annealing and genetic algorithm. As a result these two methods are hybridized and then used for missing data estimation process.

Hybrid Simulated Annealing and Genetic Algorithms

In this chapter a hybrid of simulated annealing and genetic algorithms (HSAGA) is used for the optimization process. Jwo, Liu, & Liu (1999) introduced hybrid simulated annealing and genetic algorithm method and successfully applied this to power system while Wong (2001) successfully applied HSAGA method to better manage peak system demand. Yao et al. (2003) developed the HSAGA where the crossover and mutation were the multi-adaptive annealing crossover and successfully applied this to the synthesis of ether whereas Hwang and He (2006) developed the HSAGA method and applied this to parameter estimation in auto regressive and moving average exogenous models. Furthermore, Soke and Bingul (2005) developed HSAGA and applied this to two dimensional non-guillotine rectangular packing problems while Jeong and Lee (1996) applied HSAGA to system identification problems.

Simulated annealing (SA) is an algorithm that is inspired by the natural process of crystallization that locates an approximation to the global optimum. It is in essence, a generalization to the Monte Carlo method and relies on the Metropolis algorithm (Metropolis et al, 1953). In the SA approach, a current solution is replaced with a "nearby" random solution with a probability that depends on the difference between the corresponding function values and the temperature (T). The temperature T decreases throughout the process, and when T starts approaching zero, there is less random changes in the solution. In this chapter, the temperature is made to decrease exponentially. Simulated annealing keeps moving

Figure 11.2. Autoassociative neural network

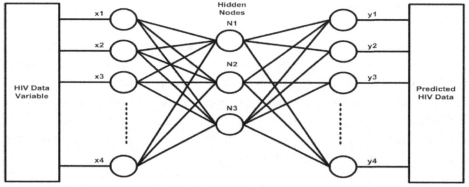

towards the best solution and it has the advantage of reversal in fitness and, therefore, it can move to a solution with a worse fitness than it currently has, but the good aspect of that is that it ensures that the solution is not found at a local optimum, but rather at a global optimum solution.

As indicated in earlier chapters, GA was inspired by Darwin's theory of natural evolution. Genetic algorithm is a simulation of natural evolution where the law of the survival of the fittest is applied to a population of individuals. This natural optimization method is used in this chapter. GA is implemented by generating a population and creating a new population by performing the following procedures: (1) crossover; (2) mutation; (3) and reproduction. The details of these procedures can be found in Holland (1975) and Goldberg(1989). Unlike in the previous chapters, a floating point genetic algorithm procedure as opposed to binary space genetic algorithm is used in this chapter.

The crossover operator mixes genetic information in the population by cutting pairs of chromosomes at random points along their length and exchanging over the cut sections. This has a potential of joining successful operators together. Arithmetic crossover technique (Goldberg, 1989) is used in this chapter. Arithmetic crossover takes two parents and performs an interpolation along the line formed by the two parents. For example if two parents *p1* and *p2* undergo crossover, then a random number *a* which lies in the interval [0,1] is generated and the new off-springs formed are *p1(a-1)* and *pa*. Mutation is a process that introduces to a population, new information. Non-uniform mutation (Goldberg, 1989) is used, in this chapter, and it changes one of the parameters of the parent based on a non-uniform probability distribution. The Gaussian distribution starts with a high variance and narrows to a point distribution as the current generation approaches the maximum generation.

Reproduction takes successful chromosomes and reproduces them in accordance to their fitness functions. In this chapter, normalized geometric selection method is used (Goldberg, 1989; Mitchell, 1999). This method is a ranking selection function which is based on the normalized geometric distribution. Using this method the least fit members of the population are gradually driven out of the population. The basic HSAGA method that is implemented in this chapter is as follows:

1. Randomly create an initial population of a certain size.
2. Evaluate all of the individuals in the population using the objective function.
3. Use the normalized geometric selection method to select a new population from the old population based on the fitness of the individuals as given by the objective function.
4. Apply some genetic operators, non-uniform mutation and arithmetic crossover, to members of the population to create new solutions.
5. Repeat steps 2-4, which is termed one generation, until a certain fixed number of generations has been achieved
6. From step 5 identify the best solution from a population
7. Introduce random changes to the solution in step 5 such that the resulting state falls within bounds.
8. Given the current temperature, compare the energy (missing data equation error in this chapter) between the previous state and the current state.
9. Accept or reject this state based on the Metropolis et al. (1953) criterion.
10. Repeat step 7 to 9 until some stopping criterion is reached.

Neural Network and HSAGA method for Missing Data

A technique that combines neural networks with genetic algorithm optimization routine was used by Abdella and Marwala to impute missing data in an industrial database (Abdella & Marwala, 2006). As derived in Chapter II, the missing data estimation error equation is written in terms of the known measurement $\{X_k\}$, unknown measurement to be estimated $\{X_u\}$, the autoassociative network model, f, and the Euclidean norm, $\| \|$, as follows:

$$\varepsilon = \left\| \left(\left\{ \begin{matrix} \{X_k\} \\ \{X_u\} \end{matrix} \right\} - f\left(\left\{ \begin{matrix} \{X_k\} \\ \{X_u\} \end{matrix} \right\}, \{W\} \right) \right) \right\| \tag{11.6}$$

The HSAGA optimization routine is then used to guess missing values that minimize the error function in equation 11.6 given the optimization bounds calculated using decision trees.

Combining C4.5 with Autoassociative Neural Network-HSAGA (AANN-HSAGA)

The AANN-HSAGA technique is a model-based data imputation technique is better suited for users that are familiar with the missing data problems, and possess the necessary expertise to apply their knowledge in building an accurate model. C4.5 on the other hand is not model-based as it does not make any assumptions on the data parameters. It has been used for data completion by treating missing value imputation of discrete valued attributes as a classification task (Lakshminarayan, Harp, & Samad, 1999) but the database in this case was largely made up of continuous attributes which C4.5 does not naturally handle.

The combination of these two procedures is shown in Figure 11.3. C4.5 is used to classify intervals of the missing continuous attributes and these intervals are then used as the bounds in which the GA searches for missing value. GA bounds are commonly set to the entire normalized range for the attribute (0 to 1) but empirical results show that by limiting the HSAGA bounds using this approach, there is a significant improvement in the accuracy of the AANN-HSAGA architecture as will be shown later.

HYBRID PRINCIPAL COMPONENT ANALYSIS, NEURAL NETWORK AND HYBRID SIMULATED ANNEALING GENETIC ALGORITHM (PCA-NN-HSAGA) APPROACH

This section describes the use of the principal component analysis (PCA) model with a neural network (NN) and HSAGA method to impute missing data. Principal component analysis is described in detail in Chapter III.

Principal Component Analysis

As indicated in earlier chapters, principal component analysis is a popular statistical technique commonly used to find patterns in high-dimensional data and to reduce these dimensions (Jollifie, 1986). It has been applied in several fields such as face recognition (Turk & Pentland, 1991), enhancing visualization of

Figure 11.3. Decision tree, auto-associative neural network and hybrid simulated annealing and genetic algorithm

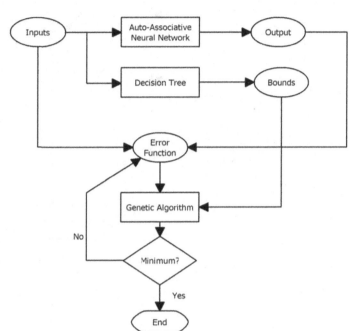

high-dimensional data (Wolfgang, Nocke, & Schumann, 2006) and image compression (Ye, Janardan, & Li, 2004), to mention but a few. By identifying the major causes of variation in a dataset, the PCA analysis provides a more compressed description of the data. Furthermore, the principal components, ordered by importance in accounting for as much of the variation in the data as possible, provide a basis for dimension reduction by selecting the major principal components and omitting the less significant ones. An overview of the PCA notation is now presented as follows.

Let $X_{M \times N}$ be the input set of M records with N dimensions after centralization (i.e. after subtracting the sample mean from each input). The aim is to derive a mapping $\psi : \chi \mapsto Z$ that maps the input features into a K-dimensional space with $K < N$. The mapping is given by:

$$Y_{K \times M} = (U_{N \times K})^T . X_{M \times N}^T \tag{11.7}$$

where $U_{N \times K}$ is a feature vector made up of K principal eigenvectors of the covariance matrix of X.

Proposed Data-Imputation Model

The proposed PCA-NN-HSAGA imputation model is shown in Figure 11.4. An MLP neural network is trained to map the input set $X_{M \times N}$ to the transpose of the reduced-dimension dataset in Equation 11.7, thereby, creating two models (PCA and NN) for the data that should produce identical outputs. The difference between the outputs of these two models is used as the error function for the HSAGA algorithm to minimize when searching for a missing value. Thus equation 11.6 is modified to become:

Figure 11.4. Proposed PCA-NN-GA data imputation model

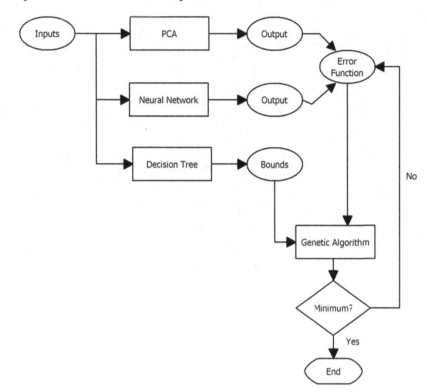

$$\varepsilon = \left\| \left(f_{PCA} \begin{Bmatrix} \{X_k\} \\ \{X_u\} \end{Bmatrix} - f \left(\begin{Bmatrix} \{X_k\} \\ \{X_u\} \end{Bmatrix}, \{W\} \right) \right) \right\|$$

(11.8)

where f_{PCA} is the PCA model's function that transforms the input vector $\{X\}$ using the feature vector of K selected eigenvectors, U_K, f_{NN} is the neural network model's function and the rest of the notation is the same as before (see equation 11.6). Once again, the C4.5 decision tree algorithm is applied to this architecture to predict HSAGA search bounds for the various missing variables.

EXPERIMENT AND RESULTS

This section presents the experiment carried out using the aforementioned procedures to predict missing values in the HIV dataset. As indicated before the methods proposed in this chapter work irrespective of the missing data mechanism (Schafer, 1997).

Experimental Data

As indicated before, the experimental data used is from an antenatal clinic survey conducted in South Africa in 2001. It consists of the following attributes:

- **Age:** Ranging from 14 to 50 years.
- **Education Level:** Where a number ranging from 0 to 13 represents the highest school grade completed by a candidate with 13 indicating tertiary level education.
- **Father's Age:** The age of the father responsible for the most recent pregnancy.
- **Gravidity:** The number of times a candidate has fallen pregnant.
- **Parity:** The number of successful pregnancies.
- **Race:** Encoded in a binary fashion to cover five possible options which are Asian, African, Mixed and European and Other.
- **Province:** Encoded in a binary fashion to cover all 9 provinces in South Africa.
- **HIV status** of the candidate with a 0 or 1 representing negative or positive status, respectively.

All records with missing fields are removed, the most prevalent being education level with 22% missing, as well as outliers and records with logical errors leaving a total of 12179 records from an initial 16743 records. These are then randomly mixed and split to form a training dataset (9745 records), validation set for use in early-stopping (1217 records) and a testing dataset consisting of previously unseen data with missing values for various attributes to be predicted (1217 records).

Experiment 1

A multi-layer perceptron neural network consisting of 13 input nodes for the aforementioned variables, 11 hidden nodes and 13 output nodes is used to build a model of the data (Nabney, 2007; MathWorks, 2007). Training is performed using the scaled conjugate gradient (SCG) supervised learning algorithm (Møller, 1993), which is explained in Chapter II. The number of hidden nodes is chosen experimentally, for a fixed number of training cycles, to minimize the root mean square error (RMSE) over the training dataset: Here y and y_0 are the true and actual network outputs, respectively, and n is the size of the dataset.

In order to minimize redundancy in the autoassociative network, the investigated number of hidden nodes is always less than the number of input nodes (Leke & Marwala, 2006). The number of training cycles (110) is determined in a similar fashion, for a fixed number of hidden nodes, using the early stopping method (Nelson & Illingworth, 1991) to prevent over-fitting of the data. The results of the network optimization are shown in Figures 11.5 and 11.6. An implementation of the HSAGA is conducted (Houck, Joines, & Kay, 2007) and when predicting missing values for all attributes, the same parameters are used namely: normal geometric selection, simple crossover, non-uniform mutation, 20 generations, a population size of 50 and exponential temperature cooling schedule. C4.5 is trained to predict intervals for various variables as shown in Table 11.1, using the training and validation datasets described in the preceding sub-section.

It is then applied to the test dataset and the predicted intervals (bounds) passed on to the HSAGA. Since the HIV attribute is binary and thus could be handled natively by C4.5, a C4.5 prediction of 0 translated to HSAGA bounds of [0, 0.5] and a prediction of 1 to [0.5, 1]. The experimental results presented in Table 11.2 are an average from three runs of the experiment.

The accuracy of imputed values is measured as a percentage of imputed values that are offset from the true value within a specified range as follows:

- **Age:** To within 2, 4, 6 and 10 years.
- **Education Level:** To within 1, 2, 3 and 5 grades.

Figure 11.5. Auto-associative neural network optimization results (Optimum Nodes)

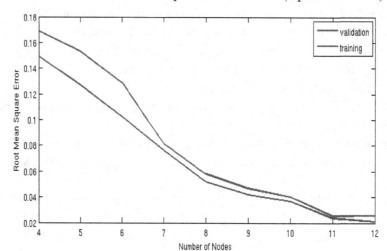

Figure 11.6. Auto-associative neural network optimization results (Optimum Training Cycles)

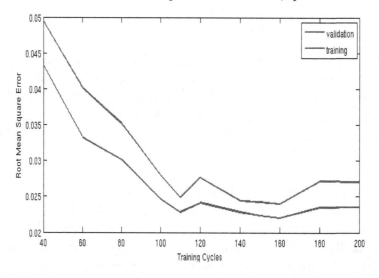

- **Father's Age:** To within 2, 4, 6 and 10 years.
- **Gravidity:** Exact number of pregnancies and to within one, three and five pregnancies.
- **Parity:** Exact number of pregnancies and to within one, three and five pregnancies.
- **HIV status:** Accuracy measured using specificity.

Experiment 2

The PCA analysis is conducted on the combined training and validation datasets and for different dimensions, the Root-Mean-Square-Error (RMSE) is calculated on the recovered data as well as the accuracy of the predicted values to within 5%. Figure 11.7 shows these results and, as can be seen, there is a significant change in accuracy between using 6 and 7 dimensions, while using 11 dimensions only alters the accuracy slightly. Thus the new reduced dimension is chosen to be 7.

Table 11.1. Attribute intervals predicted by C4.5

Attribute	Interval
Age	4 years e.g 20 – 24, 25 – 29 etc.
Education	2 grades e.g 0 – 2, 3 – 5, etc
Father's Age	4 years e.g 20 – 24, 25 – 29, etc
Gravidity	2 pregnancies e.g 0 – 2, 3 – 5, etc
Purity	pregnancies e.g 0 – 2, 3 – 5, etc

Table 11.2. Percentages of imputed data within the specified ranges for ANN-HSAGA

Method	Age	Edu	Fat	Gra	Par	HIV
AANN-HSAGA	47.7	32.5	31.4	80.4	50.9	77.0
	75.0	46.0	54.7	97.1	91.0	-
	89.0	59.7	73.0	99.6	98.5	-
	97.0	76.7	90.8	100.0	99.7	-
C4.5, AANNHSAGA	52.3	52.1	41.7	81.8	60.8	99.7
	79.4	69.5	68.6	97.8	92.9	-
	89.6	79.4	82.7	99.7	98.6	-
	97.9	91.8	93.2	100.0	99.7	-

Figure 11.7. Root Mean Square Error for varying PCA dimensions

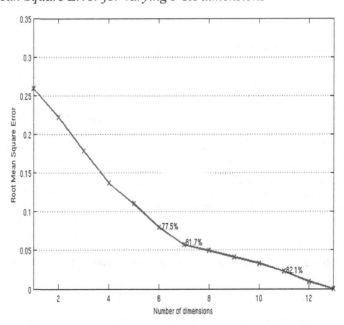

An MLP neural network consisting of 13 input nodes, 17 hidden nodes and 7 output nodes is set up and trained using 140 cycles using the same techniques detailed in Experiment 1 and these optimization results are shown in Figures 11.8 and 11.9. The GA, C4.5 and accuracy of imputed missing values

Figure 11.8. Optimization nodes for the PCA-NN-HSAGA neural network

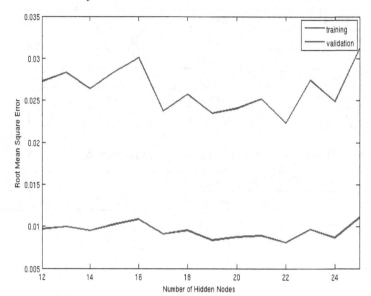

Figure 11.9. Optimization results for the PCA-NN-HSAGA neural network

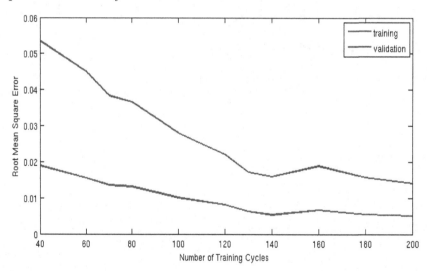

is performed using the same settings and to within the same ranges as described above. These results are presented in Table 11.3.

DISCUSSION

Impact of Bounds Selection

The most glaring insight that can be gained from thee experimentation conducted is that bounds in optimization search routines matter. Whilst it can be argued that given a big enough population size

Table 11.3. Percentages of Imputed Data within the specified ranges for PCA-NN-HSAGA

Method	Age	Educ.	Father	Grav.	Parity	HIV
PCA-NN-HSAGA	30.6	43.5	20.3	40.6	38.7	70.1
	55.0	62.6	37.2	84.1	72.8	-
	71.8	77.2	50.0	99.0	91.5	-
	91.4	91.0	69.9	99.7	98.8	-
C4.5, PCA-NN-HSAGA	50.6	51.5	43.4	53.7	47.5	99.5
	77.1	69.7	68.8	93.8	80.9	-
	88.0	81.3	81.3	99.9	98.0	-
	97.3	93.0	93.7	100	99.9	-

and number of generations HSAGA will find an optimum solution, this solution is not necessarily the global optimum. Furthermore, any experiment can have various parameters that can be tweaked and the experiment repeated until adequately accurate results are achieved, however, this is an unfair presentation of a given paradigm and it does not guarantee repeatability on another set of data. Prediction of bounds directly from given inputs, as presented in this chapter, repeatedly improves the accuracy of missing data imputation irrespective of the underlying architecture of 10.5% for the AANN-HSAGA architecture and 15.5% for the PCA-NN-HSAGA architecture.

Note on Proposed Architectures

Depending on the desired accuracy for a given variable, both architectures produce satisfactory results and can be used to complete incomplete databases. For example, a person's age group can be accurately defined to within 4 years, 20 − 24 for early twenties and 25 − 29 for late twenties say, to which the C4.5, AANN-HSAGA and C4.5, PCA-NN-HSAGA architectures would give an accuracy of 79.4% and 77.1%, respectively. Similarly, imputing the education variable (which had the largest number of missing values) to within 3 years accurately captures a person's level of schooling to say elementary school (grade 0 − 2), lower middle school (grade 3 − 5), upper middle school (grade 6 − 8), high school (grade 9−11) and (high school/college) graduate (grade 12+). The C4.5, AANN-HSAGA predicts this education level to an accuracy of 79.4% while C4.5, PCA-NN-HSAGA predicts to 81.3%. The autoassociative based architecture, however, performs better than the PCA based architecture for all variables except *Education* which could be attributed to the loss in data associated with dimensionality reduction under PCA.

CONCLUSION

In conclusion, a novel paradigm that combines decision trees with neural networks, principal component analysis and hybrid simulated annealing-genetic algorithm method is introduced to estimate missing data. Two separate architectures, one based on an autoassociative neural network and the other principal component analysis, are set up and each is combined with a decision tree as well as the genetic algorithm optimization routine. Empirical results indicate that both architectures can adequately impute missing data and the addition of a decision tree improves results for both by an average of 13%.

FURTHER WORK

This chapter dealt with the issue of the significance of optimization bounds using decision trees as the estimator of those bounds. The questions that still need to be answered are:

- What is the best approach to optimization bounds estimation within the context of missing data imputation?
- What is the best combination of machine learning infrastructure, bounds estimator and optimization method? Is the combination of multi-layer perceptron, hybrid simulated annealing genetic algorithms method and decision trees that is implemented in this chapter adequate?
- How should this infrastructure proposed in this chapter be designed to optimize computational load versus accuracy?

REFERENCES

Abdella, M., (2005). *The use of genetic algorithms and neural networks to approximate missing data in database.* Unpublished master's thesis, University of the Witwatersrand, Johannesburg.

Abdella, M., & Marwala, T. (2005). Treatment of missing data using neural net- works and genetic algorithms. In *Proceedings of the 2005 IEEE International Joint Conference on Neural Networks* 1, Mauritius (pp.598-603).

Abdella, M., & Marwala, T. (2006). The use of genetic algorithms and neural networks to approximate missing data in database. *Computing and Informatics. 24,* 1001-1013.

Abdi, H. (1994). A neural network primer. *Journal of Biological Systems. 2*(3), 247-283.

Allison, P. (2002). *Missing data.* Thousand Oaks, CA: Sage.

Aitkenhead, M. J. (2008). A co-evolving decision tree classification method. *Expert Systems with Applications. 34*(1), 18-25.

Allison, P. D. (2000). Multiple imputation for missing data: A cautionary tale. *Sociological Methods and Research. 28,* 301-309.

Anonymous, (2000). *HIV/AIDS/STD strategic plan for South Africa.* Department of Health, South Africa.

Arentze, T., & Timmermans, H. (2007). Parametric action decision trees: Incorporating continuous attribute variables into rule-based models of discrete choice. *Transportation Research Part B: Methodological, 41*(7), 772-783.

Banzhaf, W., Nordin, P., Keller, R., & Francone, F. (1998). *Genetic programming: An Introduction on the automatic evolution of computer programs and its applications*. 5th Edition. San Francisco, California: Morgan Kaufmann Publishers.

Barcena, M. J., & Tussel, F. (2002). Multivariate data imputation using trees. *Documentos de Trabajo BILTOKI. 5*. Retrieved August, 25, 2008, from http://ideas.repec.org/p/ehu/biltok/200205.html.

Beynon, M. J., Peel, M. J., & Tang, Y-C. (2004). The application of fuzzy decision tree analysis in an exposition of the antecedents of audit fees. *Omega. 32*(3), 231-244.

Bishop, C. M. (1995). *Neural networks for pattern recognition*. Oxford: Oxford University Press.

Breiman, L., Friedman, J., Olshen, R., Stone, C. (1984). *Classification and regression trees*. Belmont, MA: Wadsworth International Group.

Chi, Z., Suters, M., & Yan, H. (1996). Handwritten digit recognition using combined ID3-derived fuzzy rules and Markov chains. *Pattern Recognition. 29*(11), 1821-1833.

Dempster, A., Liard, N., Rubin, D. (1977). Maximum likelihood from incomplete data via the EM algorithm (with discussion). *Journal of the Royal Statistical Society. B39*, 1-38.

Draper, N., & Smith, H. (1998). *Applied regression analysis*. New York. John Wiley & Sons, New York.

Fogarty, D. J. (2006). *Multiple imputation as a missing data approach to reject inference on consumer credit scoring*. Retrieved January 1, 2008, from http://interstat.statjournals.net/YEAR/2006/articles/0609001.pdf

Forrest, S. (1996). Genetic algorithms. *ACM Computer Survey. 28*(1), 77.

Freeman, J., & Skapura, D. (1991). *Neural networks: Algorithms, applications and programming techniques*. Reading, MA: Addison-Wesley.

Goldberg, D. (1989). *Genetic algorithms in search, optimization, and machine learning*. Reading, MA: Addison-Wesley.

Hamzei, G. H. S., & Mulvaney, D. J. (1999). On-line learning of fuzzy decision trees for global path planning. *Engineering Applications of Artificial Intelligence, 12*(1), 93-109.

Han, J., & Kamber, M. (2000). *Data mining: Concepts and techniques*. San Mateo, CA: Morgan Kauffmann Publishers.

Hawkins, M., & Merriam V. (1991). *An over-modeled world*. New York: Direct Marketing Press.

Haykin, S. (1999). *Neural networks*. New Jersey: Prentice-Hall,.

He, Y. (2006). *Missing data imputation for tree-based models*. Doctoral dissertation, University of California, Los Angeles.

Holland, J. (1975). *Adaptation in natural and artificial systems*. Ann Arbor: University of Michigan Press.

Houck, C. R., Joines, J.A., & Kay, M. G. (1995). *A genetic algorithm for function optimisation: A Matlab implementation* (Tech. Rep. NCSU-IE TR 95-09). Chapel Hill: Carolina State University Press.

Houck, C., Joines, J., & Kay, M. (2007). *The genetic algorithm optimization toolbox (GAOT)*. Retrieved June, 2, 2008, from http://www.ise.ncsu.edu/mirage/GAToolBox/gaot/

Hu, M. S., Savucci, & Choen, M. (1998). Evaluation of some popular imputation algorithms. In *Proceedings of the Survey Research Methods Section of the American Statistical Association* (pp. 308-314).

Huisman, M. (2000). Post-stratification to correct for nonresponse: classification of zip code areas. In *Proceedings of the 14th Symposium on Computational Statistics* (pp. 325-330).

Hwang, S-F., & He, R-S. (2006). Improving real-parameter genetic algorithm with simulated annealing for engineering problems. *Advances in Engineering Software, 37*(6), 406-418.

Ichihashi, H., Shirai, T., Nagasaka, K., & Miyoshi, T. (1996). Neuro-fuzzy ID3: A method of inducing fuzzy decision trees with linear programming for maximizing entropy and an algebraic method for incremental learning. *Fuzzy Sets and Systems, 81*(1), 157-167.

Jeong, I-k., & Lee, J-j. (1996). Adaptive simulated annealing genetic algorithm for system identification. *Engineering Applications of Artificial Intelligence, 9*(5), 523-532.

Jollifie, I. T. (1986). *Principal component analysis.* New York: Springer-Verlag.

Jones, M., & Konstam, A. (1999). The use of genetic algorithms and neural networks to investigate the Baldwin effect. In *Proceedings of the 1999 ACM symposium on Applied Computing* (pp. 275-279).

Jwo, W-S., Liu, C-W., & Liu, C-C. (1999). Large-scale optimal VAR planning by hybrid simulated annealing/genetic algorithm. *International Journal of Electrical Power & Energy Systems, Volume 21*(1), 39-44.

Kim, J. (2005). *Parameters estimation in stochastic volatility models with missing data using particle methods and the EM algorithm.* Unpublished doctoral dissertation, University of Pittsburg, Pittsburg.

Kinney, E. L., & Murphy, D. D. (1987). Comparison of the ID3 algorithm versus discriminant analysis for performing feature selection. *Computers and Biomedical Research, 20*(5), 467-476.

Kirchner, K., Tölle, K-H., & Krieter, J. (2006). Optimisation of the decision tree technique applied to simulated sow herd datasets. *Computers and Electronics in Agriculture, 50*(1), 15-24.

Kolarik, T., & Rudorfer, G. (1994). Time series forecasting using neural networks. In *Proceedings of the International Conference on APL: The Language and its Applications* (pp. 86-94).

Lakshminarayan, K., Harp, S., Samad, T. (1999). Imputation of missing data in industrial databases. *Applied Intelligence, 11*(3), 259-275.

Leke, B. B., & Marwala, T. (2006). Autoencoder networks for HIV classification. *Current Science, 91*(11), 1467-1473.

Leke, B. B., Marwala, T., & Tettey, T. (2006). Autoencoder networks for HIV classification. *Current Science, 91*(11), 1467-1473.

Li, X-B. (2005). A scalable decision tree system and its application in pattern recognition and intrusion detection. *Decision Support Systems, 41*(1), 112-130,

Little, R., & Rubin, D. (2002). *Statistical analysis with missing data.* New York: John Wiley & Sons.

Liu, X., & Pedrycz, W. (2007). The development of fuzzy decision trees in the framework of axiomatic fuzzy set logic. *Applied Soft Computing, 7*(1), 325-342.

Liu, C., & Yan, Y. (2004). Robust state clustering using phonetic decision trees. *Speech Communication, 42*(3-4), 391-408.

Marx, J. L. (1982). New disease baffles medical community. *Science, 217*(4560). 618-621.

MathWorks, T. (2007). *MATLAB-The language of technical computing.* Retrieved May, 5 2008, from http://www.mathworks.com

Metropolis, N., Rosenbluth, A., Rosenbluth, M., Teller, A., & Teller, E. (1953). Equation of state calculations by fast computing machines. *The Journal of Chemical Physics, 21,* 1087–1092.

Michalewicz, Z. (1996). *Genetic algorithms + Data structures = Evolution programs.* New York: Springer-Verlag.

Mitchell, M. (1999). *An introduction to genetic algorithms.* Cambridge, MA: MIT Press.

Mohamad, H. H. (1995). *Fundamentals of artificial neural networks.* Cambridge, MA: MIT Press.

Møller, M. (1993). A scaled conjugate gradient algorithm for fast supervised learning. *Neural Networks, 6,* 525-533.

Mugambi, E. M., Hunter, A., Oatley, G., & Kennedy, L. (2004). Polynomial-fuzzy decision tree structures for classifying medical data. *Knowledge-Based Systems, 17*(2-4), 81-87.

Murthy, S., Kasif, S., Salzberg, S., Beigel, R. (1993). OC1: Randomized induction of oblique decision trees. In *Proceedings of the Eleventh National Conference on Artificial Intelligence* (pp. 322-327).

Nabney, I., (2007). Netlab neural network software. Retrieved February 28, 2008. from http://www.ncrg.aston.ac.uk/netlab/

Nelson, M., & Illingworth, W. (1991). *A practical guide to neural nets.* Reading, MA: Addison-Wesley.

Nelwamondo, F. V. (2006). *Computational intelligence techniques for missing data imputation.* Unpublished master's thesis, University of the Witwatersrand, Johannesburg.

Nelwamondo, F. V., & Marwala, T. (2007a). Rough set theory for the treatment of incomplete data. In *Proceedings of the IEEE Conference on Fuzzy Systems* (pp. 338-343).

Nelwamondo, F. V., & Marwala, T. (2007b). Techniques for handling missing data: applications to online condition monitoring. *International Journal of Innovative Computing, Information and Control* (Accepted).

Nelwamondo, F. V., Mohamed, S., & Marwala, T. (2007). Missing data: A comparison of neural network and expectation maximization techniques. *Current Science, 93 (11),* 1514-1521.

Pal, M., & Mather, P. M. (2003). An assessment of the effectiveness of decision tree methods for land cover classification, *Remote Sensing of Environment, 86 (4),* 554-565.

Pearson, R. K. (2006). The problem of disguised missing data. *ACM SIGKDD Explorations Newsletter, 8 (1),* 83-92.

Polat, K., & Güneş, S. (2007). A novel hybrid intelligent method based on C4.5 decision tree classifier and one-against-all approach for multi-class classification problems. *Expert Systems with Applications,* (in press)

Quinlan, J. (1993). *C4.5 Programs for machine learning.* San Mateo, California: Morgan Kaufmann Publishers.

Roth, P. (1994). Missing data: A conceptual overview for applied psychologists. *Personnel Psychology, 47,* 537-560.

Rumelhart, D., Hinton, G., & Williams, R. (1986). *Learning internal representations by error propagation.* Cambridge, MA: MIT Press.

Salzberg, S. (1995). Locating protein coding regions in human DNA using a decision tree algorithm. *Journal of Computational Biology, 2 (3),* 473-485.

Salzberg, S. L., Chandar, R., Ford, H., Murthy, S., & White, R. (1995). Decision trees for automated identification of cosmic ray hits in Hubble Space Telescope images. *Publications of the Astronomical Society of the Pacific, 107,* 1-10.

Schafer, J. (1997). *Analysis of incomplete multivariate data.* London, UK: Chapman and Hall.

Schafer, J., & Graham, J. (2002). Missing data: Our view of the state of the art. *Psychological Methods, 7,* 147 -177.

Shao, X., Zhang, G., Li, P., & Chen, Y. (2001). Application of ID3 algorithm in knowledge acquisition for tolerance design. *Journal of Materials Processing Technology, 117 (1-2),* 66-74

Siciliano, R., & Mola, F. (1998). Ternary classification trees: A factorial approach visualization of categorical data, In M. Greenacre and J.Blasius J. (Eds.), *Visualization of categorical data* (pp. 311-323). San Diego, CA: Academic Press.

Soke, A., & Bingul, Z. (2006). Hybrid genetic algorithm and simulated annealing for two-dimensional non-guillotine rectangular packing problems. *Engineering Applications of Artificial Intelligence, 19 (5),* 557-567.

Spurgeon, S. E. F., Hsieh, Y-C, Rivadinera, A., Beer, T. M., Mori, M., & Garzotto, M. (2006). Classification and regression tree analysis for the prediction of aggressive prostate cancer on biopsy. *The Journal of Urology, 175 (3),* 918-922.

Sun, W., Chen, J., & Li, J. (2007). Decision tree and PCA-based fault diagnosis of rotating machinery. *Mechanical Systems and Signal Processing, 21 (3),* 1300-1317.

Tan, P-N., Steinbach, M., & Kumar, V. (2005). *Introduction to data mining.* Reading, MA: Addison-Wesley.

Thompson, B. B., Marks, R. J., & Choi, J. J. (2002). Implicit learning in autoencoder novelty assessment. In *Proceedings of the IEEE International Joint Conference in Neural Networks* (pp. 2878-2883).

Ture, M., Tokatli, F., & Kurt, I. (in press). Using kaplan-meier analysis together with decision tree methods (C&RT, CHAID, QUEST, C4.5 AND ID3) in determining recurrence-free survival of breast cancer patients. *Expert Systems with Applications.*

Turk, M., & Pentland, A. (1991). Eigenfaces for recognition. *Journal of Cognitive Neuroscience, 3 (1),* 71-86.

Waheed, T, Bonnell, R. B., Prasher, S. O., & Paulet, E. (2006). Measuring performance in precision agriculture: CART—A decision tree approach. *Agricultural Water Management, 84 (1-2),* 173-185.

Wei, J-M., Wang, S-Q, Wang, M-Y, You, J-P and Liu, D-Y. (2007). Rough set based approach for inducing decision trees. *Knowledge-Based Systems, 20 (8),* 695-702.

Wolfgang, M., Nocke, T., & Schumann, H. (2006). Enhancing the visualization process with principal component analysis to support the exploration of trends. In *Proceedings of the Asia Pacific Symposium on Information Visualization,* (Vol. 60, pp. 121-130).

Wong, S. Y. W. (2001). Hybrid simulated annealing/genetic algorithm approach to short-term hydro-thermal scheduling with multiple thermal plants. *International Journal of Electrical Power & Energy Systems, 23(7),* 565-575.

Yamada, S., Tanino, T., & Inuiguchi, M. (2001). An inner approximation method incorporating a branch and bound procedure for optimization over the weakly efficient set. *European Journal of Operational Research, 13 (2),* 267-286.

Yansaneh, I. S., Wallace, L. S., & Marker, D. A. (1998). Imputation methods for large complex datasets: An application to the NEHIS. In *Proceedings of the Survey Research Methods Section,* 314-319.

Yao, R., Yang, B., Cheng, G., Tao, X., & Meng, F. (2003). Kinetics research for the synthesis of branch ether using genetic-simulated annealing algorithm with multi-pattern evolution. *Chemical Engineering Journal, 94 (2),* 113-119.

Ye, J., Janardan, R., and Li, Q. (2004). GPCA: An efficient dimension reduction scheme for image compression and retrieval, In *Proceedings of the tenth ACM SIGKDD International Conference on Knowledge Discovery and Data Mining* (pp. 354-363).

Yoon, Y., & Peterson, L. L. 1990, Artificial neural networks: An emerging new technique. In *Proceedings of the 1990 ACM SIGBDP Conference on Trends and Directions in Expert Systems* (pp. 417-422).

Yu, Z., (in press). Solving bound constrained optimization via a new nonmonotone spectral projected gradient method. *Applied Numerical Mathematics.*

Yuan, Y., (2000). Multiple imputation for missing data: Concepts and new development. *SUGI Paper* 267-25, Retrieved August 25, 2008, from http://support.sas.com/rnd/app/papers/multipleimputation.pdf

Chapter XII
Control of Biomedical System Using Missing Data Approaches

ABSTRACT

Neural networks are used in this chapter for classifying the HIV status of individuals based on socio-economic and demographic characteristics. The trained network is then used to create an error equation with one of the demographic variables as a missing input and the desired HIV status as one of the variables. The missing variable thus becomes a control variable. This control mechanism is proposed to assess the effect of education level on the HIV risk of individuals and, thereby, assist in understanding the extent to which the spread of HIV can be controlled by using the education level. An inverse neural network model and a missing data approximation model based on autoassociative neural network and genetic algorithm (ANNGA) are used for the control mechanism. Therefore, the ANNGA is used to obtain the missing input values (education level) for the first model and an inverse neural network model is then used to obtain the missing input values (education) for the second model. The two models are then compared and it is found that the proposed inverse neural network model outperforms the ANNGA model. The methodology thus shows that HIV spread can be controlled to some extent by modifying a demographic characteristic educational level.

INTRODUCTION

Acquired Immunodeficiency Syndrome (AIDS) was first defined in 1982 to depict the first cases of strange immune system failures that were observed in the preceding years. The Human Immunodeficiency Virus (HIV) was soon after identified as the origin of AIDS. From the time of the detection of the virus and the disease a great deal of effort has been done to stop the spread of HIV with very little success. AIDS is currently an epidemic, which at the end of 2003 had killed an estimated 2.9 million lives. Epidemiology examines the function of host, agent and environment to elucidate the occurrence and transmission of a disease. Risk factor epidemiology examines the individual demographic and social characteristics and attempts to identify causes that position an individual at risk of acquiring a disease (Poundstone, Strathdee, & Celectano, 2004).

In this chapter, the demographic and social characteristics of the individuals and their behavior are used to establish the risk of HIV infection and this process is called "biomedical individualism" (Poundstone, Strathdee, & Celectano, 2004; Fee & Krieger, 1993; Leke et al., 2006a&b; Leke & Marwala, 2007). Because of the quick spread of the virus in the world today, especially in Sub-Saharan Africa, there has been an increase in the need to identify strategies and mechanisms for controlling the virus. This, on the other hand, seems ineffective given that the virus spread still appears to be uncontrolled particularly in the developing world. Social factors influence the risk of exposure as well as the probability of transmission of the disease. These social factors are, therefore, essential in order to comprehend and model the disease. By identifying the individual risk factors that lead to the disease, it is feasible to adjust social conditions which give rise to the disease and, therefore, design successful HIV intervention policies (Poundstone, Strathdee, & Celectano, 2004; Leke, Marwala, & Tettey, 2006; Leke, Marwala, & Manana, 2008). Using this information, a model can be constructed and then used to control the disease using conventional adaptive control strategies (Widrow & Walach, 1993; Aoyama, Doyle III, & Venkatasubramanian, 1999; Akin, Kaya, & Karakose, 2003; Acosta & Todorovich, 2003; Altinten et al., 2003; Ahn & Kha, 2007).

Adaptive control theory provides a plausible way to solve many of complex problems. Two distinct approaches can be used to control a system adaptively; the direct adaptive control and indirect adaptive control. In the *direct control approach*, the parameters of the controller are directly adjusted to reduce a distance of the output error. In the *indirect control*, the parameters of the plant are estimated as elements of a vector at any instant k; and the parameters vector of the controller is adapted based on the estimated plant vector. A general configuration of the indirect adaptive control as a self-tuning controller is shown in Figure 12.1 (Widrow & Walach, 1993). At each sampling instant, the input and output of the generating unit are sampled and a plant model is obtained by an on-line identification algorithm to represent the dynamic behavior of the generating unit at that instant in time. The required control signal is computed based on the identified model and various control techniques can be used to compute the control signal. All control algorithms assume that the identified model is a good approximation of the system that needs to be controlled. In this chapter, control algorithm is created using computational intelligence.

Computational intelligence (CI) has been used successfully in medical informatics for decision making, clinical diagnosis, prognosis, and prediction of outcomes (Tandon, Adak, & Kaye, 2006; Alkan, Koklukaya, & Subasi, 2005; Sawa & Ohno-Machado, 2003, Szpurck et al., 2005; Tan & Pan, 2005). CI can be understood as the aptitude of a machine or object to carry out similar types of purposes that characterize the human thought (Kalogirou, 2003). Neural networks, which are a type of computational

Figure 12.1. A general adaptive controller

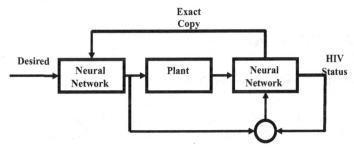

intelligence machines, are capable of non-linear pattern recognition with no necessity for an exact model based on scenarios. When applied to classification problems, neural networks learn which characteristics or combinations of characteristics are helpful for distinguishing classes. An additional purpose of a pattern recognition system is to discover a separator that partitions classes, locating as many samples into the correct classes as possible (Hudson & Cohen, 2000). Talebi, Khorasani, and Patel (1998) used neural network based control schemes for flexible-link manipulators while Aziz, Hussain, and Mujtaba (2003) implemented neural network inverse-model-based control technique for batch reactors. Other implementations of neural networks in control systems include: inverted pendulum (Wu et al., 2002), for DC motors (Nouri, Dhaouadi, & Braiek, 2008), in temperature control (Juang, Huang, & Duh, 2006); yeast fermentation control (Marwala, 2004; Kalman, 2007); in mineral grinding plants (Flament, Thibault, & Hodouin, 1993) as well as for braking control system (Ohno et al., 1994).

Altinten et al. (2007) used genetic algorithm for self-tuning PID control of jacketed batch polystyrene reactor while Arumugam, Rao, and Palaniappan (2005) introduced a new hybrid genetic operators for real coded genetic algorithm in order to identify optimal control of a class of hybrid systems. McGookin, Murray-Smith, and Fossen (2000) used genetic algorithm for ship steering control system optimization while Hu and Chen (2005) used genetic algorithm based on receding horizon control for arrival sequencing and scheduling. Other diverse applications of genetic algorithms to control include in water pollution (Rauch & Harremoës, 1999), feed-batch reactors (Nougués, Grau, & Puigjaner, 2000), decentralized control system structure selection and optimization (Lewin & Parag, 2003), pH control (Mwembeshi, Kent, & Salhi, 2004), inventory control (Disney, Naim, & Towill, 2000) as well as in high rise buildings (Pourzeynali, Lavasani, & Modarayi, 2007).

Two control mechanisms are proposed, in this chapter, to assess the demographic characteristics, particularly the educational level, required to control the HIV risk of an individual. The first model is implemented using the inverse neural networks to predict the *educational level* that yields a negative HIV status for an individual whose HIV status is predicted as positive. The second model is implemented using an autoassociative neural networks and genetic algorithm to estimate the educational level, whereby the educational level is considered a missing entry in the dataset. For this model, if an individual's status is predicted as positive, the educational level in the dataset is discarded and considered as missing input and genetic algorithm and autoassociative neural network are then used to estimate this entry in the same manner in which missing data are estimated in earlier chapters. The reason why education level is used is because of studies that have been conducted that demonstrated the reduction of HIV risk when the level of education is increased (Anonymous, 2008). In this chapter, a brief background on neural networks and biomedical individualism are explained within the context of employing missing data estimation methods for control of HIV/AIDS system (Ashhab, 2008). In addition, a summary of computational intelligence for HIV/AIDS predictions and the applications of genetic algorithms are also summarized.

THEORETICAL BACKGROUND

Multi-Layer Perceptron Neural Networks

As explained in earlier chapters, neural computation is an intelligent system that relates the input parameters to the output parameters. As indicated before, neural networks have been widely and successfully applied to adaptive control (Lee et al., 1992; Lewis, Jagannathan, & Yesildirek, 1999; Kazmierkowski, 2003; Ng & Hussain, 2004; Jaramillo et al., 2005; Salman, 2005; Arab–Alibeik & Setayeshi, 2005; Chen & Narendra, 2001; Hunt et al., 1992; Jeng, 2000). In this chapter, such an adaptive control system is based on artificial neural network, which is an inter-connected structure of processing elements called neurons and consists of three main components (Pillutla & Keyhani, 1997; Guessasma, Montavon, & Coddet, 2002): input space, hidden layer and output space. In this chapter, neural networks are used in three ways. The first one is by constructing the input output relationship with demographic characteristics as input and HIV status as output and this can be viewed in Figure 12.2.

The second one is by constructing what is termed the inverse neural network, where the demographic characteristics less educational level and HIV status are inputs and educational level is the output and this can be seen in Figure 12.3.

This is because in this chapter the main interest is on how the variable *education level* can be used to deal with the HIV problem. The third implementation of neural network is as an autoassociative network with demographic characteristics and HIV status as both inputs and outputs, and this is shown in Figure 12.4.

In this chapter, the multilayer perceptron (MLP) type of neural network, which is a feed-forward network is used and is trained using supervised learning and, therefore, requires a desired response to be trained (Bishop, 1995; Machowski & Marwala, 2004). In essence, the MLP learns how to transform

Figure 12.2. The MLP that predicts HIV status from demographic characteristics

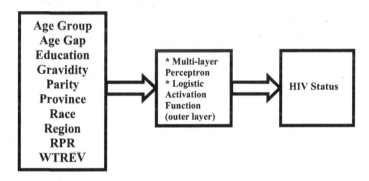

Figure 12.3. An inverse MLP that predicts education level from demographic characteristics and desired HIV status

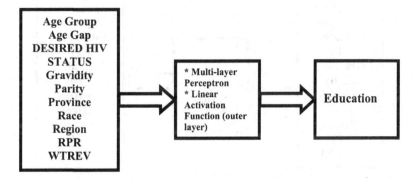

input data into a desired response. With one or more hidden layers, they can approximate virtually any input-output relationships. The MLP is considered the most widely used neural network architecture for practical applications and it has been used in areas such as option pricing in finance (Pires & Marwala, 2004), earthquake prediction (Chakraverty, Marwala, & Gupta, 2006), aircraft control (Soares, Burken, & Marwala, 2006), in braking control (Ohno et al., 1994), power system stabilization (Pillutla & Keyhani, 1997), in combustion control (Slanvetpan, Barat, & Stevens, 2003) and in fault detection in mechanical systems (Marwala & Hunt, 2001).

In this chapter, the MLP architecture, for all neural networks trained, contains a hyperbolic tangent basis function in the hidden units. For the neural network that contains the HIV status as the output, which is shown in Figure 12.2, logistic activation function is used in the outer layer because of its classification advantages while for the inverse neural network and autoassociative network used in this chapter, a linear basis function is used in the output units because of its regressive advantages (see Figures 12.3 and 12.4). The network input vector, {x}, to output {y}, relationship can be described mathematically as follows (Bishop, 1995):

$$y_k = f_{outer}\left(\sum_{j=0}^{M} w_{kj}^{(2)} \tanh\left(\sum_{i=0}^{d} w_{ji}^{(1)} x_i\right)\right)$$

(12.1)

In equation 12.1, $w_{ji}^{(1)}$ and $w_{kj}^{(2)}$ indicate weights in the first and second layer, respectively, going from input i to hidden unit j, M is the number of hidden units, f_{outer} is the activation function in the outer layer and d is the number of output units. When $j=0$ then the weight becomes a bias and x_o is set to 1.

Because of its computational efficiency, the scaled conjugate gradient method is used to train the networks (Møller, 1993). Training the neural network identifies the weights in equation 12.1. This is done by creating an objective function, which is the sum-of-squares of errors cost function as follows (Bishop, 1995):

$$E = \sum_{n=1}^{N}\sum_{k=1}^{K} \{t_{nk} - y_{nk}\}^2$$

(12.2)

Figure 12.4. The autoassociative neural network that maps demographic characteristics to themselves

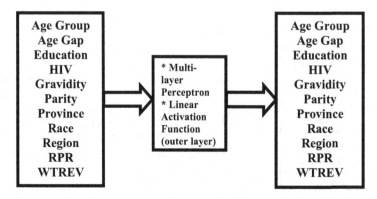

In equation 12.2 the parameter t is the target data, N is the number of training examples and K is the number of outputs. In this chapter, the scaled conjugate gradient method (Møller, 1993) is used in conjunction with the back-propagation method (Bishop 1995) to minimize equation 12.2.

Biomedical Individualism

Biomedical individualism (Poundstone, Strathdee, & Celectano, 2004) is defined as the root of risk factor epidemiology and this is dissimilar to social epidemiology. In social epidemiology, social conditions are viewed as the basic causes of diseases while in biomedical individualism, demographic and behavioral characteristics are considered to be the causes of diseases. Poundstone, Strathdee, and Celectano (2004) linked the demographic properties to the risk of HIV. They, as well, related both demographic and social characteristics, for example structural violence and discrimination, race/ethnicity and racism (Laumann & Youm, 1999) stigma and collective denial, legal structures, demographic change as well as policy environment, to the spread of HIV. This, consequently, justifies the utilization of such socio-demographic parameters in creating a model to predict the HIV status of individuals as is done in this chapter.

Computational Intelligence in HIV/AIDS Predictions

Artificial neural networks have been used to classify and predict the status of HIV/AIDS patients from symptoms (Laumann & Youm, 1999). In their study, Laumann and Youm used data that had complete entries from a publicly available AIDS Cost and Services Utilization Survey performed in the United States of America, to create the multi-layer perceptron architecture, with 15 linear inputs and 3 hidden logistic nodes and one output being the HIV status. The results they obtained from this implementation showed accuracy of 88%. A study was also performed to classify the functional health status of HIV and AIDS patients into *in good health* or *not in good health* classes by means of neural networks (Takano, Nakamura, & Watanabe, 2002). Other applications of neural networks in HIV/AIDS research have been in bioinformatics where modeling of the HIV has been conducted at a molecular level, such as the prediction of HIV-1 Protease Cleavage Sites. The above results show that artificial neural network can perform well in pattern recognition and signal processing. The methodology presented here is aimed at using other demographic and social factors to predict the HIV status of an individual.

Genetic Algorithms (GA)

This chapter uses genetic algorithm to control the risk of HIV. As indicated in earlier chapters, GA is a machine learning model that derives its behavior from the processes of evolution and is inspired by Darwin's theory of natural evolution (Holland, 1975; Goldberg, 1989; Aparisi & García-Díaz, 2004). The evolution process is achieved by creating within a machine or computer a population of individuals represented by chromosomes (Arfiadi & Hadi, 2001; Cheng, 1999) and in this chapter these chromosomes indicate the education levels. Therefore, in this chapter a demographic parameter education is used to control the risk of HIV by reducing this risk. In this chapter, GA essentially evolves a population of education levels that given the relationships between demographic parameters and HIV statuses, which are identified by the neural network model, is able to ensure a reduction in the HIV risk. As discussed in earlier chapters, the processes that contribute to this evolution are crossover, mutation and selection or reproduction (Onnen et al., 1997).

In this chapter, GA is used for two reasons and these are: (1) to identify optimal architectures for models indicated in Figures 12.3 and 12.4; and (2) for the missing data process where an input is identified to ensure the desired outcome and this is the control process (Park & Koh, 2004; Perkoz et al., 2007). Genetic algorithm has been used widely and successfully for control. Cho, Cho, and Wang (1997) successfully used GA in fuzzy-PID hybrid control by applying it to automatic rule identification.

Further applications of GA to control include in spacecraft control (Karr & Freeman, 1997), in chemical reactions (Karr et al., 1993; Sarkar & Modak, 2003; Sarkar & Modak, 2004), in helicopters (Phillips, Karr, & Walker, 1996), in communications (San José-Revuelta, 2005), in optimal control (Kundu & Kawata, 1996), in production control (Kurian & Reddy (1999) and many more applications (Lennon & Passino, 1999; Li et al., 2000). In many applications of genetic algorithms to fuzzy control, GA is normally used for identification of fuzzy rules as was the case by Herrera, Lozano, and Verdegay (1998). As indicated earlier, GA is population based approach and, therefore, is less subjected to becoming trapped to a local minimum than traditional optimization methods (Hanna et al., 2008) an advantage that is observed in Chapter X. Even though not assured to offer the globally optimum solution, GA has been observed in earlier chapters and in the literature to be highly efficient at reaching a very near optimum solution in a computationally efficient manner (Moore, Musacchio, & Passino, 2001). In general, in the implementation of GA, parameters that need to be specified by the user are: population size, crossover rate; mutation rate and chromosome type (San José-Revuelta, 2007). The implementation of GA is as follows; firstly, the fitness of all the individuals in the population is evaluated, then a new population is created by performing operations such as crossover, mutation and fitness-proportionate reproduction on the individuals whose fitness has just been measured; finally, the old population is discarded and a new iteration is performed on the new population. Each step (iteration) in this loop is referred to as a generation. More details on GA can be found in Davis (1991) and Michalewicz (1996). In this chapter, on implementing genetic algorithm, arithmetic crossover, non-uniform mutation and normalized geometric selection are chosen (Yuzgec, Becerikli, & Turker, 2006).

METHODOLOGIES FOLLOWED

In this chapter the process of designing the adaptive controller adopted is as follows:

1. **Generate HIV Prediction Model:** Use a mathematical model to appropriately represent the HIV prediction from demographic properties as is demonstrated in Figure 3.2.
2. **Generate Inverse Neural Model:** Use a model to predict an input given the output and other inputs, thus modeling output-input relationship as is shown in Figure 3.3.
3. **Generate Missing Data Model:** Use the autoassociative network model and genetic algorithms to predict missing data given all the other inputs and an output as was done in earlier chapters.

Generating the Model for HIV Prediction

Data Source

Demographic and medical data come from the South African ante-natal sero-prevalence survey of 2001, which was described in earlier chapter. This is a national survey, and any pregnant women attending selected public health care clinics participating for the first time in the survey are eligible to participate. The ante-natal sero-prevalence surveys are used as the main source of HIV prevalence data worldwide, reasons for this are that ante-natal clinics are found throughout the world, and pregnant women are ideal candidates for the study as they are sexually active.

Missing Data

Out of the total dataset cases, 12945 complete cases are selected, out of 13087 cases (98.91%) and the incomplete entries 142 cases (1.09%) are set aside. This is because to design the missing data estimation method implemented, in this chapter, a complete dataset is required.

Variables

The variables used in this chapter are: *race, region, age* of the *mother, age* of the *father, education level* of the mother, *gravidity, parity, province* of origin, *race, region* of origin and *HIV* status (Rauner & Brandeau, 2001). The qualitative variables such as race and region are converted to integer values. The age of mother and father are represented in years. The integer value representing education level represents the highest grade successfully completed, with 13 representing tertiary education. *Gravidity* is the number of pregnancies, complete or incomplete, experienced by a female, and this variable is represented by an integer between 0 and 11. *Parity* is the number of times the individual has given birth, (for example, multiple births are counted as one). Both these quantities are important, as they show the reproductive activity as well as the reproductive health state of the women. The HIV status is binary coded; a 1 represents positive status, while a 0 represents negative status. Thus the final number of input variables is 9. There is one output.

Dataset Used

The dataset is divided into three sets; training, validation and testing sets. The sets are created by dividing the huge dataset into three equivalent small datasets of 1988 entries each. The inputs used are; age of the mother, age gap, education level of the mother, gravidity, parity, province of origin, race and region of origin.

Model

Three datasets are used for training, validation and testing of the required models which are shown in Figures 12.2. Because the determination of network architecture involves the optimization of the number of hidden nodes, it is incorrect to use the testing set to compare results before determining the final number of hidden nodes, and thus the final architecture. In addition, genetic algorithm is used to optimize the network architectures through the optimal selection of the number of hidden nodes. The optimal feed-forward network obtained has 9 inputs representing demographic characteristics, 7 hidden units and 1 output representing the HIV status. This network has hyperbolic tangent function in the inner layer and logistic activation function in the outer layer. More details on this framework may be viewed in Figure 12.2.

Generating the Inverse Neural Network Model to Predict Missing Input

The datasets presented in the previous section are used to create a neural network model. In this model, one of the input variables (*educational level*) is replaced by the output (*HIV*) status; meanwhile the output becomes the replaced input (*education*), as is shown in Figure 12.2. Thus, there still exist 9 inputs and one output. This is known as an inverse neural network, since it relates the output to an input. A genetic algorithm is then used to optimize the network parameters (number of hidden nodes) and the optimal network is made up of 9 inputs (one of the inputs being replaced by the output), 8 hidden nodes and 1 output (in this case education). Using genetic algorithm in inverse predictive models has been studied before by researchers such as Mannino and Koushik (2000). The combined model for the above system is as shown in Figure 12.5.

In Figure 12.5 the inputs are used in the feed-forward neural network represented by "neural networks" in Figure 12.3 to predict the HIV status. If the HIV status is positive then the inverse neural network model is used to predict the input parameter value (*education*) required to make the status negative, by replacing the HIV input value, in the input dataset by 0 which is the desired outcome, indicating the HIV negative status. This is then sent to the forward neural network again to make sure the prediction yields a negative HIV status.

Generating the Missing Data Model

The aim of this model is to assess the educational level that will yield a change in an individual's HIV status from positive to negative, thereby, reducing the risk of HIV infection by that individual. Initially, the prediction model is used to predict the status of individuals from the input dataset. For every set of inputs, if the prediction yields negative status, then that educational level (one of the inputs) is kept as the right level for that individual. If the output from the prediction model, however, yields a positive status, then the educational level (one of the input in the dataset) is discarded and considered as missing data. A missing data approximation method that uses neural networks is then used to obtain the missing input described above that predicts a negative HIV status for a given individual. The missing data approximation model is designed using neural networks and genetic algorithm. The proposed method is as shown in Figure 12.6.

Figure 12.5. Inverse neural network model

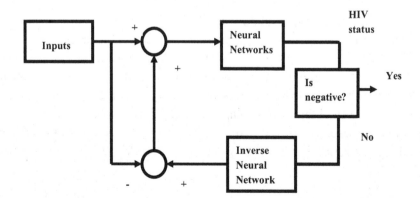

Figure 12.6. A schematic diagram indicating the implementation of the missing input

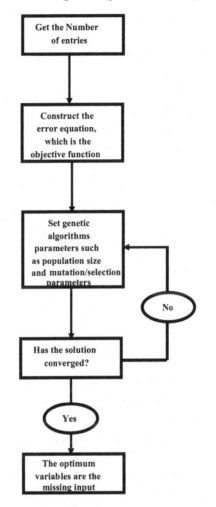

First the number of missing entries (inputs) is determined, and in this chapter this is the educational level. The missing data equation which is explained in the previous chapters is then developed as the difference between the expected output variables, calculated using the autoassociative MLP, and the actual output variables squared as is shown in equation 12.3.

Generating the Model for HIV Control

The datasets are then used to generate the overall model. This first model uses genetic algorithms and neural networks to predict the educational level using a missing data model. The second model uses inverse neural networks to predict the input required to obtain HIV negative status.

Both these models are implemented in Matlab© Simulink (Matlab, 2004). The structure implemented is as shown Figure 12.7. Hence, we can categorize the input vector $\{X\}$ elements into $\{X\}$ known vector represented by $\{X_k\}$ and $\{X\}$ unknown scalar representing education level represented by X_{Ed}. Rewriting equation 12.3 in terms of $\{X_k\}$ and X_{Ed} we then have:

Figure 12.7. Simulink model of the system

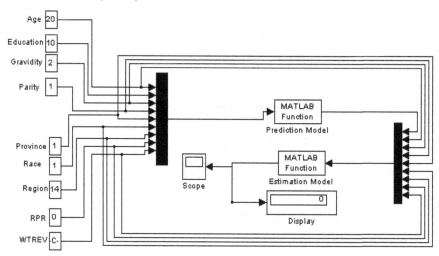

$$e = \left\| \left(\left\{ \begin{matrix} \{X_k\} \\ X_{Ed} \end{matrix} \right\} - f\left(\left\{ \begin{matrix} \{X_k\} \\ X_{Ed} \end{matrix} \right\}, \{W\} \right) \right) \right\|$$

(12.3)

Here ‖ ‖ is the Euclidean norm. In this book, equation 12.3 is called a missing data estimation error equation. A genetic algorithm is used to minimize the missing entry objective function and the optimum solution yielded by the GA is the estimated value of the missing variable, which in this chapter is the *education level* that will change the HIV status from positive to negative.

The above models are used to generate inputs that will obtain the educational level required to yield a negative status for an individual whose status is predicted as positive, thereby finding the parameters which are needed for an individual to be less prone or susceptible to contracting HIV. For the above models, an input dataset is sent into a prediction model, which is implemented. The prediction of the network at this stage is then verified and if the result yielded is 1 (positive) the estimation model which is the inverse model implemented or the missing data model is then used. The input thus required for the output to be zero is then obtained. The results obtained are presented to the inverse neural network model (output-input relationship) and the genetic algorithm estimation model (missing data model) in the next section.

TESTING THE PROPOSED PROCEDURE

The data described earlier in the chapter is used to train a feed-forward network of 9 inputs representing the demographic characteristics, 7 hidden units and 1 output representing the HIV status. This network has hyperbolic tangent function in the inner layer and logistic activation function in the outer layer. More details on this framework may be viewed in Figure 12.2. Furthermore, an inverse neural network with 9 inputs which are demographic characteristics and desired HIV status which is negative, 8 hidden units and 1 output unit representing education level is constructed using hyperbolic tangent function in the inner layer and linear activation function in the outer layer. More details on this inverse neural network are shown in Figure 12.3. These two sets of neural networks (i.e. Figure 12.2 and 12.3) are combined to form a control system that is illustrated in Figure 12.5.

In Figure 12.5 the demographic characteristics are given as inputs into the feed-forward neural network system and an HIV prediction is made. If the HIV prediction is negative then the system terminates because this is the desired outcome. If, however, the outcome is positive, which is not the desired outcome, then the system treats the controllable variables as missing values that need to be estimated. In this chapter, we assume that the controllable variable is the education level. In this case, therefore, the inverse neural network takes the demographic variables without the education level and with the HIV status assumed to be negative now as an input and predicts the appropriate 'missing value' which is the education level. As indicated before, both the feed-forward and inverse neural networks are trained using the scaled conjugate gradient method with error back-propagation algorithms (Bishop, 1995).

The first experiment uses the dataset for HIV modeling, predicting the HIV status of individuals from demographic characteristics, as is shown in Figure 12.2. The performance analysis for the HIV prediction model is based on classification accuracy and training times. The optimal number of hidden nodes for this feed-forward neural network is 7 hence the structure is 9-7-1 and this network gives an accuracy of 84% on the test datasets. The confusion matrix obtained for the above network is as shown in Table 12.1.

The second experiment uses the HIV dataset described in earlier sections and one of the inputs, and in this chapter education level, is replaced by the desired output which is the HIV status of negative. The educational level is used as the output variables while the HIV status is used as one of the inputs and set to negative because this is the desired outcome. The inverse neural network model is thus created and the optimal number of hidden nodes for the inverse neural network model obtained is 8 thus yielding a 9–8–1 network structure. This network gives a mean square error of 5.9956 on the training dataset, thus about 94% accuracy and 11.573 on the validating dataset thus about 88% accuracy. The training time is 46.9980s.

The third experiment consists of cases where one of the input (educational level) to the neural network is assumed to be unknown and then estimated using genetic algorithm method the same way a missing data estimation procedure described in earlier chapters operates. This involves first constructing an autoassociative network with demographics characteristics and HIV status as both inputs and outputs. Then genetic algorithm is implemented as an optimization approach. On implementing genetic algorithm, arithmetic crossover, non-uniform mutation and normalized geometric selection are chosen. Optimization bounds of genetic algorithm are set based on maximum and minimum values observed in the data. The genetic algorithm parameters are set based on background reading as follows: probability of crossover is chosen to be 0.75 as suggested in Marwala and Chakraverty (2006) and probability of

Table 12.1. Classifier confusion matrix of the two algorithms

Confusion Matrix	Predicted Positive	Predicted Negative
Actual Positive	680	313
Actual Negative	10	983

mutation is chosen to be 0.0333. Genetic algorithm has a population of 20 and is run for 150 generations. This network gives a mean square error of 18.875 on the training dataset, and 23.175 on the validating dataset. The adaptive controller is then implemented using both the inverse neural network model and the genetic algorithm missing data model. The inverse neural network model converges faster and yields better results compared to the genetic algorithm model. The inverse neural network model, however, has a problem of singularity of a solution due to the fact that the possible outcomes might be above the bounds of the input set. The genetic algorithm model yields a solution within the bounds due to the fact that it is a bounded optimization method. The estimated missing values, however, are more realistic as they are based on the possible values of the input unlike the inverse neural network model which is strictly based on the correctness of the model. When both models are used, the status of the individuals could be controlled adaptively. A table of the predictions obtained for the inverse neural network model as well as genetic algorithm model for the educational level is as shown in Table 12.2.

The sources of errors in the experiment are mainly dataset related errors due to the data being biased towards one class. To minimize the dataset errors, reliable data are replicated for the class with less data. To minimize the neural network training errors, standard procedures are used for training, generalization and testing of neural networks. The effects of these errors are, however, minimal on the over-all predictability.

Table 12.2. Educational level obtained from the two methods. Key: INVNN= Inverse Neural Networks, GA Pred= GA Prediction

Actual	GA Prediction	INVNN	Actual	GA Pred	INVNN	Actual	GA Pred	INVNN
0.9231	0.4515	0.7308	0.7692	0.6608	0.4573	0.6154	0.0850	0.5527
0.5385	0.1661	0.8107	0.8462	0.3717	0.8000	0.7692	0.6150	0.6102
0.6923	0.8924	0.3695	0.9231	0.8228	0.8000	0.6923	0.357	0.6116
0.6154	0.2730	0.6328	0.8462	0.505	0.5078	0.6923	0.7578	0.5740
0.9231	0.0008	0.7343	0.7692	0.1026	0.3923	0.4615	0.0192	0.5337
0.1539	0.1473	0.4895	0.7692	0.3011	0.4804	0.6923	0.8241	0.7307
0.6923	0.6313	0.7775	0.7692	0.4040	0.5273	0.7692	0.3551	0.8000
0.9231	0.5362	0.527	0.1539	0.1225	0.4923	0.8462	0.6022	0.8106
0.6923	0.0067	0.6648	0.7692	0.4304	0.6935	0.6923	0.6449	0.8106

CONCLUSION

In this chapter, a method based on neural networks and genetic algorithms is proposed to understand how educational level can be used to control the HIV risk. The model is aimed at obtaining the educational level, which will make individuals less prone and susceptible to HIV contraction using demographic characteristics. A classifier is first developed which maps demographic characteristics to HIV status and has a classification accuracy of 84%. An adaptive control module is then generated and implemented. This module is tested on a set of demographic characteristics.

The proposed method is able to estimate the education level needed to reduce the risk of HIV using GA and inverse neural networks. The inverse neural network model yields faster results compared to the GA model. A higher accuracy is obtained using the inverse neural network model than when using the autoassociative network and genetic algorithm missing data model. The low accuracy of combined autoassociative and genetic algorithm model may be attributed to the fact that genetic algorithms converge slowly, hence the global optimum may not have been obtained. A model can thus be developed using inverse neural networks, to effectively assess the educational level required by individuals, in order to control the HIV status of an individual.

FURTHER WORK

In this chapter two control mechanisms are created to understand the impact of demographic variable education on the risk of HIV. For future work more demographic variables should be used. As a further work, a much clearer link between the control parameters and HIV policy instruments should be established.

REFERENCES

Acosta, G., & Todorovich, E. (2003). Genetic algorithms and fuzzy control: A practical synergism for industrial applications. *Computers in Industry, 52*(2), 183-195.

Ahn, K. K., & Kha, N. B. (in press). Modeling and control of shape memory alloy actuators using Preisach model, genetic algorithm and fuzzy logic. *Mechatronics.*

Akin, E., Kaya, M., & Karakose, M. (2003). A robust integrator algorithm with genetic based fuzzy controller feedback for direct vector control. *Computers & Electrical Engineering, 29*(3), 379-394.

Alkan, A., Koklukaya, E., & Subasi, A. (2005). Automatic seizure detection in EGG using logistic regression and artificial neural network. *Journal of Neuroscience Methods, 148*(2), 167-176.

Altinten, A., Erdoan, S., Hapolu, H., & Alpbaz, M. (2003). Control of a polymerization reactor by fuzzy control method with genetic algorithm. *Computers & Chemical Engineering, 27*(7), 1031-1040.

Altinten, A., Ketevanlioğlu, F., Erdoğan, S., Hapoğlu, H., & Alpbaz, M. (in press). Self-tuning PID control of jacketed batch polystyrene reactor using genetic algorithm. *Chemical Engineering Journal.*

Anonymous. (2008). *School attendance lowers HIV risk*. Retrieved March 1, 2008, from http://iafrica. com/aidswise/news/798770.htm

Aoyama, A., Doyle III, F. J., & Venkatasubramanian, V. (1999). Fuzzy neural network systems techniques and their applications to nonlinear chemical process control systems. In C. T. Leondes (Ed.), *Fuzzy theory systems* (pp. 485-526). New York: Academic Press.

Aparisi, F., & García-Díaz, J. C. (2004). Optimization of univariate and multivariate exponentially weighted moving-average control charts using genetic algorithms. *Computers & Operations Research, 31*(9), 1437-1454.

Arab–Alibeik, H., & Setayeshi, S. (2005). Adaptive control of a PWR core power using neural networks. *Annals of Nuclear Energy, 32*(6), 588–605.

Arfiadi, Y., & Hadi, M. N. S. (2001). Optimal direct (static) output feedback controller using real coded genetic algorithms. *Computers & Structures, 79*(17), 1625-1634.

Arumugam, M. S., Rao, M. V. C., & Palaniappan, R. (2005). New hybrid genetic operators for real coded genetic algorithm to compute optimal control of a class of hybrid systems. *Applied Soft Computing, 6*(1), 38-52.

Ashhab, M. S. (2008). Fuel economy and torque tracking in camless engines through optimization of neural networks. *Energy Conversion and Management, 49*(2), 365-372.

Aziz, N., Hussain, M. A., & Mujtaba, I. M. (2003). Implementation of neural network inverse-model-based control (NN-IMBC) strategy in batch reactors. *Computer Aided Chemical Engineering, 15*(2), 708-713.

Bishop, C. M. (1995). *Neural networks for pattern recognition*. Oxford: Oxford University Press.

Chakraverty, S., Marwala, T., & Gupta, P. (2006). Response prediction of structural system subject to earthquake motions using artificial neural network. *Asian Journal of Civil Engineering, 7*(3), 301-308.

Chen, L., & Narendra, S. K. (2001). Nonlinear adaptive control using neural networks and multiple models. *Automatica, 37*(8), 1245–1255.

Cheng, F. Y. (1999). Multiobjective optimum design of structures with genetic algorithm and game theory: Application to life-cycle cost design. *Computational Mechanics in Structural Engineering* (pp. 1-16).

Cho, H-J., Cho K-B., & Wang, B-H. (1997). Fuzzy-PID hybrid control: Automatic rule generation using genetic algorithms. *Fuzzy Sets and Systems, 92*(3), 305-316.

Davis, L. (1991). *Handbook of genetic algorithms*. New York: Van Nostrand.

Disney, S. M., Naim, M. M., & Towill, D. R. (2000). Genetic algorithm optimisation of a class of inventory control systems. *International Journal of Production Economics, 68*(3), 259-278.

Fee, E., & Krieger, N. (1993). Understanding AIDS: Historical interpretations and limits of biomedical individualism. *American Journal of Public Health, 83*, 1477–1488.

Flament, F., Thibault, J., & Hodouin, D. (1993). Neural network based control of mineral grinding plants. *Minerals Engineering, 6*(3), 235-249.

Goldberg, D. E. (1989). *Genetic algorithms in search optimization and machine learning*. Reading, MA: Addison-Wesley.

Guessasma, S., Montavon, G., & Coddet, C. (2002). On the neural network concept to describe the thermal spray deposition process: An introduction. In *Proceedings of the International Thermal Spray Conference and Exposition*, Düsseldorf, DVS-Verlag GmbH (pp. 435–439).

Hanna, J., Upreti, S. R., Lohi, A., & Ein-Mozaffari, F. (2008). Constrained minimum variance control using hybrid genetic algorithm – An industrial experience. *Journal of Process Control, 18*(1), 36-44.

Herrera, F., Lozano, M., & Verdegay, J. L. (1998). A learning process for fuzzy control rules using genetic algorithms. *Fuzzy Sets and Systems, 100*(1-3), 143-158.

Holland, J. (1975). *Adaptation in natural and artificial systems*. Ann Arbor: University of Michigan Press.

Hu, X-B., & Chen, W-H. (2005). Genetic algorithm based on receding horizon control for arrival sequencing and scheduling. *Engineering Applications of Artificial Intelligence, 18*(5), 633-642.

Hudson, D. L., & Cohen, M. E. (2000). *Neural networks and artificial intelligence for biomedical engineering*. Piscataway, NJ: IEEE Press.

Hunt, K. J., Sbarbaro, D., Zbikowski, R., & Gawthrop, P. J. (1992). Neural networks for control systems: A survey. *Automatica, 28*(6), 1083 – 1112.

Jaramillo, M. A., Peguero, J. C., de Salazar, E. M., &del Valle, M. G. (2005). Neural network control in a wastewater treatment plant. In *Recent Advances in Multidisciplinary Applied Physics* (pp. 241-245).

Jeng, J-T. (2000). Nonlinear adaptive inverse control for the magnetic bearing system. *Journal of Magnetism and Magnetic Materials, 209*(1–3), 186–188.

Juang, C-F., Huang, S-T., & Duh, F-B. (2006). Mold temperature control of a rubber injection-molding machine by TSK-type recurrent neural fuzzy network. *Neurocomputing, 70*(1-3), 559-567.

Kalman, Z. (2007). Model based control of a yeast fermentation bioreactor using optimally designed artificial neural networks. *Chemical Engineering Journal, 127*(1-3), 95-109.

Kalogirou, S. A. (2003). Artificial intelligence for the modeling and control of combustion processes: A review. *Progress in Energy and Combustion Science, 29*, 515–566.

Karr, C. L., & Freeman, L. M. (1997). Genetic-algorithm-based fuzzy control of spacecraft autonomous rendezvous. *Engineering Applications of Artificial Intelligence, 10*(3), 293-300.

Karr, C. L., Sharma, S. K., Hatcher, W. J., & Harper, T. R. (1993). Fuzzy control of an exothermic chemical reaction using genetic algorithms. *Engineering Applications of Artificial Intelligence, 6*(6), 575-582.

Kazmierkowski, M. P. (2003). Neural networks and fuzzy logic control in power electronics. In M. P. Kazmierkowski, L. Kazmierkowskl, R. Krishnan and F. Blaabjerg (Eds.), *Control in power electronics* (pp. 351-418). New York: Academic Press.

Kundu, S., & Kawata, S. (1996). Genetic algorithms for optimal feedback control design. *Engineering Applications of Artificial Intelligence, 9*(4), 403-411.

Kurian, T. K., & Reddy Ch. V. K. (1999). On-line production control using a genetic algorithm. *Computers & Industrial Engineering, 37*(1-2), 101-104.

Laumann, E. O., & Youm, Y. (1999). Racial/ethnic group differences in the prevalence of sexually transmitted diseases in the United States: A network explanation. *Sexually Transmitted Diseases, 26*, 250– 61.

Lee, T. H., Hang, C. C., Lian, L. L., & Lim, B. C. (1992). An approach to inverse nonlinear control using neural networks. *Mechatronics, 2*(6), 595-611.

Leke, B. B., & Marwala, T. (2007). Using inverse neural network for HIV adaptive control. *International Journal of Computational Intelligence Research, 3*(1), 11-15.

Leke, B. B., Marwala, T., & Manana, J. V. (2008). Computational intelligence for HIV modelling. In *Proceedings of the IEEE Conference on Intelligent Engineering Systems* (pp. 127-132).

Leke, B. B., Marwala, T., & Tettey, T. (2006). Autoencoder networks for HIV classification. *Current Science, 91*(11), 1467-1473.

Leke, B. B., Marwala, T., Tim, T., & Lagazio, M. (2006a). Using genetic algorithms versus line search optimization for HIV predictions. *WSEAS Transactions on Information Science and Applications, 4*(3), 684-690

Leke, B. B., Marwala, T., Tim, T., & Lagazio, M. (2006b). Prediction of HIV status from demographic data using neural networks. In *Proceedings of the IEEE International Conference on Systems, Man and Cybernetics, Taiwan* (pp. 2339-2344).

Lennon, W. K., & Passino, K. M. (1999). Techniques for genetic adaptive control. In M. Gupta and N. Sinha (Eds.), *Soft computing and intelligent systems: Theory and applications* (pp. 257-278). New York: Academic Press.

Lewin, D. R., & Parag, A. (2003). A constrained genetic algorithm for decentralized control system structure selection and optimization. *Automatica, 39*(10), 1801-1807.

Lewis, F. L., Jagannathan, S., & Yesildirek, A. (1999). Neural network control of robot arms and non-linear systems. London: Taylor and Francis.

Li, S., Liu, D. K., Fang, J. Q., & Tam, C. M. (2000). Multi-level optimal design of buildings with active control under winds using genetic algorithms. *Journal of Wind Engineering and Industrial Aerodynamics, 86*(1), 65-86.

Machowski, L. A., & Marwala, T. (2004). Representing and matching 2D shapes of natural objects using neural networks. In *Proceedings of the IEEE International Conference on Systems, Man and Cybernetics, The Hague, Nederland* (pp. 6366-6372).

Mannino, M. V., & Koushik, M. V. (2000). The cost-minimizing inverse classification problem: A genetic algorithm approach. *Decision Support Systems, 29*(3), 283-300.

Marwala, T. (2004). Control of complex systems using Bayesian neural networks and genetic algorithm. *International Journal of Engineering Simulation, 5*(2), 28-37.

Marwala, T., & Chakraverty, S. (2006). Fault classification in structures with incomplete measured data using autoassociative neural networks and genetic algorithm. 2006, *Current Science Journal, 90*(4), 542–549.

Marwala, T., & Hunt, H. E. M. (2001). Detection and classification of faults using maximum-likelihood and Bayesian approach. In *Proceedings of SPIE: The International Society for Optical Engineering, 4359*(1), 207-213.

MATLAB 7.1 Manual. (2004). *Matlab and Simulink for Technical Computing (Release 13)*. New York: Mathworks Press.

McGookin, E. W., Murray-Smith, D. J., Li, Y., & Fossen, T. I. (2000). Ship steering control system optimisation using genetic algorithms. *Control Engineering Practice, 8*(4), 429-443.

Michalewicz, Z. (1996). *Genetic algorithms + Data structures = Evolution programs*. 3rd Edition, Berlin: Springer.

Møller, M. (1993). A scaled conjugate gradient algorithm for fast supervised learning. *Neural Networks, 6*(4), 525-533.

Moore, M. L., Musacchio, J. T., & Passino, K. M. (2001). Genetic adaptive control for an inverted wedge: Experiments and comparative analyses. *Engineering Applications of Artificial Intelligence, 14*(1), 1-14.

Mwembeshi, M. M., Kent, C. A., & Salhi, S. (2004). A genetic algorithm based approach to intelligent modelling and control of pH in reactors. *Computers & Chemical Engineering, 28*(9), 1743 -1757.

Ng, C. W., & Hussain, M. A. (2004). Hybrid neural network—prior knowledge model in temperature control of a semi-batch polymerization process. *Chemical Engineering and Processing, 43*(4), 559-570.

Nougués, J. M., Grau, M. D., & Puigjaner, L. (2002). Parameter estimation with genetic algorithm in control of fed-batch reactors, *Chemical Engineering and Processing, 41*(4), 303-309.

Nouri, K., Dhaouadi, R., & Braiek, N. B. (2008). Adaptive control of a nonlinear DC motor drive using recurrent neural networks. *Applied Soft Computing, 8*(1), 371-382.

Ohno, H., Suzuki, T., Aoki, K., Takahasi, A., & Sugimoto, G. (1994). Neural network control for automatic braking control system. *Neural Networks, 7*(8), 1303-1312.

Onnen, C., Babuka, R., Kaymak, U., Sousa, J. M., Verbruggen, H. B., & Isermann, R. (1997). Genetic algorithms for optimization in predictive control. *Control Engineering Practice, 5*(10), 1363-1372.

Park, K-S., & Koh, H-M. (2004). Preference-based optimum design of an integrated structural control system using genetic algorithms. *Advances in Engineering Software, 35*(2), 85-94.

Perkgoz, C., Azaron, A., Katagiri, H., Kato, K., & Sakawa, M. (2007). A multi-objective lead time control problem in multi-stage assembly systems using genetic algorithms. *European Journal of Operational Research, 180*(1), 292-308.

Phillips, C., Karr, C. L., & Walker, G. (1996). Helicopter flight control with fuzzy logic and genetic algorithms. *Engineering Applications of Artificial Intelligence, 9*(2), 175-184.

Pillutla, S., & Keyhani, A. (1997). Power system stabilization based on modular neural network architecture. *International Journal of Electrical Power & Energy Systems, 19*(6), 411-418.

Pires, M. M., & Marwala, T. (2004). Option pricing using neural networks and support vector machines. In *Proceedings of the IEEE International Conference on Systems, Man and Cybernetics, The Hague, Nederland,* (pp. 1279-1285).

Poundstone, K., Strathdee, S., & Celectano, D. (2004). The social epidemiology of human immunodeficiency virus/acquired Immunodeficiency syndrome. *Epidemiologic Reviews, 26,* 22–35.

Pourzeynali, S., Lavasani, H. H., & Modarayi, A. H. (2007). Active control of high rise building structures using fuzzy logic and genetic algorithms. *Engineering Structures, 29*(3), 346-357.

Rauch, W., & Harremoës, P. (1999). Genetic algorithms in real time control applied to minimize transient pollution from urban wastewater systems. *Water Research, 33*(5), 1265-1277.

Rauner, M., & Brandeau, M. (2001). AIDS policy modeling for the 21st century: An overview of key issues. *Health Care Management Science, 4,* 165–180.

Salman, R. (2005). Neural networks of adaptive inverse control systems. *Applied Mathematics and Computation, 163*(2), 931–939.

San José-Revuelta, L. M. (2005). Entropy-guided micro-genetic algorithm for multiuser detection in CDMA communications. *Signal Processing, 85*(8), 1572-1587.

San José-Revuelta, L. M. (2007). A new adaptive genetic algorithm for fixed channel assignment. *Information Sciences, 177*(13), 2655-2678.

Sarkar, D., & Modak, J. M. (2003). Optimisation of fed-batch bioreactors using genetic algorithms. *Chemical Engineering Science, 58*(11), 2283-2296.

Sarkar, D., & Modak, J. M. (2004). Optimization of fed-batch bioreactors using genetic algorithm: Multiple control variables. *Computers & Chemical Engineering, 28*(5), 789-798.

Sawa, T., & Ohno-Machado, L. (2003). A neural network-based similarity index for clustering DNA microarray data. *Computers in Biology and Medicine, 33*(1), 1-15.

Slanvetpan, T., Barat, R. B., & Stevens, J. G. (2003). Process control of a laboratory combustor using artificial neural networks. *Computers & Chemical Engineering, 27*(11), 1605-1616.

Soares, F., Burken, J., & Marwala, T. (2006). Neural network applications in advanced aircraft flight control system, a hybrid system, a flight test demonstration. *Lecture Notes in Computer Science, 4234,* 684-691.

Szpurek, D., Moszynski, R., Smolen, A., & Sajdak, S. (2005). Artificial neural network computer prediction of ovarian malignancy in women with adnexal masses. *International Journal of Gynecology & Obstetrics, 89*(2), 108-113.

Takano, T., Nakamura, K., & Watanabe, M. (2002). Urban residential environments and senior citizens; longevity in megacity areas: The importance of walkable green spaces. *Journal of Epidemiology Community Health, 56,* 913–918.

Talebi, H. A., Khorasani, K., & Patel, R. V. (1998). Neural network based control schemes for flexible-link manipulators: Simulations and experiments. *Neural Networks, 11*(7-8), 1357-1377.

Tan, A-H., & Pan, H. (2005). Predictive neural network for gene expression data analysis. *Neural Networks, 18*(3), 297-306.

Tandon, R., Adak, S., & Kaye, J. A. (2006). Neural network for longitudinal studies in Alzheimer's disease. *Artificial Intelligence in Medicine, 36*(3), 245-255.

Widrow, B., & Walach, E. (1993). *Adaptive inverse control.* Upper Saddle River, New York: Prentice-Hall.

Wu, Q., Sepehri, N., & He, S. (2002). Neural inverse modeling and control of a base-excited inverted pendulum. *Engineering Applications of Artificial Intelligence, 15*(3-4), 261-272.

Yuzgec, U., Becerikli, Y., & Turker, M. (2006). Nonlinear predictive control of a drying process using genetic algorithms. *ISA Transactions, 45*(4), 589-602.

Chapter XIII
Emerging Missing Data Estimation Problems:
Heteroskedasticity; Dynamic Programming and Impact of Missing Data

ABSTRACT

This chapter is divided into three parts: The first part presents a computational intelligence approach for predicting missing data in the presence of concept drift using an ensemble of multi-layered feed-forward neural networks. An algorithm that detects concept drift by measuring heteroskedasticity is proposed. Six instances prior to the occurrence of missing data are used to approximate the missing values. The algorithm is applied to simulated time series data sets resembling non-stationary data from a sensor. Results show that the prediction of missing data in non-stationary time series data is possible but is still a challenge. In the second part, an algorithm that uses dynamic programming and neural networks to solve the problem of missing data imputation is presented. A model that uses autoassociative neural networks and genetic algorithms is used as a basis; however, the neural networks are not trained using the entire data set. Data are broken up into granules and various models are created. The models are tested on a real dataset and the results show that the proposed method is effective in missing data estimation. In the third part of this chapter, a study of the impact of missing data estimation on fault classification in mechanical systems is undertaken. The fault classification task is implemented using the extension network as well as Gaussian mixture models. When the imputed values are used in the classification of faults using the extension networks, the fault classification accuracy of 95% is observed for single-missing-entry cases and 92% for two-missing-entry cases while the full database set is able to give classification accuracy of 97%. On the other hand, the Gaussian mixture model gives 94% for single-missing-entry cases and 92% for two-missing-entry cases while the full database set is able to give classification accuracy of 96%.

INTRODUCTION: HETEROSKEDASTICITY

The problem of missing data has intensively been researched but continues to be mainly unsettled. One of the causes for this is that the complexity of approximating missing variables is exceedingly reliant on the problem domain. This complexity, moreover, increases when data are missing in an on-line application where data have to be used as soon as they are obtained. A difficult characteristic of the missing data problem is when data are missing from a time series that exhibit non-stationarity. Most machine learning techniques and algorithms that have been developed thus far assume that data will continuously be obtainable. In addition, they assume that data conform to a stationary distribution.

Non-stationarity of a data essentially means that the character or the nature of the data is actually changing as a function of time. There are lots of non-stationary quantities in the natural world that fluctuate with time. Familiar examples include the stock market, weather, heartbeats, seismic waves as well as animal populations. There are some engineering and measurement systems that have been developed to detect and to quantify non-stationary quantities. Such instruments are not resistant to failures. These instruments include the wavelet methods which are time-frequency analysis methods (Marwala, 2002; Bujurke et al., 2007) and fractals methods (Lunga & Marwala, 2006a; Sadana, 2003&2005; Reiter, 1994). In this chapter, a procedure known as heteroskedasticity (Nelwamondo & Marwala, 2007a) is used to analyze concept drift with the aim of ensuring that the deployed missing data estimation method remains relevant even in the presence of the concept drift.

Computational intelligence techniques have previously been employed for analyzing non-stationary data such as the stock-market, nevertheless, the volatility of the data render the problem too complex to easily analyze. The 2003 Nobel Prize Laureates in Economics, Granger (2003) and Engle (1982) made an exceptional contribution to non-linear data analysis. Granger showed that long-established statistical methods could be deceiving if applied to variables that wander over time without returning to some long-run resting position. Engle (1982) on the other hand contributed a pioneering innovation of an Autoregressive Conditional Heteroskedasticity (ARCH), a technique to analyze and understand unpredictable movements in financial market prices. This method is, moreover, applicable to risk assessment. Dufour et al. (2004) introduced simulation-based finite-sample tests for Heteroskedasticity and ARCH effects. Hafner and Herwartz (2001) proposed option pricing under linear autoregressive dynamics, heteroskedasticity, and conditional leptokurtosis whereas Khalaf, Saphores, and Bilodeau (2003) introduced simulation-based exact jump tests in models with conditional heteroskedasticity and Inkmann (2000) introduced mis-specified heteroskedasticity in the panel probit model and made a comparison between Gaussian mixture models (GMM) and simulated maximum likelihood. Other work on the heteroskedasticity include its use on analyzing the performance of bootstrap neural tests for conditional heteroskedasticity in ARCH models (Siani & Peretti, 2007), pooling of cross-sectional and time-series data in the presence of heteroskedasticity as well as analyzing auto-correlation- and heteroskedasticity-consistent t-values with trending data (Krämer & Michels, 1997).

Numerous techniques for solving missing data problems have been developed and discussed at length in the literature (Little & Rubin, 1987). However, limited attempt has been made to approximate missing data in strictly non-stationary processes, where concepts change with time. The challenge with missing data problems in this application is that the approximation process must be complete before the next sample is taken. Moreover, more than one technique may be required to approximate the missing data due to drifting of concepts. As a result, the computational time needed, the amount of computational memory required and the model complexity may grow indefinitely as new data continually arrive (Last, 2002).

This chapter confronts the above-mentioned problems by proposing a computational intelligence technique that makes use of an ensemble of multi-layer perceptrons feed-forward neural networks to deal with missing data estimation in the presence of concepts that are drifting. In the proposed algorithm, concept drift is detected by measuring heteroskedasticity and based on this detection an appropriate missing data estimation model is then utilized. The proposed technique learns new concepts incrementally and, therefore, on-line. Incremental learning methods are ideally suited for on-line learning and methods that have been used thus far include Learn++ (Polikar, 2007; Lunga & Marwala, 2006b; Vilakazi & Marwala, 2007a,b&c), genetic programming based techniques (Hulley & Marwala, 2007) and others. The resulting state of knowledge is then used in predicting the missing values. A brief definition of non-stationary and chaotic systems will be discussed, followed by a discussion on the principle of concept drift and its detection techniques. A formal algorithm of approximating missing data in the presence of concept drift will then be presented, followed by empirical evaluation and results.

DEFINITION OF NONSTATIONARY AND CHAOTIC SYSTEMS

Non-stationarity is a common property to many macro-economic and financial time series data (Engle, 2003). Non-stationarity signifies that a variable has no clear tendency to return to a constant value or a linear trend. Frequently, stationarity is described in terms of the mean and the auto-covariance. If data are randomly sampled and they are established to have a constant mean, and the auto-covariance is a function that depends only on the distance in placement, the data are considered to be stationary, or more properly, wide-sense stationary. Chaotic systems on the other hand are non-stationary systems that are sensitive to initial conditions. This renders such systems very difficult to predict. The reason for this difficulty is that concepts drift before reaching a particular observation.

Some work has been conducted for dealing with missing data in non-stationary time series (Stefanakos & Athanassoulis, 2001). In most cases, attempts are made to first make the data stationary using differencing procedures (Stefanakos & Athanassoulis, 2001). In cases of missing data, applying the differencing techniques proposed in the literature is not feasible for solving the problem of missing data (Ljung, 1989). A fascinating technique was proposed by Stefanakos and Anthnassoulis (2001), which functions by completing missing values at the level of uncorrelated residuals subsequent to removing any systematic trends such as periodic components. Their scheme is complex and only works well with seasonal data. It, therefore, would not be precise in situations where the concept being predicted or learned varies with time.

THE PRINCIPLE OF CONCEPT DRIFT

The principle of concept drift implies that the concept about which data are acquired may shift from time to time. Predicting values of a rapidly drifting concept is not possible if the concept changes each time step without restriction (Last, 2002). The rate of concept drift is defined as the probability that the target function is different over two successive examples (Helmbold & Long, 1991). There are two types of concept drifts that have been reported in the literature and these types are categorized by the rate of the drift and are referred to as: (1) sudden and (2) gradual concept drift.

One drawback of concept drift is that for a high volume of non-stationary data streams, the time it takes to predict may grow indefinitely (Last, 2002). In all cases of concept drift, incremental learning methods that continuously revise and refine the approximation model need to be devised and these methods need to incorporate new data as they arrive. This can be achieved by continually using recent data while not forgetting the past data. However, in some cases, past data might be invalid and may need to be forgotten. Harries and Sammut (1988) have developed an off-line method for partitioning data streams into a set of time-dependent conceptual clusters. Their approach was, however, aimed at detecting concept drift in off-line systems. This work looks at a technique of detecting concept drift in an on-line application.

CONCEPT DRIFT DETECTION USING HETEROSKEDASTICITY

Techniques of detecting concept drift are quite essential in time series data. The biggest challenge to this task is due to data being collected over time. Ways of detecting concept drift may vary in accordance to the pattern at which the concept is drifting. In most cases, the use of a window, where old examples are forgotten has proven to be sufficient (Last, 2002; Helmbold & Long, 1991). Known examples of window based algorithms include Time-Window Forgetting, FLORA and FRANN (Widmer & Kubat, 1993). A cyclically drifting concept exhibits a tendency to return to previously visited states. However, there are many algorithms such as STAGGER (Allison, 2002) and FLORA 3 (Widmer & Kubat, 1993) that have been developed to deal with cyclic concept drift. In this kind of drift, old examples need not be forgotten as they may re-appear at a later stage. An effective missing data estimator must be able to track such changes and to rapidly adapt to them. In the light of this challenge, this chapter proposes the use of heteroskedasticity as a means of detecting concept drift.

Heteroskedasticity occurs when the variables in a sequence have differing variances. Heteroskedasticity can arise in a variety of ways such as changes in behaviors of data under different conditions. Heteroskedasticity has been modeled as an ARCH or Generalized Autoregressive Conditional Heteroskedasticity (GARCH). Only in recent times, a new model has been developed and this model is Non-stationary Non-linear Heteroskedasticity (NNH) that pre-supposes stochastic volatility (Park, 2002). For a volatile NNH model, this chapter considers the sample auto-correlations of the squared processes of obtaining the data from the sensor. The sample auto-correlations are defined as (Park, 2002):

$$R_{nk}^2 = \frac{\sum_{k+1}^n \left(y_t^2 - y_n^2\right)\left(y_{t-k}^2 - y_n^2\right)}{\sum_1^n \left(y_t^2 - y_n^2\right)^2}$$

(13.1)

In equation 13.1, y_n^2 designates the sample mean of y_t^2 and $y_t^2 = \sigma_t \varepsilon_t$. In this chapter, it is assumed that ε is independently identically distributed (0,1) and is updated using filtration γ denoting information accessible at time t whereas σ on the other hand is adapted to γ_{t-1}.

The NNH model, therefore, specifies the conditional heteroskedasticity as a function of some explanatory variables, completely in parallel with the conventional approach. This work considers an aspect of NNH that the variable affecting the conditional heteroskedasticity is non-stationary and typically follows a random walk (Park, 2002).

MISSING DATA APPROXIMATION IN THE PRESENCE OF CONCEPT DRIFT

This section presents an algorithm used in this investigation. In missing data estimation, it is important to understand the mechanism by which data variables become missing. This allows one to be able to know what approach to pursue in order to deal with the missing values. As stated in earlier chapters, three main mechanisms for missing data have been identified in the past and these are Missing Completely At Random (MCAR), Missing At Random (MAR) and Missing Not At Random (MNAR) (Little & Rubin, 1987). As discussed before the MCAR is the type where the probability of missing value for variable X is not related to the value X itself or to any other variable in the data set. MAR takes place if the probability of missing data on a particular variable X depends on other variables, but not on X itself and the MNAR case comes up if the probability of missing data, X, is related to the value of X itself even if the other variables in the analysis are controlled.

Learning and Forgetting

The most common technique for on-line learning is through the use of a window and this technique only trusts the most recent examples. Examples are added to the window as they arrive and older ones are removed from the window. In the simplest case, the window will be of a fixed size; however, adaptive windows have also been reported in the literature (Kubat & Widmer, 1996). It is very important to choose the window size very well as small windows will help in fast adaptation to new concepts, meanwhile, bigger windows will offer a good generalization. In this case, the choice of the window size is a compromise between fast adaptability and good generalization (Scholz & Klinkenberg, 2005).

The idea of forgetting an example has been criticized for weakening of the existing description items (Kubat & Widmer, 1996). This kind of forgetting also assumes that only the latest examples are relevant, which might not be the case. Helmbold and Long (1991) have, however, shown that it is sufficient to use a fixed number of previous examples. An algorithm that removes inconsistent examples more efficiently will manage to track concept sequences that change more rapidly (Kubat & Widmer, 1996). In this chapter, the window is selected such that it is not too narrow to accommodate a sufficient number of examples. Again, the window size is selected to avoid slowing down the reaction to concept drift.

Algorithm Description

The proposed algorithm makes use of an ensemble of regressors and circumvents discarding old knowledge by discarding old networks. Instead, networks are stored and ranked according to a particular concept which is derived from the heteroskedasticity calculations. The algorithm is divided into three sections, namely, training, validation and testing as described below.

Training: Batch learning is initially used. In this training approach, each missing datum is predicted by means of the past i instances where $i = 6$ in this chapter. This implies that the window size is fixed at six samples. While sliding this window all the way through the data, the heteroskedasticity of each window is calculated. All vectors are then grouped according to their heteroskedasticity range. This process results in disordering the sequence of the data. An ensemble of neural networks that predict data for a particular heteroskedasticity is then trained. A multi-layer perceptron (MLP) neural network which is described in earlier chapters is used.

In the complete range of heteroskedasticity [0, 1], a sub-range of length 0.05 is used and various neural networks are trained. This practice leads to 20 sub-ranges and as a result, 20 trained neural networks. In a case where data do not have a heteroskedasticity value of a particular sub-range, no network will be trained for such a sub-range. In this chapter, each network is assigned to a sub-range and is optimized for such range. The objective here is to have at least one neural network designed for each individual sub-range. However, this does not imply that only one network will be used in a particular sub-range as it is needed to add diversity to the system. The next step is on validation.

Validation: All networks produced in the training section above are subjected to a validation set containing all the groupings of the data. Each neural network regressor is tested on all groups and weights are then assigned accordingly. The weight is assigned using the weighted majority scheme given as (Merz, 1997):

$$\sigma_k = \frac{1 - E_K}{\sum_{j=1}^{N}\left(1 - E_K\right)} \qquad (13.2)$$

In equation 13.2, E_k is the estimate of neural network model k's error on the validation set. This leads to each network having 20 weights, forming a neural network vector of weights.

To implement this procedure a testing process is conducted on a pre-determined range and a vector of instances is created. The heteroskedasticity of this vector is evaluated. From all the networks created, only those networks that have high weights assigned to them in the validation set for the same range of heteroskedasticity are chosen. In this application, all available networks are used and the missing values are approximated as shown below:

$$f(x) = y \equiv \sum_{k=1}^{k=N} \alpha_i f_k(x) \qquad (13.3)$$

In equation 13.3, α is the weight assigned during the validation stage when no data are missing and N is the total number of neural networks used. For a given network, the weights are normalized such that $\sum_{k=1}^{i=N} \alpha_i \approx 1$. After enough new instances have been sampled, the training process is repeated. The next section reports on the evaluation of the algorithm on two data sets.

Table 13.1. Illustration of how weights are assigned to each neural network

Network	Range 1	Range 2	Range 3	Range 4	...	Range 20
1	$\alpha_1^{(1)}$	$\alpha_2^{(1)}$	$\alpha_3^{(1)}$	$\alpha_4^{(1)}$	...	$\alpha_{20}^{(1)}$
2	$\alpha_1^{(2)}$	$\alpha_2^{(2)}$	$\alpha_3^{(2)}$	$\alpha_4^{(2)}$	...	$\alpha_{20}^{(2)}$
3	$\alpha_1^{(3)}$	$\alpha_2^{(3)}$	$\alpha_3^{(3)}$	$\alpha_4^{(3)}$	...	$\alpha_{20}^{(3)}$
...	...	...	...	...	...	...
20	$\alpha_1^{(20)}$	$\alpha_2^{(20)}$	$\alpha_3^{(20)}$	$\alpha_4^{(20)}$	...	$\alpha_{20}^{(3)}$

EMPIRICAL EVALUATION: SIMULATED DATA SET

Case Study 1: Firstly the algorithm proposed above is evaluated on the time series data produced by numerical simulation. A sequence of un-correlated Gaussian random variables is generated with zero mean and variance of 0.108 as prepared by Stephanos and Anthanassoulis (2001). In this chapter, data are simulated as if they are coming from a sensor that measures some variable that exhibit non-stationary characteristics. The data are made to show some cyclic behavior that simulates a cyclic concept drift. Figure 13.1(A) shows a sample of the simulated data.

Case Study 2: The second test data is created using the Dow Jones stock market data. The stock market is well known for being difficult to predict as it exhibits non-stationarity. The opening price of the Dow Jones stock is also simulated as some data collected from some sensor and sampled at a constant interval. Samples of these data are shown in Figure 13.1(B).

The relative performance of the algorithm is measured by how close the prediction is to the actual data. The results obtained with this data are summarized in the next section.

EXPERIMENTAL RESULTS

Firstly, the effect of the number of regressors on the error is evaluated. Mean Square Error is computed for each prediction and results are presented in Figure 13.2. Performance in terms of accuracy is shown in Figure 13.3 for the study in *Case 1* and this figure evaluates predictability of missing sensor values within 10% tolerance.

It can be observed from both figures that as the number of estimators is increased, the Mean Squared Error is reduced. On the other hand, there is a point at which increasing the number of regressors does not significantly affect the error. For each case study, the algorithm is tested with 500 missing points and the same algorithm. Performance is calculated based on how many missing points are estimated within a given percentage tolerance as follows:

Figure 13.1. Sample data with (A) cyclic concept drift and (B) gradual and sudden concept drift

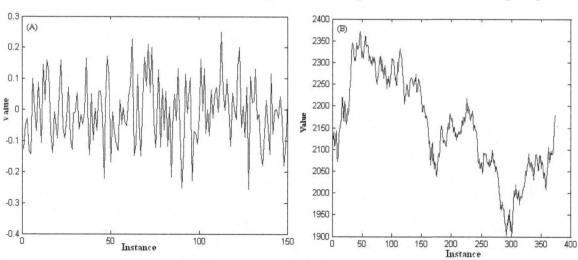

Figure 13.2. Effect of the number of regressors on the Mean Square Error for (a) the simulated data and (b) the real data of the from the stock market

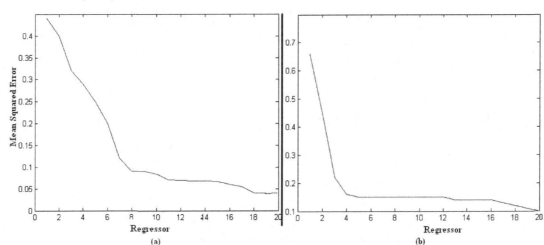

Figure 13.3. Effect of the number of regressors on the prediction accuracy for Case study

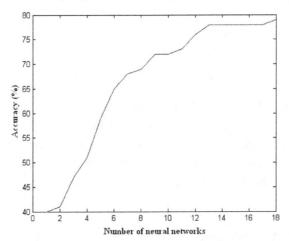

$$Accuracy = \frac{n_T}{N} \times 100\% \tag{13.4}$$

In equation 13.4, n_T is the number of predictions within a 10% tolerance and N is the total number of instances being evaluated. The results obtained are summarized in Table 13.2. In addition, the correlation coefficients between the missing data and the estimated data are computed and the results are shown in Table 13.2.

Results shown in Table 13.2 prove that prediction of missing data when there is a large concept drift is not very accurate. It can be seen that there is poor correlation between the estimated data and the actual data for *Case study 2*. This position is further investigated in this study, paying a particular attention to the data set of *Case study 2*.

Table 13.2. Results obtained from both case studies and the Correlation coefficients between the estimated and the actual values for prediction within 10%

Case study	Estimate within 10%	Estimate within 10%	Corr Coefficient
1	78%	41%	0.78
2	46%	25%	0.26

Figure 13.4. Best results obtained with the data set of case study 2

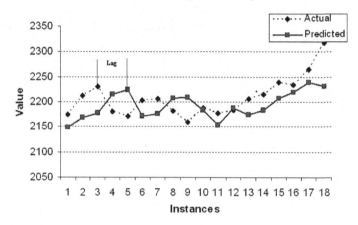

The best results obtained in the estimation of missing data in that case are shown in Figure 13.4.

It is observed that there is a time lag of approximately two instances. Pan, Tilakaratne, and Yearwood (2005) also found a lag in their stock market prediction. The findings in this chapter show that this lag is responsible for the poor correlation coefficient reported in Table 13.2. Results obtained here give some insight into the employment of heteroskedasticity as a measure of concept drift.

DISCUSSION AND CONCLUSION

This part of the chapter presented an algorithm to approximate missing data in non-stationary time series that may also possess concept drift. An ensemble of estimators is used and the final output is computed using the weighted approach of combining regression machines that are constructed using multi-layer perceptron neural networks. Results demonstrate that the predictability increases as the number of neural networks used to form an ensemble is increased. This is seemingly caused by the concept drift. The concept simply drifts from a region of one mastered by one neural network to a region mastered by another. The approach used in this work can, therefore, function as it covers the entire range of possible outcomes. However, the major shortcoming is that data covering the entire range need to be there in the beginning. In addition, an ensemble with a large number of neural networks slows down the prediction which then reduces the usability of the method in fast sampled data. This technique can be computationally expensive as it requires a large memory to store all the networks and their assigned weights.

INTRODUCTION: DYNAMIC PROGRAMMING FOR MISSING DATA ESTIMATION

The problem of missing data causes problems to a variety of fields from sensor readings in machine operation to risk analysis. A good number of models built to run off a specific number of inputs will breakdown when one or more inputs are not available. In many such applications, merely ignoring or deleting the incomplete record, a situation known as case deletion, is not an alternative, as it may carry a great deal of harm than good (Allison, 2002). In a statistical model, case deletion can lead to biased results and in practical applications of such as statistical models the consequences may be severe (Roth & Switzer, 1995).

Many techniques, to impute missing data that are intended to minimize the bias or output error of a model, have been researched extensively. A good number of these methods are statistically based techniques. One of the most successfully used of these techniques is the Bayesian multiple imputation. Methods that use computational intelligence techniques such as neural networks like the one that was proposed by Abdella and Marwala (2006) have also revealed excellent results. Nevertheless, most of these techniques do not run at optimal manner. As a consequence, a lot of processing power and time is wasted in repeated calculations.

In this chapter, dynamic programming can be viewed as a stage-wise search technique with the main features being a sequence of decisions that exploits the duplications as well as the arrangement of the data for missing data imputation. For the duration of the search for the optimal solution, early decisions solutions that can not possibly give optimal results are discarded. The fundamental concept behind this procedure is to keep away from performing the same calculations more than once and this is achieved by storing the results obtained in each sub-problem. Dynamic programming uses the concept of optimality that can be translated to optimization in stages. It follows that, for an optimal sequence of decisions, each sub-sequence must be optimal (Bellman, 1957).

The estimation of missing data requires a system that possesses the knowledge of certain characteristics such as the correlations between variables, which are inherent in the input space. Computational intelligence techniques and maximum likelihood techniques do possess such characteristics and as a result are important for imputation of missing data (Nelwamondo & Marwala, 2007). The concept of dynamic programming can be a useful tool to the problem of missing data that optimizes all the sub-steps in the solution. By using the concept of optimality, to obtain the best estimate of missing data, all steps leading to the solution need to be optimized. This concept has several advantages that can improve the method proposed by (Abdella & Marwala, 2006), which shall be used as a baseline method in this paper.

Therefore, missing data is estimated, in this section, by solving the following missing data estimation equation (Abdella & Marwala, 2006):

$$\varepsilon = \left\| \left(\left\{ \begin{matrix} \{X_k\} \\ \{X_u\} \end{matrix} \right\} - f\left(\left\{ \begin{matrix} \{X_k\} \\ \{X_u\} \end{matrix} \right\}, \{w\} \right) \right) \right\| \tag{13.5}$$

In equation 13.5 vectors $\{X_k\}$ and $\{X_u\}$ are the known and unknown measurements, respectively, while the vector $\{w\}$ is the mapping weight vector that maps the input to the output and in this chapter using a two-layered autoassociative multi-layered perceptron neural network, which is defined by the function

$f(*)$ (in equation 13.5) with the hyperbolic tangent function in the inner layer and linear function in the outer layer (Marwala, 2007), $\| \|$ is the Euclidean norm and ε is the error. The missing data vector $\{X_u\}$ is, therefore, obtained by minimizing the error ε using genetic algorithm. The autoassociative neural network is trained using the NETLAB Toolbox (Nabney, 2002) and genetic algorithm is implemented using the GAOT toolbox (Houck, Joines, & Kay, 1995).

This element, of this chapter, puts forward a novel technique to estimate missing data using the principle of dynamic programming using autoassociative neural networks and genetic algorithms.

DYNAMIC PROGRAMMING

Mathematical Background

Dynamic programming is a technique of finding solutions to problems showing the characteristics of overlapping sub-problems and optimal sub-structure. In finding this solution dynamic programming is computationally efficient when compared to conventional techniques. Dynamic programming was initially coined by Richard Bellman to depict the procedure to work out problems where it is desired to uncover the best resolution one after another (Bellman, 1953).

The expression "programming" in "dynamic programming" possesses no specific relationship to computer programming and in its place emanates from the word "mathematical programming", another word for the expression optimization. Consequently, the "program" is the optimal map for a process that is generated. As an example, decided program of events at a trade fair is usually called a program. Therefore, in this chapter the word programming, means identifying an acceptable program of action also known as a feasible route to a solution.

In this chapter, as in dynamic programming, the word 'optimal sub-structure' implies that optimal solutions of sub-problems can be utilized to identify the optimal solutions of the entire problem. Generally, problems that display optimal sub-structure can be solved by utilizing a three-step procedure (Bertsekas, 2000):

- Divide the problem into sub-problems.
- Solve each sub-problem optimally by utilizing this three-step procedure recursively.
- Utilize the optimal solutions from these sub-problems to come up with an optimal solution for the entire problem.

The sub-problems should also be solved by breaking them into sub-sub-problems, and so forth, until a simple case that can be easily solved is identified. This dynamic programming procedure has been successfully utilized in numerous disciplines ranging from engineering to economics. Chen and Tanaka (2002) successfully applied dynamic programming to discover solution to inverse heat conduction problems while Kwatny et al. (2006) effectively used dynamic programming for system management in power plants. Additional successful applications of dynamic programming include process control (Wang, 2001; Lee, Kaisare, & Lee, 2006; Lee & Lee, 2006), telecommunications (Flippo et al., 2000), transportation (Secomandi, 2000; Kang & Jung, 2003), asset allocation (Kung, 2007), speech synthesis (Yang & Stone, 2002; Salor & Demirekler, 2006), face recognition (Yao & Li, 2006), ecological management (Grüne, Kato, & Semmler, 2006) and economics (Genc, Reynolds, & Sen, 2007). Despite all these impressive successes, dynamic programming has not yet been widely applied in the missing data estimation problem. Some of the related applications to missing data estimation include input estimation application in linear time-variant system by Nordström (2006).

As explained earlier, the problem is broken into sub-problems and each sub-problem is broken into sub-sub-problems until the problem becomes manageable. This can be very computationally taxing and to steer clear of this, solutions to the sub-problems that were previously solved are stored. Subsequently, if solutions to similar problems afterwards are required, then it becomes possible to recover and use the stored solution again. This technique is known as memoization. If it becomes clear that the stored solution will not be required any longer, then such solution can then be discarded. In other situations, the solutions to sub-problems can be calculated in anticipation of its need in the future. Consequently, dynamic programming makes use of the following attributes which are described above:

- Overlapping sub-problems
- Optimal substructure
- Memoization

On implementing dynamic programming, the following procedures are typically pursued:

- *Top-down approach*: The problem is divided into sub-problems, and the solutions to these sub-problems are obtained and memorized, for the situation when these solutions may be required sometimes in the future. This process, therefore, merges recursion and memoization.
- *Bottom-up approach*: Every sub-problem that is anticipated to be required in the future is solved beforehand and then utilized to construct solutions to the bigger problem at hand and in this chapter the missing data estimation. However, it is occasionally not instinctive to outline every sub-problem required for solving the problem at hand in advance.

The necessary condition for optimality of the global problem given the optimal solutions for sub-problems in dynamic programming is ensured by the use of the Bellman equation (Bellman, 1957). The Bellman equation, within the context of the problem of missing data imputation problem, essentially states that given an appropriate initial estimation of the missing data x_0, the infinite horizon dynamic programming problem can be written as follows (Bellman, 1957):

$$V(x) = \max_{y \in \Im(x)} \left[F(x, x') + \beta V(x') \right], \forall x \in X \tag{13.6}$$

In equation 13.6, the function V solves the Bellman equation and is known as the value function and is described by the missing value error estimation equation described in equation 13.6. The function $F(x,x')$ is the policy function while x and x' are the current state and state to be chosen. From theoretical perspective, if the policy function is optimal for the infinite summation, then it has to be true that irrespective of the initial estimate of the missing data, the remaining decisions must characterize an optimal policy concerning the missing data estimate resulting from that first decision, as articulated by the Bellman equation. The principle of optimality is connected to the theory of optimal substructure.

Base Model for Imputation

The model, in equation 13.5, proposed by Abdella and Marwala (2006) is utilized as a baseline technique. This method as implemented in this chapter combines the use of multi-layer perceptron autoassociative neural networks with genetic algorithms to approximate missing data. A genetic algorithm is used to estimate the missing values by optimizing an objective function as presented in equation 13.5. The complete vector combining the estimated and the observed values is fed into the autoassociative network as input and as shown in Figure 13.5.

Implementation of Dynamic Programming for Missing Data Problem

The concept of dynamic programming is added to the base model. The method implemented does not train its neural networks on the full set of complete data, where complete data refers to a record that has no missing fields. Instead, the data are broken up into categories and a neural network is trained for each category using only the data from that category. This can be viewed as breaking the global problem into over-lapping sub-problem as per the first step in dynamic programming schedule. The rationale as to why this method could improve performance of the base model is that, the base model assumes that data variables are somehow related to one another. From statistical analysis on the HIV data that is used in this chapter, it can be figured out that the parameters such as income of an individual have an influence on the other variables such as HIV risk. It, therefore, follows that; those parameters from one category might not have an effect on known parameters from other categories. Following the fundamental concept of dynamic programming, it is a necessity to optimize each and every step of the solution. To achieve this, different models have to be created and all extraneous data should not be used to create such a model. Separate neural networks should be trained for each combination of granular data. Looking at this from a behavioral point, a person is likely to act more similarly to someone close to their age than someone much older or younger than them. Optimal models need to be created for each granule. Figure 13.6 is a flow chart representation of the proposed algorithm.

Figure 13.5. Auto-associative neural network and genetic algorithm model for missing data imputation as explained by equation 13.5.

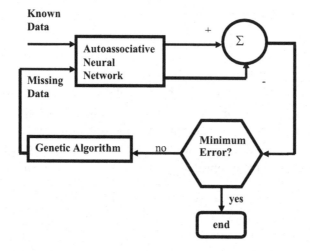

After implementation it is found that the algorithm shown in Figure 13.6 breaks down when there is a group of categories with no complete records. In this case the algorithm supplies no training data for the neural network. Two approaches are taken to solve this problem. The first approach is to broaden the granule in the hope of finding data that would then fit the record in question. If a record is alone in its category then the algorithm creates a data set for training ignoring numerical categories. In other words, it uses all data with the same population group and province. In an unlikely event that there is no complete data in that combination of population group and province then the record is passed un-altered (i.e. no imputation is done) and its number is logged as an error. Two approaches can be taken from there. They are as follows:

1.   Case deletion is an option if the percentage of records that will not be imputed is negligible.
2.   A network can be trained using the entire data set as done in Nelwamondo & Marwala (2007b).

The second way to prevent these "lone categories" is to use previously imputed data to train the networks. Thus the first time a lone category appears it is imputed using a broader category. A lone category is defined as a granule in which there are no complete records. There is then one record that can be used for training. This is not an ideal implementation as it means a network is trained using only one record but serves as the basis for future extension. A suggested extension of this is to retrain such a network after every imputation it performs.

There are two explicit cases where dynamic programming has been used to improve the efficiency of the program:

Figure 13.6. Auto-associative neural network and genetic algorithm model for missing data imputation using dynamic programming

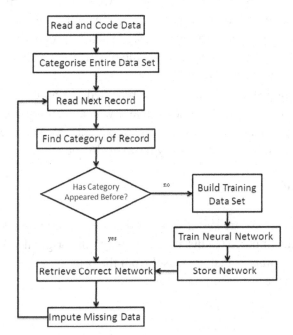

1.   The entire data set is categorized at the start of the program. This way when the program builds a dataset for training in a certain category it does not have to redundantly find the category of every record.

2.   The first time a neural network is trained in a category it is stored in a multi-dimensional array. When future data from this category needs to be imputed, instead of retraining a network, the stored one is used. This way two networks in the same category are never trained as this would be redundant and networks for categories that never require imputations are not trained as would be the case if the training was done during initialization.

EXPERIMENTAL EVALUATION

Data and Analysis

The data used in this test are obtained from the South African antenatal sero-prevalence survey of 2001. The data for this survey are obtained from questionnaires answered by pregnant women visiting selected public clinics in South Africa. Only women participating for the first time in the survey were eligible to answer the questionnaire.

Data attributes used in this study are the *HIV status, education level, gravidity, parity, age, age of the father, population group* and *region* and are described in detail in earlier chapters. The HIV status is the decision and is represented in a binary form, where 0 and 1 represent negative and positive, respectively. Population group is measured on the scale 1 to 4 where 1, 2, 3, and 4 represent African, Mixed, European and Asian, respectively and these are expressed in binary form. The data used is obtained in three regions and are referred to as region A, B and C in this investigation. The education level is measured using integers representing the highest grade successfully completed, with 13 representing tertiary education. Gravidity is the number of pregnancies, complete or incomplete, experienced by a female, and this variable is represented by an integer between 0 and 11. Parity is the number of times the individual has given birth and multiple births are counted as one. Both parity and gravidity are important, as they show the reproductive activity as well as the reproductive health state of the women. Age gap is a measure of the age difference between the pregnant woman and the prospective father of the child. A sample of this data set is shown in Table 13.3.

During the pre-processing stage, all outliers are removed. As an example, all cases where parity exceeds the gravidity are removed as it is known that a woman can not give birth more than she has been pregnant if multiple pregnancy is countered as one pregnancy. All nine provinces of the country are represented and the age is found to be mainly between 16 and 45.

Testing Methodology

Elements of the complete data set are systematically removed. The program is then run on these now incomplete data sets and the imputed results are compared to the original values. The accuracy is measured as the percentage of numbers that are offset from the original value within a certain range. The following are the ranges that are measured for each field:

Table 13.3. Extract of the HIV database used, with missing values

Population Group	Region	Education	Gravidity	Parity	Age	Father	HIV
1	C	?	1	2	35	41	0
2	B	13	1	0	20	22	0
3	?	10	2	0	?	27	1
2	C	12	1	?	20	33	1
3	B	9	?	2	25	28	0
?	C	9	2	1	26	27	0
2	A	7	1	0	15	?	0
1	C	?	4	?	25	28	0
4	A	7	1	0	15	29	1
1	B	11	1	0	20	22	1

- Age, to within 2, 5 and 10 years
- Education Level, to within 1, 3 and 5 grades
- Gravidity, to within 1, 3, and 5 pregnancies
- Parity, to within 1, 2 and 4 children
- Father's Age, to within 3, 7 and 15 years
- HIV status, the correctly predicted percentage, the percentage of false positives and the percentage of false negatives

Testing is first done for single missing fields and then for multiple missing fields. The next sub-section presents the results.

Results

Table 13.4 shows the average percentages of the missing data to within each range for the tests done on one missing fields. The vertical axis differentiates between the ranges listed for each field respectively.

These results illustrate the varying degree of predictability of the data. If one is trying to predict gravidity to within one pregnancy then this method will give an accurate answer 96% of the time. However, one should not use it to predict education level to within one grade as it only succeeds 38% of the time. In the case of two missing data points, the combination of missing age and all the other fields are tested. Table 13.5 presents the results as above. The bottom three rows of each column correspond to age, whilst the top three correspond to the field that heads the column.

It can be seen from this table that removing two fields has no noticeable difference to removing one except with the combination of age and father's age. Both of these significantly lose predictability when they are both absent. This indicates a correlation between the two. A few tests are done on three and four missing fields and there are no noticeable differences, apart from when both age and father's age are missing.

Table 13.4. Percentages of imputed data that fell within the ranges for records missing only one field

Ranges	Age (%)	Education (%)	Gravidity (%)	Parity (%)	Father (%)	HIV (%)
1	48	38	96	95	45	66
2	79	66	99.6	99	75	20
3	95	81	100	99.9	93	14

Table 13.5. Percentages of imputed data that falls within the ranges for records missing only one field

Tolerance	Education	Gravidity	Parity	Father	HIV
1	38	96	96	37	67
2	65	99.5	99	68	17
3	80	100	99.9	92	16
1 (age)	50	47	50	35	48
2 (age)	81	78	79	67	79
3 (age)	96	95	95	90	96

CONCLUSION AND DISCUSSION

A model using autoassociative neural networks, genetic algorithm and data categorization is built to impute missing data in an HIV database. Dynamic programming technique is used to improve the efficiency of the code. The results from testing shows varying degree of predictability of each field, with gravidity being the most predictable and education level the least. The results also show that some fields are independent of each other and removing both has no effect on the model's ability to predict them. However, fields that are dependant on each other yield much lower predictability when both are removed, such as a combination of age and father's age.

INTRODUCTION: IMPACT OF MISSING DATA ESTIMATION ON DECISION MAKING

In the previous chapters, methods for imputing missing data are described. One question that still needs to be answered is: What is really the impact of estimating missing data on a decision making process? This question of course assumes that the missing data estimation process is aimed at decision making. In this chapter, a problem of fault classification in mechanical systems (Marwala, 2000) is considered. In particular, decision making process in the presence of missing data where missing data are imputed is studied. A particular attention is paid to how different classification procedures perform in the presence of missing data imputation. The method of imputing data is based on autoassociative multi-layer neural networks combined with genetic algorithm. This proposed study is, therefore, evaluated on a fault classification problem in a population of cylindrical shells by using the estimated values and observed values for fault classification. For the task of fault identification the extension network as well as Gaussian mixture models are used and these are described below.

EXTENSION NEURAL NETWORK (ENN)

ENN is an emerging pattern classification system based on concepts from neural networks and extension theory (Wang & Hung, 2003). The extension theory uses a novel distance measurement for classification processes, and the neural network can embed the salient features of parallel computation and learning capability and is shown in Figure 13.7. The classifier is well suited to classification problems where there exist patterns with a wide range of continuous inputs and a discrete output indicating which class the pattern belongs to. ENN has been successfully used for condition monitoring in electrical components (Vilakazi & Marwala, 2006) and in mechanical systems (Mohamed, Tettey, & Marwala, 2006; Marwala & Vilakazi, 2007a). ENN comprises of an input layer and an output layer. The input layer nodes receive an input feature pattern and use a set of weighted parameters to generate an image of the input pattern. There are two connection weights between input nodes and output nodes; one connection represents the lower bound for this classical domain of features and the other represents the upper bound. The entire network is thus represented by a matrix of weights for the upper and lower limits of the features for each class W_U and W_L. A third matrix representing the cluster centers is also defined as (Wang & Hung, 2003):

$$z = \frac{W_u + W_l}{2}$$

(13.7)

ENN uses supervised learning, which tunes the weights of the ENN to achieve a good clustering performance or to minimize the clustering error. The network is trained by adjusting the network weights and recalculating the network centers for each training pattern depending on the extension distance of that pattern to its labeled cluster.

Figure 13.7. Extension neural network architecture

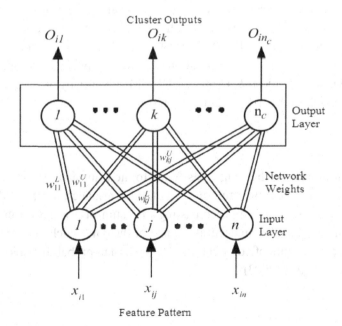

Each training pattern adjusts the network weights and the centers by amounts that depend on the learning rate. In general, the weight update for a variable x_i is (Wang & Hung, 2003):

$$w^{new} = w^{old} - \eta(x_i - w^{old})$$

(13.8)

where η is the learning rate and w can either be the upper or the lower weight matrices of the network centers. It can be shown that for t training patterns for a particular class C, the weight is given by (Mohamed, Tettey, & Marwala ,2006):

$$w^c(t) = (1-\eta)w^c(0) - \eta \sum (1-\eta)^{t-1} x_i^c$$

(13.9)

This equation demonstrates how each training pattern reinforces the learning in the network by having the most recent signal determines only a fraction of the current value. The equation indicates that there is no convergence of the weight values since the learning process is adaptive and reinforcing. Equation 13.9 also highlights the importance of the learning rate, η. Small values of η require many training epochs, whereas large values may results in oscillatory behavior of the network weights, resulting in poor classification performance.

GAUSSIAN MIXTURE MODELS (GMM)

Gaussian Mixture Models have been a reliable classification tools in many applications of pattern recognition, particularly in speech and face recognition. GMM have proved to perform better than HMM in text independent speaker recognition (Reynolds, Quatieri, & Dunn, 2003). The success of GMM in classification of dynamic signals has also been demonstrated by many researchers such as Cardinaux, Cardinaux, and Marcel (2003) who compared GMM and multi-layer perceptron in face recognition and found that the GMM approach easily outperforms the MLP approach for high resolution faces and is significantly more robust in imperfectly located faces. Other applications of GMM include in fault detection of mechanical systems (Nelwamondo & Marwala, 2006; Nelwamondo, Marwala, & Mahola, 2006; Marwala, Mahola, & Nelwamondo, 2006; Marwala, Mahola, & Chakraverty, 2007). Some of the advantages of using GMM are that it is computationally inexpensive and is based on well understood statistical models (Reynolds, Quatieri, & Dunn, 2003). GMM works by creating a model of each fault which is written as (Reynolds, Quatieri, & Dunn, 2003):

$$\lambda = (w, \mu, \Sigma)$$

(13.10)

where λ is the model, w represents the weights assigned to the Gaussian means, μ is the diagonal variance of the features used to model the fault and Σ is the covariance matrix. GMM contains a probability density function of the observation consisting of a sum of normal observations. A weighted sum of Gaussians normally provides an accurate model of the data. Each Gaussian comprises a mean and a covariance, hence, a mixture of components. The Gaussian probability density function is given by (Reynolds, Quatieri, & Dunn, 2003):

$$p(\{x\}) = \frac{1}{\sigma\sqrt{2}x} e^{\frac{-(x-\mu)^2}{2\sigma^2}}$$

(13.11)

where μ is the mean and σ is the standard deviation of the distribution of a variable x. For a case where $\{x\}$ is a vector of features, which is the case in this chapter, equation 13.11 becomes (Reynolds, Quatieri, & Dunn, 2003):

$$p(\{x\}) = \frac{1}{\sqrt{(2\pi)^n |\Sigma|}} e^{-\frac{1}{2}\left[(\{x\}-\{\mu\})'\Sigma^{-1}(\{x\}-\{\mu\})\right]}$$

(13.12)

where n is the size of the vector feature, $\{x\}$ and Σ is the covariance. The log-likelihood is then computed as (Reynolds, Quatieri, & Dunn, 2003):

$$\hat{s} = \arg \max_{1 \le f \le F} \sum_{k=1}^{K} \log\ p(x_k|\lambda_f)$$

(13.13)

where f represents the index of the type of fault, whereas F is the total number of known fault conditions and $x = \{x_1, x_2, ..., x_K\}$ is the unknown fault vibration segment. $P(x_k|\lambda_f)$ is the mixture density function. An arbitrary probability density of a sample vector x can be approximated by a mixture of Gaussian densities (Bishop, 1995) as:

$$p(\{x\}|\lambda) = \sum_{i=1}^{M} w_i p(\{x\})$$

(13.14)

where all mixtures weights, w_i, are adjusted to satisfy the constrains and $0 \le w_i \le 1$. Training GMM is a fast and straight-forward process which estimates the mean and covariance parameters from the training data. The training procedure estimates the model parameters from a set of observations using the Expectation Maximization (EM) algorithm which is described in detail in Chapter IV. The EM algorithm tries to increase the expected log-likelihood of the complete data $\{x\}$ given the partially observed data and finds the optimum model parameters by iteratively refining GMM parameters for a given bearing fault feature vector.

MISSING ENTRY ESTIMATION METHODOLOGY

As described earlier, the missing entry methodology here combines the autoassociative neural networks and optimization method, viz. the genetic algorithm. The missing data are estimated by minimizing the missing data estimation error equation (equation 13.5) by using genetic algorithm. Since the impact of a neural-network-genetic-algorithm missing data estimation procedure is evaluated on condition monitoring in mechanical systems using vibration data in this chapter, the next section describes this case study.

EXAMPLE: MECHANICAL SYSTEM

In this section the impact of the missing data on fault identification in mechanical system is studied. In particular a population of cylinders is used and this is explained in detail in Marwala (2000). These sets of cylinders are sustained by inserting a sponge rested on a *bubble-wrap*, to simulate a 'free-free' environment. The details of this experiment are found in Marwala (2000). The sponge is included within the cylinders to control boundary conditions. The impulse hammer test is performed on each of the 20 steel seam-welded cylindrical shells (1.75 ± 0.02 mm thickness, 101.86 ± 0.29 mm diameter and of height 101.50 ± 0.20 mm). The impulse is applied at 19 different locations. The sponge is inserted inside the cylinder to control boundary conditions and by rotating it every time a measurement is taken.

From the data measured in the previous section, 10 parameters are selected. The autoassociative network with 10 inputs and 10 outputs is constructed and several numbers of hidden units are tried and it is found that 10 hidden units is the optimal network that gives the best prediction of the input data. The first experiment consists of cases where one of the inputs to the neural network is assumed to be unknown and then estimated using genetic algorithm method as is done in Chapter II. On implementing genetic algorithm, the arithmetic crossover, non-uniform mutation and normalized geometric selection as described before are used. On implementing arithmetic crossover several parameters needed to be chosen and these are bounds and the probability of crossover. The bounds are determined from the method explained in Chapter XI while the probability of crossover is chosen to be 0.75. On implementing mutation the parameters that need to be chosen are the bounds, and these are chosen as is done for crossover, and the probability of mutation, that is chosen to be 0.0333. Genetic algorithm that has a population of 20 and 25 generations is implemented.

The proposed method for the case of one missing data per input set, estimated the missing value to the accuracy of 93%. When the proposed method is tested for the case with two missing data per input set, the accuracy of the estimated values is 91%. The estimated values together with the accurate values are also indicated in Figure 13.8. This figure illustrates that the proposed missing data estimator gives the results that are consistent and accurate. In fact the data in this figure shows the correlation between the estimated data and the correct data to be 0.9. In many cases the estimated values are intended for a particular reason, and in this chapter they are intended to fulfill the goal of fault classification in a population of cylinders. The estimated values are then used in the classification of faults using the extension networks and the fault classification accuracy of 95% is observed for single-missing-entry cases and 92% for two-missing-entry cases while the full database set is able to give classification accuracy of 97%. On the other hand the Gaussian mixture models give 94% for single-missing-entry cases and 92% for two-missing-entry cases while the full database set is able to give classification accuracy of 96%. These results indicate that the impact of missing data estimation on the ENN is marginally better than on GMM.

CONCLUSION

In this study, a method based on autoassociative neural networks and genetic algorithms is proposed to estimate missing entries in data. This procedure is tested on a population of cylindrical shells. The proposed method is able to estimate single-missing-entries to the accuracy of 93% and two-missing-entries to the accuracy of 91%. The estimated values are then used in the classification of faults using the extension networks and the fault classification accuracy of 95% is observed for single-missing-entry cases and 92% for two-missing-entry cases while the full database set is able to give classification accuracy of 97%. On the other hand the Gaussian mixture model gives 94% for single-missing-entry cases and 92% for two-missing-entry cases while the full database set is able to give classification accuracy of 96%. The extension neural network is, therefore, found to be marginally more resistant to missing data that the Gaussian mixture models.

FUTURE WORK

This chapter presents two emerging areas in the field of missing data estimation and these are heteroskedasticity and dynamic programming. For future work, an investigation should be conducted as to how heteroskedasticity compares with other non-stationarity detection methods such as fractals. Furthermore, an investigation should be done on how the incremental learning method proposed in this chapter compares with other methods such as Learn++ and genetic programming. On the dynamic programming side, an investigation should be carried out on how dynamic programming performs compared to other methods such as genetic programming. Furthermore, additional examples should be included to further test the robustness of dynamic programming procedure. In this chapter, a method based on autoassociative neural networks and genetic algorithms is proposed for missing data estimation. The proposed method is tested on how well the estimated data would be able to predict faults in a population of cylinders. For future work, a study should be conducted on the sensitivity of the choice of the global optimization method on the prediction of faults.

REFERENCES

Abdella, M., & Marwala, T. (2006). The use of genetic algorithms and neural networks to approximate missing data in database, *Computing and Informatics, 24,* 1001–1013.

Allison, P. D. (2002). *Missing Data: Quantitative Applications in the Social Sciences.* Thousand Oaks, CA: Sage.

Bellman, R. (1957). *Dynamic programming.* Princeton: Princeton University Press.

Bertsekas, D. P., (2000). *Dynamic programming and optimal control.* New York: Athena Scientific.

Bishop, C. M. (1995). *Neural networks for pattern recognition.* Oxford: Oxford University Press.

Bujurke, N. M., Salimath, C. S., Kudenatti, R. B., & Shiralashetti, S. C. (2007). A fast wavelet-multigrid method to solve elliptic partial differential equations. *Applied Mathematics and Computation, 185,* 667-680.

Cardinaux, F., Cardinaux C., and Marcel, S. (2003). Comparison of MLP and GMM classifiers for face verification on XM2VTS. *Springer-Verlag GmbH, 2688,* 911-920.

Chen, W., & Tanaka, M. (2002). Solution of some inverse heat conduction problems by the dynamic programming filter and BEM. In *International Symposium on Inverse Problems in Engineering Mechanics,* Nagano, Japan (pp. 23-28).

Dufour, J-M, Khalaf, L., Bernard, J-T., & Genest, I. (2004). Simulation-based finite-sample tests for heteroskedasticity and ARCH effects. *Journal of Econometrics, 122*, 317-347.

Engle, R. (1982). Autoregressive conditional heteroskedasticity with estimates of the variance of UK inflation. *Econometrica, 50*, 987–1008.

Engle, R. F. (2003). Time-series econometrics: Co-integration and autoregressive conditional heteroskedasticity. *Advanced Information on the Bank of Sweden Prize in Economic Sciences in Memory of Alfred Nobel* (pp. 1–30).

Flippo, O. E., Kolen, A. W. J., Koster, A. M. C. A., & van de Leensel, R. L. M. J., (2000). A dynamic programming algorithm for the local access telecommunication network expansion problem. *European Journal of Operational Research, 127*, 189-202.

Genc, T. S., Reynolds, S. S., & Sen, S. (2007). Dynamic oligopolistic games under uncertainty: A stochastic programming approach. *Journal of Economic Dynamics and Control, 31*, 55-80.

Granger, C. W. J. (2003). Time series analysis, co-integration and applications. *Nobel Price Lecture* (pp. 360–366).

Grüne, L., Kato, M., & Semmler, W. (2005). Solving ecological management problems using dynamic programming. *Journal of Economic Behavior & Organization, 57*, 448-473.

Hafner, C. M., & Herwartz, H. (2001). Option pricing under linear autoregressive dynamics, heteroskedasticity, and conditional leptokurtosis. *Journal of Empirical Finance, 8*, 1-34.

Harries, M., & Sammut, C. (1988). Extracting hidden context. *Machine Learning 32*, 101–126.

Helmbold, D. P., & Long, P. M. (1991). Tracking drifting concepts using random examples. In *Proceedings of the Fourth Annual Workshop on Computational Learning Theory* (pp. 13–23).

Houck, C. R., Joines, J. A., & Kay, M. G. (1995). *A genetic algorithm for function optimisation: A MATLAB implementation* (Tech. Rep. NCSU-IE TR 95-09). Chapel Hill: North Carolina State University.

Hulley, G., & Marwala, T. (2007). Genetic algorithm based incremental learning for optimal weight and classifier selection. *Computational Models for Life Sciences. 952*, 258-267.

Inkmann, J. (2000). Misspecified heteroskedasticity in the panel probit model: A small sample comparison of GMM and SML estimators. *Journal of Econometrics, 97*, 227-259.

Kang, D-J., & Jung, M-H. (2003). Road lane segmentation using dynamic programming for active safety vehicles. *Pattern Recognition Letters, 24*, 3177-3185

Khalaf, L., Saphores, J-D., & Bilodeau, J-F. (2003). Simulation-based exact jump tests in models with conditional heteroskedasticity. *Journal of Economic Dynamics and Control, 28*, 531-553.

Krämer, W., & Michels, S. (1997). Autocorrelation- and heteroskedasticity-consistent t-values with trending data. *Journal of Econometrics, 76*, 141-147.

Kubat, M., & Widmer, G. (1996), Learning in the presence of concept drift and hidden contexts. *Machine Learning, 23*, 69–101.

Kung, J. J. (2007) (in press). Multi-period asset al.location by stochastic dynamic programming. *Applied Mathematics and Computation.*

Kwatny, H. G., Mensah, E., Niebur, D., & Teolis, C. (2006). Optimal power system management via mixed integer dynamic programming. In *Proceedings of the 2006 IFAC Symposium on Power Plants and Power Systems Control,* Kananaskis, Canada (pp. 353-358).

Last, M. (2002). On-line classification of non-stationary data streams. *Intelligent Data Analysis 6,* 129–147.

Lee, J. M., Kaisare, N. S., & Lee, J. H. (2006). Choice of approximator and design of penalty function for an approximate dynamic programming based control approach. *Journal of Process Control, 16,* 135-156.

Lee, J. H., & Lee, J. M., (2006). Approximate dynamic programming based approach to process control and scheduling. *Computers & Chemical Engineering, 30,* 1603-1618.

Little, R. J. A., & Rubin, D. B. (1987). *Statistical analysis with missing data.* New York: John Wiley and Sons.

Ljung, G. M. (1989). A note on the estimation of missing values in time series. *Communication of Statistics, 18*(2), 459–465.

Lunga, D., & Marwala, T. (2006a). Time series analysis using fractal theory and on-line ensemble classifiers. *Lectures Notes in Artificial Intelligence, 4304/2006,* 312-321.

Lunga, D., & Marwala, T. (2006b). On-line forecasting of stock market movement direction using the improved incremental algorithm. *Lecture Notes in Computer Science, Volume, 4234,* 440-449.

Marwala, T. (2000). *Fault identification using neural networks and vibration data.* Unpublished doctoral dissertation, University of Cambridge, Cambridge.

Marwala, T. (2002). Finite element updating using wavelet data and genetic algorithm. *American Institute of Aeronautics and Astronautics, Journal of Aircraft, 39,* 709-711.

Marwala, T. (2007). *Computational intelligence for modelling complex systems.* New Delhi: Research India Publications.

Marwala, T., Mahola, U., & Chakraverty, S. (2007). Fault classification in cylinders using multi-layer perceptrons, support vector machines and Gaussian mixture models. *Computer Assisted Mechanics and Engineering Sciences, 14*(2), 307-316.

Marwala, T., Mahola, U., & Nelwamondo, F. V. (2006). Hidden Markov models and Gaussian mixture models for bearing fault detection using fractals. In *Proceedings of the IEEE International Joint Conference on Neural Networks,* British Columbia, Canada (pp. 5876-5881).

Merz, C .J. (1999). Using correspondence analysis to combine classifiers. *Machine Learning 36,* 33-58.

Mohamed, S., Tettey, T., & Marwala, T. (2006). An extension neural network and genetic algorithm for bearing fault classification. In *Proceedings of the IEEE International Joint Conference on Neural Networks,* British Columbia, Canada (pp. 7673-7679).

Nabney, I. T. (2002). *NETLAB: Algorithms for pattern recognition*. Berlin: Springer-Verlag.

Nelwamondo, F. V., & Marwala, T. (2006). Fault detection using Gaussian mixture models, mel-frequency ceptral coefficient and kurtosis. In *Proceedings of the IEEE International Conference on Systems, Man and Cybernetics,* Taiwan (pp. 290-295).

Nelwamondo, F. V., & Marwala, T. (2007a). Handling missing data from heteroskedastic and nonstationary data. *Lecture Notes in Computer Science*, vol. *4491*, 1297-1306.

Nelwamondo, F. V., & Marwala, T. (2007b). Rough set theory for the treatment of incomplete data. In *Proceedings of the IEEE International Conference on Fuzzy Systems* London, UK (pp. 338–343).

Nelwamondo, F. V., Marwala, T., & Mahola, U. (2006). Early classifications of bearing faults using hidden Markov models, Gaussian mixture models, mel-frequency cepstral coefficients and fractals. *International Journal of Innovative Computing, Information and Control, 2*(6), 1281-1299.

Nordström, L. J. L. (2006). IC programming algorithm for input estimation on linear time-variant systems. *Computer Methods in Applied Mechanics and Engineering, 195*, 6407-6427.

Pan, H., Tilakaratne, C., &Yearwood, J. (2005). Predicting Australian stock market index using neural networks, exploiting dynamical swings and inter-market influences. *Journal of Research and Practice in Information Technology 37*, 43–55.

Park, J.Y. (2002). Non-linear non-stationary heteroskedasticity. *Journal of Econometrics 110,* 383–415.

Polikar, R. (2007), Bootstrap inspired techniques in computational intelligence: ensemble of classifiers, incremental learning, data fusion and missing features. *IEEE Signal Processing Magazine, 24*, 59-72.

Reiter, C. A. (1994). Sierpinski fractals and GCDs. *Computers & Graphics, 18*(6), 885-891.

Reynolds, D. A., Quatieri, T. F., & Dunn, R. B. (2000). Speaker verification using adapted Gaussian mixture models. *Digital Signal Processing, 10*(1), 19–41.

Roth P. L., & Switzer III, F. S. (1995), A Monte Carlo analysis of missing data techniques in a HRM setting. *Journal of Management, 21*, 1003–1023.

Sadana, A. (2003). *Biosensors: Kinetics of binding and dissociation using fractals*. London: Elsevier.

Sadana, A. (2005). *Fractal analysis of pathogen detection on biosensors*. London: Elsevier.

Salor, Ö., & Demirekler, M. (2006). Dynamic programming approach to voice transformation. *Speech Communication, 48*, 1262-1272.

Scholz, M., & Klinkenberg, R. (2005). *An ensemble classifier for drifting concepts*. Paper presented at the Second International Workshop on Knowledge Discovery in Data Streams. Porto, Portugal.

Secomandi, N. (2000). Comparing neuro-dynamic programming algorithms for the vehicle routing problem with stochastic demands. *Computers & Operations Research, 27*, 1201-1225.

Siani, C., & de Peretti, C. (2007). Analysing the performance of bootstrap neural tests for conditional heteroskedasticity in ARCH-M models. *Computational Statistics & Data Analysis, 51*, 2442-2460.

Stefanakos, C., & Athanassoulis, G.A. (2001). A unified methodology for analysis, completion and simulation of non-stationary time series with missing values, with application to wave data. *Applied Ocean Research 23*, 207–220.

Vilakazi, C. B., & Marwala, T. (2006). Bushing fault detection and diagnosis using extension neural network. In *Proceedings of the 10ᵗʰ IEEE International Conference on Intelligent Engineering Systems* (pp. 170-174).

Vilakazi, B. C., & Marwala, T. (2007). Condition monitoring using computational intelligence. In D. Laha and P. Mandal (Eds.), *Handbook on Computational Intelligence in Manufacturing and Production Management* (pp. 106-143). New York: Idea Group Inc (IGI).

Vilakazi, B. C., & Marwala, T. (2007b). On-line incremental learning for high voltage bushing condition monitoring. In *Proceedings of the International Joint Conference on Neural Networks* (pp. 2521 – 2526).

Vilakazi, B. C., & Marwala, T. (2007c). Incremental learning and its application to bushing condition monitoring. *Lecture Notes in Computer Science, 4491*, 1241-1250.

Wang, S. (2001). A hybrid threshold curve model for optimal yield management: neural networks and dynamic programming. *Computers & Industrial Engineering, 40*, 161-173

Wang, M. H., & Hung, C. P. (2003). Extension neural network and its application. *Neural Network, 16*(5-6), 779-784.

Widmer, G., & Kubat, M. (1993). Effective learning in dynamic environments by explicit context tracking. In *Proceedings of the European Conference on Machine Learning* (pp. 227-243).

Yang, C., & Stone, M. (2002). Dynamic programming method for temporal registration of three-dimensional tongue surface motion from multiple utterances. *Speech Communication, 38*, 201-209.

Yao, Z., & Li. H. (2006). Tracking a detected face with dynamic programming. *Image and Vision Computing, 24*, 573-580.

About the Author

Tshilidzi Marwala holds a chair of systems engineering at the School of Electrical and Information Engineering at the University of the Witwatersrand. He is the youngest recipient of the Order of Mapungubwe (whose other recipients are Nobel Prize Winners Sydney Brenner and J.M. Coetzee) and was awarded the President Award by the National Research Foundation. He holds a Bachelor of Science in mechanical engineering (*Magna Cum Laude*) from Case Western Reserve University, a Master of Engineering from the University of Pretoria, PhD in engineering from University of Cambridge (St John's College) and attended a Program for Leadership Development at Harvard Business School. He was a post-doctoral research associate at the Imperial College of Science, Technology and Medicine and in year 2006 to 2007 was a visiting fellow at Harvard University. His research interests include theory and application of computational intelligence to engineering, computer science, finance, social science and medicine. He has published over 150 papers in journals, proceedings and book chapters and has supervised 30 master and PhD theses. His book *Computational Intelligence for Modelling Complex Systems* is published by Research India Publications. He is the associate editor of the *International Journal of Systems Science*. His work has appeared in publications such as the *New Scientist* and *Time Magazine*. He was a chair of the Local Loop Unbundling Committee, is a deputy chair of the Limpopo Business Support Agency and has been on boards of EOH (Pty) Ltd, City Power (Pty) Ltd, State Information Technology Agency (Pty) Ltd, Statistics South Africa and the National Advisory Council on Innovation. He is a trustee of the Bradlow Foundation as well as the Carl and Emily Fuchs Foundation. He is a senior member of the IEEE and a member of the ACM.

Index

A

AANN-PSO 71, 72
adaptation speed 168
adaptive boosting 176
auto-associative neural network 19, 29, 40, 45, 46, 47, 49, 53, 55, 57, 59, 65, 73, 77, 82, 87, 88, 95, 119, 120, 135, 178, 184, 209, 243
auto-associative neural networks with evolutionary optimization (AANN-EO) 71

B

back-propagation 24, 25, 26, 49, 53, 58, 74, 191, 192, 205, 232, 234, 236, 261, 267
Bayesian auto-associative neural network 45, 47, 55, 57, 65
Bayesian MLP 57, 63, 65
Bellman equation 287
biomedical individualism 188, 205, 257, 258, 261, 270
boundary mutation 221
boundary region 103, 104

C

C4.5 decision tree 234, 238, 244, 254
canonical 49, 50, 51, 53, 138, 196
canonical distribution 49, 50, 51
case deletion 1, 3, 6, 18, 285
classification and regression trees (CART) 236, 255
cold-deck imputation 6, 8

committee of networks

committee of networks 142, 143, 144, 145, 146, 149, 150, 154, 155, 156, 157, 158, 167, 177
concept drift 276, 277, 278, 279, 280, 282, 283, 284, 298
cross-over 32, 33, 34, 54, 61, 77, 93, 125, 225
cross-over operator 32
cultural algorithms 124, 125, 138
cultural genetic algorithm 117, 125, 127, 128, 129, 133, 135

D

decision tree based imputation 9
decision trees 1, 18, 115, 117, 234, 235, 236, 237, 238, 239, 242, 249, 250, 251, 252, 253, 255
direct control approach 257
diverse committee 146
domain specific knowledge 125
dynamic programming 15, 216, 276, 285, 286, 287, 288, 289, 297, 298, 299, 300, 301

E

Euclidean norm 26, 30, 56, 78, 121, 129, 143, 177, 192, 212, 242, 266, 286
evolutionary algorithms 212
evolutionary committee 156, 157, 182
evolutionary optimization 71, 88
expectation maximization 3, 14, 71, 72, 78, 79, 152, 154, 235, 295